Advances in Optical Networks and Components

Advances in Optical Networks and Components

Partha Pratim Sahu

CRC Press
Taylor & Francis Group
Boca Raton London New York

CRC Press is an imprint of the
Taylor & Francis Group, an **informa** business

First edition published 2020
by CRC Press
6000 Broken Sound Parkway NW, Suite 300, Boca Raton, FL 33487-2742

and by CRC Press
2 Park Square, Milton Park, Abingdon, Oxon, OX14 4RN

ISBN: 978-0-367-26565-6 (hbk)
ISBN: 978-0-429-29396-2 (ebk)

Typeset in Times
by codeMantra

To my family:

*My grand mother Sushila Sahu and
my parents Harekrishna Sahu and Jyotsana Sahu,*

*My wife Arpita Sahu, and
my daughters Prakriti and Ritushree Sahu.*

Contents

Preface

As discussed earlier, volume 1 consists of nine chapters (Chapters 1–9) on hardware components used in optical networks and basics of data transmission and optical access technology used in optical networks. In volume 2, we try to focus on wavelength router using wavelength converter, traffic grooming and priority assignment, and survivability and security of optical networks. The chapters in volume 2 are as follows.

Chapter 1: Optical Ring Metropolitan Area Network
>   This chapter describes mainly optical ring topology used in metropolitan area networks.

Chapter 2: Queuing System and Its Interconnection with Other Networks
>   This chapter discusses queuing theories used to analyze the performance of computer communication networks along with optical networks.

Chapter 3: Routing and Wavelength Assignment
>   This chapter discusses different static and dynamic routing and wavelength assignment approaches used in optical networks.

Chapter 4: Virtual Topology
>   This chapter addresses the formulation of virtual topology and its design for optical networks.

Chapter 5: Wavelength Conversion in WDM Networks
>   This chapter mentions various aspects and benefits of wavelength conversion in the networks and its incorporation in a wavelength-routed network design for efficient routing and wavelength assignment.

Chapter 6: Traffic Grooming in Optical Networks
>   This chapter addresses different schemes of static and dynamic traffic grooming in optical networks.

Chapter 7: Survivability of Optical Networks
>   This chapter presents fault management schemes such as protection deployed in survivable networks for SONET/SDH ring and mesh optical networks.

Chapter 8: Restoration Schemes in Survivability of Optical Networks
>   This chapter discusses different restoration schemes in survivable optical networks.

Chapter 9: Network Reliability and Security
>   This chapter discusses optical signal security schemes used in WDM optical network apart from basic theory of network security.

Chapter 10: FTTH Standards, Deployments, and Issues
>   This chapter presents different FTTH standards, deployments, and research issues.

Chapter 11: Math Lab Codes of Optical Communication
>   This chapter discusses mathematical simulation codes of optical fiber link design and the codes that help in designing physical link in optical networks.

IMPORTANCE OF THIS BOOK

The main purpose of volume 2 is to provide students, researchers, and practicing engineers with an expert guide to the fundamental concepts, issues, and state-of-the-art developments in optical networks. This book discusses the concepts of these devices along with their fabrication processes, optical encryption, etc. Since optical networks mostly handle data communication, this book also discusses data transmission control and protocols to interlink data communication with optical networks.

1. The examples mentioned in volume 1 are very helpful for undergraduates and postgraduate students to understand basic problems and their solution.
2. These examples are helpful in solving problems in the development of high-speed optical networks for network software development companies/vendors.
3. The practices mentioned in this book are useful in learning the formulation and modeling of conventional network topologies and nationwide mesh network topologies.

    Apart from research students, UG and PG students (who have started learning and initiated research on this topic), the following persons will gain knowledge from this book:
    • service providers of optical backbone such as telecommunication industries (BSNL, India), telecommunication agencies of different countries, and experts/engineers in software industries
    • scientific advisors of government organizations/laboratories working on high-speed networks.

    The audience may need to understand the issues and challenges in designing such networks. It is anticipated that the next generation of the Internet will employ WDM-based optical backbones. In all chapters, we discuss these issues and challenges.

INTENDED AUDIENCE

The intended audiences of this book are researchers, industry practitioners, and graduate students. (It is both a graduate textbook and a reference for doctoral students' research.) Many electrical engineering, electronics engineering, communication engineering, and computer engineering programs over the years in the world have been offering a graduate course on this topic. Each chapter is typically organized in a stand-alone and modular fashion, so any of them can be skipped if desired.

I also hope that industry professionals will find this book useful as a reference. There are large groups of people involved in physical-layer optics and having interest in network architectures, protocols, and the corresponding engineering problems to design new state-of-the-art optical-networking products.

Acknowledgments

Though my name is visible on the cover page, a lot of people have worked to produce this book. First and foremost, I wish to thank my research and project students for their effort in getting this book to its current form. Much of the book's content is based on research that I have conducted over the years with my graduate students, research scientists, and my research group member visiting my laboratory, and I would like to acknowledge them: Dr. Bijoy Chand Chatterjee for some portion of wavelength-routing content (Chapter 3); Dr. Rabindra Pradhan for content on wavelength routing, traffic grooming and protection (Chapters 3–6); and Dr. Mahipal Singh for content on queuing system and network security (Chapters 2 and 9).

I would like to acknowledge Prof Alok Kumar Das, Prof Mrinal Kanti Naskar, Prof Debasish Datta, and Prof Utpal Biswas who guided and discussed with me over the years.

I wish to acknowledge Marc Gutierrez and Nick Mould of CRC Press who corresponded with me for their assistance during the book's production.

Finally, I wish to thank my family members for their constant encouragement: my father Harekrishna Sahu, my mother Jyotsana Sahu, my wife Arpita, and my daughters Prakriti and Ritushree. Without their support, it would not have been possible to complete this project.

Author

P. P. Sahu received his M.Tech. degree from the Indian Institute of Technology Delhi and his Ph.D. degree in engineering from Jadavpur University, India. In 1991, he joined Haryana State Electronics Development Corporation Limited, where he has been engaged in R&D works related to optical fiber components and telecommunication instruments. In 1996, he joined Northeastern Regional Institute of Science and Technology as a faculty member. At present, he is working as a professor in the Department of Electronics and Communication Engineering, Tezpur Central University, India. His field of interest is integrated optic and electronic circuits, wireless and optical communication, clinical instrumentation, green energy, etc. He has received an INSA teacher award (instituted by the highest academic body Indian National Science Academy) for high level of teaching and research He has published more than 90 papers in peer-reviewed international journals, 60 papers in international conference and has written five books published by Springer Nature, McGraw-Hill, India. Dr Sahu is a Fellow of the Optical Society of India, Life Member of Indian Society for Technical Education and Senior Member of the IEEE.

1 Optical Ring Metropolitan Area Networks

Communications networks are represented as three-tier hierarchical networks having access to a long-haul wide area network. At the bottom of the hierarchy, there is an access network connecting to customers/users within close proximity averaging regions between 10–100 km and interconnecting the access and long-haul networks. The access networks have already been discussed in Chapter 9 of volume 1. Long-haul optical backbone networks provide a coverage of interregional/global distances (2000 km or more) for transmission of signals. Metropolitan area networks (MANs) use synchronous digital hierarchy (SDH)/synchronous optical network (SONET) architecture [1–3] (as discussed in Chapter 1 of volume 1) and are placed in between the access network and long-haul optical backbone [4–29]. For example, the smaller rings having OC-3/STM-1 (155 Mbps) or OC-12/STM-4 (622 Mbps) traffic are combined into larger core inter-office (IOF) rings that interconnect central office (CO) locations at higher bit rates; e.g., OC-48/STM-16 (2.5 Gbps) Synchronous Transport Signal (STS) uses transmission of frame structure in electrical domain wave for end-user connectivity, namely voice. Internet data traffic growth is basically considered for metropolitan optical network. As the number of users and traffic/services increases, the networks require higher bandwidth for the Internet users using optical wavelength division multiplexing (WDM) technology.

1.1 DIFFERENT MANs

WDM technology is used in the backbone networks to enhance the capacity to the range of terabits. The access technology does the sharing of a large number of signals. Residential and digital subscriber line (DSL) modems provide access rates ranging from kbps to Mbps and other advanced technologies such as passive optical networks (PONs) (Chapter 9 of volume 1). Also, customers employ advanced switching/routing capable of direct line-rate inputs to the metro core having OC-48/192 and 10 Gbps Ethernet interfaces. Many SONET/SDH rings operate with the capacities of OC-48/192 rates. These backbones comprise IP (Internet Protocol), ATM (Asynchronous Transfer Mode), SONET/SDH, Ethernet (10/100 Mbps, 1.0/10 Gbps), multiplexed time division multiplexers (TDMs) voice, and other more specialized data protocols such as ESCON (Enterprise System Connectivity), FICON (Fiber Connectivity Channel) [4], Fiber Channel, cable video, etc.

The new MANs offer alternative networks for wide areas using SONET/SDH expansion [3]. The MANs should provide high bandwidth capability and use multiple protocols over a common infrastructure to increase link utilization. Intelligent service provisioning and survivability [1] are essential for advanced service-level agreements (SLAs) [5]. Moreover, new schemes offer backward compatibility for a more cost-effective network. Tunable optical sources, optical fibers, receivers, and tunable switches are required for enabling network-level wavelength routing and protection over rings or meshes. These capabilities with intelligent control architectures make the operators provide more capacity to a large number of services [6].

WDM technology offers many benefits with its deployment within MAN, but its use in a network is complex [7]. For WDM use in a larger metro area backbone, scalable "lambda" provisioning is required in Gbps range. In a MAN, operators interface with increased protocol heterogeneities, namely, interfaces and bit rates, and a MAN needs to cost-effectively handle finer "sub-wavelength" capacity increments. In long-haul networks, the input signals comprise a few protocols (SONET/SDH) [3] and interface bit rates (2.5, 10, and possibly 40 Gbps) which perform multi-protocol aggregation/grooming onto larger WDM branches, with a particular focus on data protocol efficiency. Various signal solutions are driven by advances in high-density electronic integrated circuit (IC) technology, e.g., SONET/SDH multi-service provisioning platforms (MSPPs) [8] and IP packet rings.

1.2   METRO WDM NETWORKS

The growth of optical technology requires electronics for the MANs in case of bottlenecks in the network. WDM is used in long-haul backbones providing a high-speed/high-bandwidth network. The terms "transparent" and "all-optical" represent entities (nodes, networks). WDM evolves to support intelligent, rapid provisioning of large, interconnection capacities. Optics blends in with advanced IC technologies to form an intelligent, opto-electronic metro edge.

1.2.1   WDM Ring Networks for MAN

WDM rings are made with SONET/SDH concepts [8] making time slots within wavelengths and performing optically equivalent channel operations such as add-drop, pass-through, protection, etc. These rings provide very good bandwidth scalability, data transparency, and multiple data rates. Optical bypass removes the need for complicated electronic access to client signals and gives significant cost savings over traditional add–drop multiplexer (ADM)/optical cross-connect (OXC) nodes. WDM ring architectures contribute from simple static setups (i.e., fixed nodes) to advanced sharing schemes (i.e., dynamic nodes). The basic building blocks of an optical WDM are optical ADM (OADM) nodes. Fixed OADMs operate on static or pre-fixed tuned wavelengths used in static rings for static traffic. With increasing customer dynamics, static rings can provide services, but they require complex manual wavelength planning and yield reduced wavelength. Different optical WDM optical rings are mentioned later in this chapter.

Here, re-configurability is one of the key requirements apart from scalability, in dynamic OADM rings having reconfigurable OADMs (ROADMs).

1.2.2 Metro-Edge Technology

The edge technology in a metro network provides an integration of the core inter-office and the client access points [7]. In this direction, WDM with electronic multiplexing access for sub-gigabit line rates is required to collective diverse end-user protocols onto large-granularity optical (WDM) arms in metro-edge technology [9]. Many metro-edge solutions include next-generation SONET/multi-service provisioning and IP routing of packet rings.

Next-Generation SONET/Multi-Service Provisioning Paradigms

Although a hierarchical TDM device has many issues, it plays a significant role in the convergence of data and optical networks at the metro edge [7] because of short-haul SONET/SDH. A large part of it comprises of larger OC-48/STM-16 and OC-192/STM-64 systems, and at the same time, smaller OC-3/STM-1 and OC-12/STM-4 systems are also used for deployments in metro edge [1]. The network having existing routers/switches has SONET/SDH interfaces and requires a broader generic framing protocol (GFP) (for mapping "non-standard" data protocols). There are many reported approaches to enhance SONET/SDH suited to data traffic needs. All these have two main features: efficient data-tributary mappings and integrated higher-layer protocol functionalities.

SONET/SDH mapping of smaller packet interfaces of 10 and 100 Mbps Ethernet is formed in "coarse" STM-1, and 10 Mbps Ether net allocated a full STS-1 yields 80% unused capacity. Burst data profiles further worsen bandwidth inefficiencies. By using virtual concatenation (VCAT) [10], native packet interface rates provide 1-Gbps Ethernet via seven STS-3c's. Multiple, "matched" STM-3c's are packed into existing STM standards.

Packet-Based Access Edge

Recently for metro-edge solution [7], packet rings with TDM having packet switching (statistical multiplexing) play an active role in resilient packet ring (RPR, IEEE 802.17). Specifically, a new Ethernet-layer media-access control (MAC) protocol [11] has been developed to multiplex multiple IP packets statistically onto Ethernet frames. The MAC protocol operates over various underlying network infrastructures, including SONET/SDH, WDM, and optical fiber. The RPR contributes separate layer-2 packet by passing at coarser granularities, and this enhances packet loads at the IP (layer-3) routing level, improving the quality of service (QOS). In addition, packet rings can also provide a rapid layer-2 protection signaling protocol to match the 50-ms protection-switching time of SONET/SDH.

The latest RPR framework [1] provides dual counter-propagating rings accommodating working traffic with no reserved protection bandwidth for bandwidth efficiency. The reuse of bandwidth (multicast and broadcast) still requires source stripping. The above characteristics significantly improve ring capacity/utilization/throughput. Currently, a spatial reuse protocol (SRP) framework is employed for standardization and is commercially available for all RPR features.

1.2.3   TRAFFIC GROOMING IN SONET RING NETWORKS

Traffic grooming is found to be an important framework in SONET ring network to accommodate heavy traffic.

1.2.3.1   Node Architecture

SONE/SDH ring is used in an optical network infrastructure today for a MAN [12,13]. In a SONET ring network, WDM contributes a point-to-point bandwidth transmission technology. The SONET system's hierarchical TDM schemes permit a high-speed OC-N channel to accommodate multiple OC-M channels (where M is smaller than or equal to N). The ratio of N to the smallest value of M carried by the network is defined as grooming ratio. Electronically controlled add–drop multiplexers (ADMs) are used to add/drop traffic at intermediate nodes to/from the high-speed channels. In a SONET ring network, an ADM is required for a wavelength at every node to add/drop traffic in that wavelength. The use of ADM is expensive as the same number of ADMs is needed at every network node since traffic assigned to a wavelength is only by passing an intermediate node. With the emerging optical components such as OADM (also referred to as wavelength ADM (WADM)), a node bypasses most of the wavelength channels optically and drops only the wavelengths having the traffic destined to the node. Figure 1.1 shows the architecture of a typical node in a SONET/WDM ring network. Since there is no need to add or drop any of its time slots, they are optically by passed at the node. For other wavelengths (λ_2 and λ_3) where at least one time slot needs to be added or dropped, an electronic ADM is used.

Carefully arranging these optical by passes reduces a large part of the network cost. Using OADMs reduces the number of SONET ADMs needed in the network. Then, the problems is, for a given low-speed set of traffic demands, which low-speed demands should be groomed together. The wavelengths should be used to carry the traffic, and a number of ADMs are needed at a particular node for these wavelengths.

1.2.3.2   Single-Hop Grooming in SONET/WDM Ring

As shown in Figure 1.1, ADMs do not operate the time slot interchange function, and since there is no wavelength conversion, time slot-continuity and wavelength-continuity constraints are considered at nodes where only ADMs are used [14]. These rings are built only with ADMs as single-hop rings since all the connections are "direct" connections. Low-speed OC-M connections are groomed on to OC-N wavelength channels. Based on this network model, for a given traffic matrix, satisfying

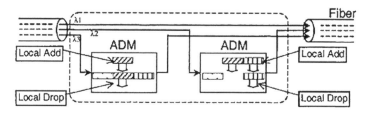

FIGURE 1.1   Typical node architecture of a SONET ring [1].

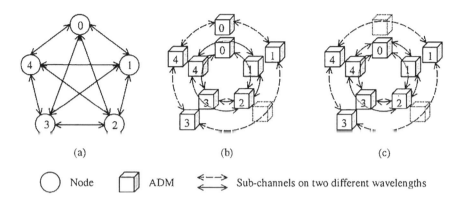

(a)   (b)   (c)

◯ Node   ⬛ ADM   ⟷ Sub-channels on two different wavelengths

FIGURE 1.2 Five-node network with uniform traffic requests. In this network, a bidirectional ring is considered with grooming ratio 2. (a) Five-node SONET ring having all the 10 requests. (b) and (c) Two ways of systematizing the connections on two wavelengths (one and half are actually in use). In (b), there is only one by passing traffic at node 2 of the second ring (dashed line) where nine ADMs are needed to carry all the requests. The number of ADMs is reduced by simply rearranging in the rings as shown in (c) in which the connections between nodes 0 or 1 and 4 on the second ring (1 ↔ 4) and part of the first ring (4 ↔ 0, 0 ↔ 1) are exchanged.

all the traffic demands as well as minimizing the total number of ADMs is a network design optimization problem and has been studied extensively in the literature (Figure 1.2).

The traffic-grooming problem is NP-complete [15]. The optimization of this problem is formulated [9] by using an integer linear program (ILP). The formulation [15] can be used for both uniform and non-uniform traffic demands in unidirectional and bidirectional SONET/WDM rings. The ILP approach considers the limited numbers of variables and equations. Hence, it is difficult to be considered for a large-sized network. In this case, by relaxing some of the constraints, the ILP formulation can solve these equations to get optimal results. Some lower bounds are analyzed for different traffic (uniform and non-uniform) and network models (unidirectional ring and bidirectional ring). These lower-bound results estimate the performance of traffic-grooming heuristic algorithms. In most of the heuristic approaches, traffic-grooming problems are solved by greedy approach, approximation approach, and simulated-annealing approach in heuristic algorithms.

1.2.3.3 Multi-Hop Grooming in SONET/WDM Ring

In single-light path-hop grooming, traffic is switched between different wavelengths. Figure 1.3a shows a single-light path hop network configuration. Figure 1.3b shows a network architecture [16,17] in which some nodes have digital cross-connects (DXCs) installed in node 3 only and traffic is transmitted from one wavelength/time slot to another wavelength/time slot at node 3. Node 3 is also called as hub node because the traffic needs to be converted from optical to electronic at the hub node when wavelength/time slot exchange occurs. This grooming approach is known as multi-hop (multi-light path hops) grooming. Depending on the implementation, there

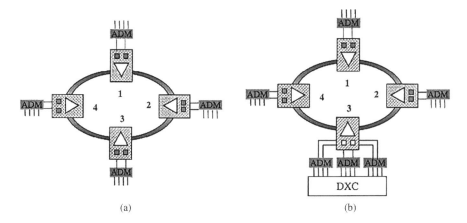

(a)   (b)

FIGURE 1.3   SONET-WDM ring network [1] (a) with hub node (b) without hub node.

is a single hub node or multiple hub nodes in the network. This kind of network is called point-to-point WDM ring network (PPWDM ring) [18].

Detailed theoretical analysis is needed for comparison between PPWDM ring, a SONET/WDM ring without hub nodes, a SONET/WDM ring with one or multiple hub nodes, etc. The results indicate that, when the grooming ratio is large, the multi-hop approach reduces the number of ADMs, but when the grooming ratio is small, the single-hop approach reduces the number of ADMs, and in general, the multi-hop approach requires more wavelengths than the single-hop approach in ring topology.

Static traffic: Traffic pattern will not change with time.

Dynamic traffic: Traffic will change with time.

Uniform traffic: Traffic demands of any two nodes are the same.

Non-uniform traffic: Traffic demands of any two nodes are not the same.

Single-hop ring: No virtual connection on the ring will be ended electronically at any intermediate node.

Multi-hop ring: Some or all of the virtual connections on the ring are remade electronically at some intermediate nodes.

Unidirectional ring: All the traffic on the ring can go along one (both) direction.

Bidirectional ring: Traffic can flow in both directions.

Virtual ring: It is established on one time slot of a wavelength. It has the capacity of one unit.

PPWDM: Point-to-point (opaque) WDM ring in which signals get cross-connected and regenerated at every node.

Distance-dependent traffic [16]: The nodes farthest apart exchange one unit of traffic, and the inter-node traffic demand increases by one unit at the same time; inter-node distance decreases by one link.

Egress node: A special node, like the central office of an access network, on the ring where all the traffic originates from or ends.

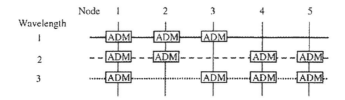

FIGURE 1.4 Five-node network design for three-wavelength traffic [1].

1.2.4 DYNAMIC GROOMING IN SONET/WDM RING

For the study of dynamic grooming, it is required to consider a set of traffic matrices which change from one matrix to another for a period of time which may be throughout a day or a month. The network needs to be reconfigured while transmitting the traffic pattern from one matrix to another matrix in the matrix set. The network design supports the accommodating of any traffic matrix in the matrix set in a non-blocking manner as well as minimizing the overall cost. This is known as a dynamic-grooming problem in a SONET/WDM ring [19]. Dynamic traffic grooming [19] is formulated in a SONET/WDM ring as a bipartite graph-matching problem, and several methods are used to reduce the number of ADMs. For this, the necessary and sufficient conditions are that a network should support such a traffic pattern. There is an allowable traffic pattern used for dynamic grooming analysis. If each node can be a source in most t duplex circuits, we call this traffic matrix a t-allowable traffic matrix. A traffic matrix set consists of t-allowable traffic matrices.

Figure 1.4 shows a five-node SONET/WDM ring network having three-wavelength transmission. Each wavelength can support two low-speed circuits. The network configuration in the figure is a 2-allowable configuration, i.e., it can support any 2-allowable traffic matrix (set). For instance, consider a traffic matrix with request streams 1–2, 1–3, 2–3, 2–4, 3–4, 4–5, and 4–5. The traffic matrix can be maintained by allotting 1–3, 2–3 on wavelength 1; allotting 1–2, 2–4, 4–5, 4–5 on wavelength 2; and allotting 3-4 on wavelength 3. However, the design is able to maintain other t-allowable traffic matrices.

1.2.5 GROOMING IN INTERCONNECTED SONET/WDM RINGS

Most of traffic-grooming studies in SONET/WDM ring networks use a single-ring network topology in which the problem of an interconnected-ring topology is considered [13]. The recent backbone networks are mainly constructed as networks of interconnected rings. The study of traffic grooming from a single-ring topology to the interconnected-ring topology is important for a network operator to design their network and to engineer the network traffic.

1.3 TRAFFIC GROOMING IN WDM RING NETWORKS

In this section, we first consider mathematical definitions for various traffic-grooming problems, which turn out to be integer linear programs (ILPs). Then, a simulated-annealing-based heuristic approach is reported for solving these optimization problems. This algorithm has achieved the best result in comparison to other approaches. Finally, the multi-hop case (with a single hub) is considered where a hub node is used to cross-connect traffic between different wavelengths and time slots. Finally a comparison of the single-hop and multi-hop approaches is provided.

1.3.1 PROBLEM DEFINITION

Here the following common notations are used throughout this study.

N = total number of nodes in the network.

W = Number of wavelengths per link in the network (each wavelength can transmit several circles in time-division fashion).

C = Grooming ratio, which is the number of circles a wavelength can carry.

$T(t_{ij})$ = Non-uniform traffic matrix, in which t_{ij} represents the traffic from node i to j. The traffic elements are carried by the ring at minimum cost.

${}^d V_{ij}^{cw}$ = Virtual connection from node i to node j on circle c, wavelength w. d = the direction of a connection, and it can be either clockwise or counter-clockwise.

O_i = In the multi-hop case, O_i represents the virtual connection that starts from node i and terminates at the hub node.

Similarly, I_i is the virtual connection that starts from the hub node and terminates at node i.

ADM_i^w = Number of ADMs at node i on wavelength w.

e is a link on the physical ring.

1.3.2 MATHEMATICAL FORMULATION OF SINGLE-HOP CONNECTIONS

The traffic-grooming problem can be mathematically formulated for a single-hop bidirectional ring [13,14]. This formulation shows whether there is a virtual connection (one unit of capacity) from node i to node j along direction d (which can be clockwise or counter-clockwise) on circle c and wavelength w.

The ADM_i^w has values either 0 or 1 representing whether there is an ADM on wavelength w at node i. There will be zero if the distance from node i to j in the specified direction exceeds $N/2$; otherwise it will be 1. There are four constraints in single-hop connections – traffic-load constraint, channel-capacity constraint, transmitter constraint, and receiver constraint. The traffic-load constraint simply states that the number of links from node i to node j on all circles is equal to the traffic specified in the traffic matrix. The ${}^d e^{cw}$ indicates a d-directional link on wavelength w and sub-channel c. In the virtual connection ${}^d V_{ij}^{cw}$, the channel-capacity constraint shows that for ${}^d e^{cw} \in {}^d V_{ij}^{cw}$, a circle carries only one connection on any given link. The transmitter and receiver constraints state that the number of connections that start and terminate at a node of a ring is limited by the capacity of the electronic ADM at that node.

Objective function

$$\text{Minimize} \sum_i \sum_w \text{ADM}_i^w$$

Subject to
Traffic-load constraint

$$\sum_w \sum_c \sum_d {}^d V_{ij}^{cw} = t_{ij} \quad \forall i,j$$

Channel-capacity constraint

$$\sum_{d\, e^w \in {}^d V_{ij}^{cw}} {}^d V_{ij}^{cw} \le 1 \quad \forall d,e,c,w$$

Transmitter constraint

$$\sum_c \sum_i {}^d V_{ij}^{cw} \le C.\text{ADM}_i^w \quad \forall d,i,w$$

Receiver constraint

$$\sum_c \sum_j {}^d V_{ij}^{cw} \le C.\text{ADM}_j^w \quad \forall d,j,w$$

For an ADM, C connections begin and end there; otherwise, no add/drop can happen. The unidirectional-ring system is taken as a special case of the formulation, where direction d can only be either clockwise or counter-clockwise. Sometimes, shortest-path routing is needed in case of bidirectional rings.

1.3.3 MATHEMATICAL FORMULATION OF MULTI-HOP METHOD

The traffic-grooming problem on a multi-hop ring is mathematically formulated [15]. A multi-hop ring uses a cross-connect to do sub-channel consolidation or segregation at a hub node. The number of ADMs at a hub node is same as that of the wavelengths. All connections that are passed through it are either ended or switched to any wavelength and time slot. A connection goes through the hub node. Bidirectional formulation for the single-hub-based multi-hop network is made here [1].

Objective function

$$\text{Minimize} \sum_i \sum_w \text{ADM}_i^w$$

Subject to
Traffic-load constraint

$$\sum_w \sum_c \left(\sum_{j(j>i)} V_{ij}^{cw} + O_i^w \right) = t_{ij} \quad \forall i$$

$$\sum_w \sum_c \left(\sum_{i(j>i)} V_{ij}^{cw} + I_j^w \right) = t_{ij} \quad \forall j$$

Channel-capacity constraint

$$\sum_{e^w \in V_{ij}^{cw}} V_{ij}^{cw} + \sum_{i<e^{cw}} O_{ij}^{cw} + \sum_{j<e^{cw}} I_j^{cw} \leq 1 \quad \forall d,e,c,w$$

Transmitter constraint

$$\sum_c \sum_j V_{ij}^{cw} + \sum_c O_i^{cw} \leq C.\text{ADM}_i^w \quad \forall d,i,w$$

Receiver constraint

$$\sum_c \sum_j V_{ij}^{cw} + \sum_c I_j^{cw} \leq C.\text{ADM}_j^w \quad \forall d,j,w$$

Bounds:
    Values of I_j^{cw}, O_i^{cw}, V_{ij}^{cw}, and ADM_j^w are either 0 or 1.
    In the above formulation, O_i^{cw} represents the virtual connection from node i to other hub node on circle c, wavelength w. Similarly, the notation I_j^{cw} indicates the connection that starts from the hub node and ends at node i. The condition $(j > i)$ states that the virtual connection originates from node i and ends at node j without going through the hub node. The condition $i < e^{cw}$ states that if s is the start node of link e^{cw} and t is the end node, then i is upstream of t. Similarly, $j > e^{cw}$ denotes that j is downstream of s. The traffic-load constraint needs to be broken into two parts for the multi-hop case. The first part in the multi-hop formulation states that any virtual connection that starts from node i will either end before it reaches the hub node or will end at the hub node. Similarly, the second traffic-load constraint shows that any virtual connection that terminates at node i is either coming from the hub node or from some node downstream of the hub node. The statements of the other constraints are similar to those in the single-hop case. The standard ILP solver CPLEX is considered for a SONET ring with a maximum size of 16 nodes. It also solves for non-uniform traffic. For the cases $N = 4$, 5 and $C = 12$ (shaded cells), no grooming is needed since all the traffic is carried by one wavelength. The simulated-annealing algorithm provides better results than the greedy algorithm, even reaching the lower bound sometimes.

1.3.4 HEURISTICS-BASED SIMULATED ANNEALING ALGORITHM FOR SINGLE HOP

The general traffic-grooming problem becomes NP-complete [12]. The solution obtained by using the ILP requires a long computation time. A simple greedy heuristic, time complexity as well as computation time is not decreased. In this direction, the simulated-annealing algorithm is employed to decrease the computation time for single-hop connections and multi-hop connections using a simple heuristic. To reduce its computational complexity, the traffic-grooming problem usually has two components [16,17] – the first step to assign the traffic demands as "circles" and second step having a traffic-grooming algorithm deployed to reorganize the circles or connections on wavelengths. The same strategy and our first-step heuristic are used in the wavelength-assignment algorithm for reducing the wavelength consumption. It is seen [20] that the number of wavelengths and the number of ADMs may not be decreased simultaneously. Both the ILPs and heuristic estimate the optimal number of ADMs and wavelength channels. The second-step heuristic randomly chooses some virtual connections in the network and changes their position by using the simulated-annealing technique to accelerate the process of branch and bound to find an optimal solution. The algorithm is given below [20].

Algorithm 1 Simulated Annealing for Single-Hop Traffic Grooming

```
        (where Δc= cost difference
   repeat {iterate around all states}
     repeat {accept ANN_CONST x C times)
       dcost <= perturb() {randomly pick two circles, swap part or all of them)
       if Δc <0 or (Δc > 0 and exp(−Δc /control)> rand[0, 1])
       then
         accept_change () {accepted the change)
         chain <= chain + 1
       else
         reject_change ()
       end if
     until chain < ANN_CONST x C
     temp<= temp x DEC_CONST
     until temp > END
```

In the above algorithm, the "perturb" term indicates to choose two circles randomly on different wavelengths exchanged their part. Whenever the principle for doing partial exchange is satisfied, it is checked if there are segments in the two circles that are swapped. If there is no segment, we simply exchange the whole circles. We will calculate exp (−Δc/control) and compare it with a random number. The perturb will still be accepted if the random number is smaller. After repeating the above process for a certain number of times, we consider that the system has reached the equilibrium state, and then we go on to decrease the control variable (the temperature). The process will be terminated when the control variable satisfies some predefined criterion.

In the above implementation, as the computation goes ahead and the control variable "temp" goes down, the chance of having a "good perturb" and the chance of accepting a "bad perturb" will reduce, so the time spent in lower temperature is much longer than that in higher temperature. The constants ANN-CONST and DEC-CONST are critical for the algorithm's performance. ANN-CONST estimates how long it is required for the system to reach to its equilibrium. DEC-CONST estimates how fast the temperature is decreased. After experimenting with these parameter values, ANN-CONST is required to be between 4 and 20 depending on the size of the search space, and DEC_CONST = 0.95 for best results for our numerical examples. We may not consider very high starting temperature. The circle-assignment heuristic in the first step serves as a pre-grooming heuristic, which puts similar connections as close to one another as possible; thus, a very high starting temperature can counterbalance the effort of the first step and prolong the convergence process. Some special techniques are also needed when implementing the perturb() function, details of which are skipped here. For the greedy heuristic, the second step can only regroup the circles but not change the existing circles. One of the major causes is getting a sub-optimal solution in [21]. The second-step heuristic using simulated-annealing algorithm gives a chance to change the circles so that it can skip out of "traps."

Heuristic for the Multimode Method

A greedy heuristic [22] algorithm based on the wavelength-combining function checks two wavelengths link by link. If the total load on any link does not exceed the wavelength capacity, then it combines them. The segment-swapping function finds the under utilized links in different wavelengths and combines them into one wavelength through segment swapping [1].

Algorithm 2 Traffic-Grooming Algorithm for Multi-Hop Method

while the number of ADMs and the number of wavelengths can be reduced do
    Establish connection on the shortest path
    Wavelength_Combining ()
    Segment_Swapping()
end while

This algorithm is applied on uniform and non-uniform traffic patterns for both single-hop and multi-hop approaches

1.4   INTERCONNECTED WDM RING NETWORKS

Multiple SONET rings can be interconnected together to increase geographical coverage area. The optical cross-connect (OXC) can be used for these interconnections. There are two problems: how to interconnect WDM rings with OXCs and how to groom traffic in interconnected rings.

1.4.1 INTERCONNECTED RINGS

Node Architecture

In an interconnected SONET/WDM ring, there are non-intersection nodes and intersection nodes. A non-intersection node has two interfaces to connect it with its two neighbors and a local interface bank for adding or dropping traffic. Figure 1.5 shows the architecture of a non-intersection node in a unidirectional ring. A bidirectional ring comprises of two unidirectional rings, and this requires extra hardware. Most of the studies will consider bidirectional rings and bidirectional SONET ADMs (SADMs) (back-to-back double SADMs for both directions assembled as one unit).

The intersection node comprises OADMs and SADMs. A digital cross-connect (DXC) is used to connect low-speed streams between two rings. Figure 1.6a shows the operation of s mechanism. Traffic, either sent to local ports or to other rings, is dropped by OADMs and SADMs, which are relayed by the DXC to their desired destination. Newly developed hardware technologies permit OXCs to switch an entire wavelength (not sub-channels as in the case of DXC). The transparent and opaque technologies build these OXCs as shown in Figure 1.6b,c, respectively. Transparent here indicates all-optical switching without the use of wavelength converter. Here, all-optical wavelength converters are used in transparent cross-connects. In Figure 1.6c, electronic switching is used in an opaque cross-connect. The signal has to be converted from optical to electronic domain before it is switched and converted back to optical. Opaque technology implies full wavelength conversion, so it potentially has an advantage in delivering higher network utilization when compared with a wavelength-continuous switch. Figure 1.6b,c also compare the operation of these two technologies. Conceptually, a transparent OXC may not be fully non-blocking although it can be made non-blocking on a single-wavelength plane. An OXC has

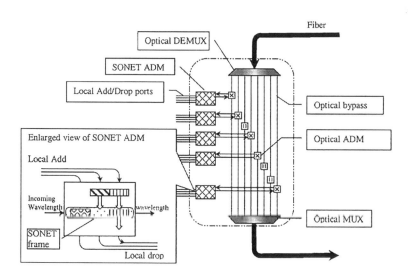

FIGURE 1.5 Simplified node architecture of a SONET/WDM ring network.

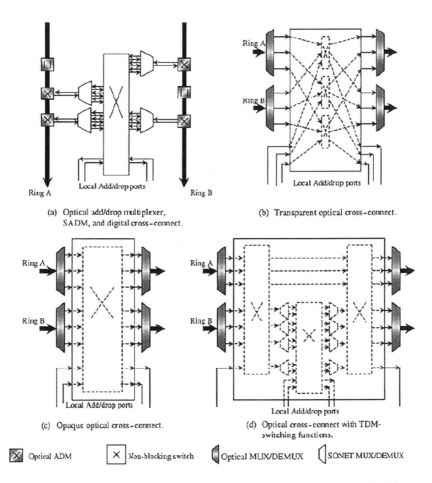

(a) Optical add/drop multiplexer, SADM, and digital cross-connect.

(b) Transparent optical cross-connect.

(c) Opaque optical cross-connect.

(d) Optical cross-connect with TDM-switching functions.

Optical ADM   Non-blocking switch   Optical MUX/DEMUX   SONET MUX/DEMUX

FIGURE 1.6 Simplified architectures of different optical cross-connects: (a) digital cross-connect (b) transparent optical cross-connect (c) Opaque optical cross-connect (d) Optical cross-connect with TDM switching.

built-in TDM-switching capability. Figure 1.6d shows another implementation of an OXC where limited sub-wavelength switching is possible.

Interconnection Strategies

Two rings may inter connect at either one or multiple points for the fault-recovery concern (when a node failure occurs at one intersection node, the rest of the nodes should still be connected so that the auto-recovery mechanism is performed for the traffic flow) [1]. SONET-ring-based protection mechanism has become known for the protection of signal [8]. A hop is defined as any connection segment that is entirely in optical domain. If traffic is dropped from one ring to an OXC and then is added onto another ring, this is counted as two hops; if traffic is switched to another ring through an optical switch, then the connection is counted as one hop. There are four strategies for connection of rings: strategy 1 connecting two rings entirely at the

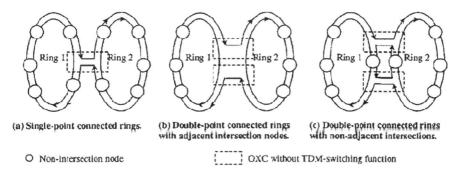

(a) Single-point connected rings.

(b) Double-point connected rings with adjacent intersection nodes.

(c) Double-point connected rings with non-adjacent intersections.

O Non-intersection node   OXC without TDM-switching function

FIGURE 1.7 Three forms of interconnections that can be formed by using strategies 2 and 3: (a) single point connected rings (b) double point connected rings with intersection adjacent nodes (c) double point connected rings with inter section nonadjacent nodes.

SONET level, strategies 2 and 3 connecting two rings at wavelength level, and strategy 4 providing a mixed connection with maximum flexibility [1]. Figure 1.7 shows three architectures for single-point (one intersection node between the two rings) and double-point (two intersection nodes) connected rings. These three interconnections operate on strategies 2 and 3. This is a scalability problem since no more than 16 nodes can be on a single SONET ring. Strategies 2 and 3 are used with other ones for interconnecting multiple rings [8].

1.4.2 TRAFFIC GROOMING IN INTERCONNECTED RINGS

Interconnected Double Rings

The formulation of the single-ring traffic-grooming problem is done mathematically using an ILP, and the heuristic algorithm improved for general traffic pattern is used. An arbitrary (static) traffic pattern is represented by a $N \times N$ traffic matrix, where N is the total number of nodes in the rings [8]. The traffic matrix is denoted by T, and each of its element t_{ij} represents the total amount of traffic demand from node i to node j measured in units of sub-channel capacity. The unidirectional formulations are less complex and shown here, but the formulations for bidirectional rings are complex. The formulation also considers that all wavelengths have same capacity; i.e., an equal number of sub-channels are considered and represented as C. The following formulation is considered for single-point-connected double rings. In the formulation for single-point-connected double rings, a minimum number of wavelengths is required to achieve maximum SADM reserves. Actually, finding this number is challenging. The formulation is based on a heuristic [8] for the cross-connect architecture shown in Figure 1.6a for strategy 1. In this case, an OXC is to interconnect two rings, and all the inter-ring traffic is added/dropped at the intersection. To each ring, all traffic to or from the other ring is the traffic to or from the intersection node. The unified traffic matrix is divided into two parts, each part belonging to a ring.

The notation $T' = \{t'_{ij}\}$ is used to represent the decomposed traffic matrix on each ring. The mathematical formulation for the traffic-grooming problem on each ring is similar to that for a single hop ring.

The $^d V_{ij}^{cw}$ represents whether there is a virtual connection (one unit of capacity) from node i to node j along direction d (which can be clockwise or counter-clockwise) on circle c and wavelength w. The t'_{ij} represents the total amount of traffic from node i to node j on a ring. $t'_{ij} = t_{ij}$ when neither i nor j is the intersection node. When one of the two is the intersection node, t'_{ij}, it will include both the intra-ring and inter-ring traffic. The intra-ring traffic is included in t_{ij}, and the inter-ring traffic part is the summation of all traffic from the nodes on another ring to the non-intersection node. The ADM_i^w represents whether there is an SADM on wavelength w at node i. Both $^d V_{ij}^{cw}$ and ADM_i^w can be either 0 or 1.

The traffic-load constraint states that the number of virtual links from node i to node j is equal to the traffic mentioned in the traffic matrix T'. For each TDM sub-channel of a wavelength on a certain link, only one connection is accommodated. In the channel-capacity constraint, $^d e^{cw}$ represents a link d (d can be either 0 or 1) on wavelength w and sub-channel c. If the virtual connection $^d V_{ij}^{cw}$ is used, then $^d e^{cw} \in {}^d V_{ij}^{cw}$. The transmitter and receiver constraints state that the number of virtual connections (that start or terminate) at any node is surrounded by the electronic multiplexing (de-multiplexing) capacity, which is same as the number of transmitters or receivers. Sometimes, shortest-path routing is required in a bidirectional ring. This requirement is also accommodated in the formulation considering $^d V_{ij}^{cw}$ to be zero if the distance from node i to j in a specified direction exceeds half the ring size. Some redundant constraints were added to the basic formulation shown in single-hop formulation.

Like strategy 1, strategy 2 has one more logical ring covering all the non-common nodes. The traffic matrix is broken down into three rings, and this decomposition leads to a sub-optimal solution. The new logical ring carries all the inter-ring traffic, as well as part of the local traffic. A local ring takes care of all or part of the local traffic. The notation $\overline{T} = \{\overline{t_{ij}}\}$ indicates traffic on the logical ring, and $T' = \{t'_{ij}\}$ for the traffic on the local rings. Now the objective function is to minimize the summation of all SADMs on all the three rings. The following additional traffic-load constraints for transparent-OXC-connected double rings (strategy 2) are considered.

$$\overline{t_{ik}} = t_{ik} \qquad (i, k \text{ are in different rings})$$

$$t'_{ij} + \overline{t_{ij}} = t_{ij} \qquad (i, k \text{ are in different rings})$$

The strategy 4's additional traffic-load constraints for OXC with TDM-switching capability are taken as [8]

$$\overline{t_{ik}} \le t_{ik} \qquad (i, k \text{ are in different rings})$$

$$t'_{ij} + \overline{t_{ij}} = t_{ij} + \sum_k \left(t_{ik} - \overline{t_{ik}} \right) \quad (i, k \text{ are in different rings, } i \text{ is the interconnection node})$$

$$t'_{ij} + \overline{t_{ij}} = t_{ij} + \sum_k \left(t_{kj} - \overline{t_{kj}} \right) \quad (i, j \text{ belong to the same rings. } j \text{ is the interconnection node})$$

$$t'_{ij} + \overline{t_{ij}} = t_{ij} \quad (i, j \text{ belong to the same rings. None of these is the interconnection node})$$

The cross-connect architecture is shown in Figure 1.6c, having wavelength conversion available for interacting traffic at the intersection node. The formulation of the problem is to represent all possible wavelength-conversion patterns and to use the same formulation of strategy 2 for each pattern. It replaces $^d V_{ij}^{cw}$ with $^d V_{ij}^{cw \to w'}$ for the inclusion of wavelength conversion. A program is written to figure out the $^d V_{ij}^{cw \to w'}$ (whether it will turn out to be 0 or 1 eventually) when vector (i, j) goes across the intersection node for a given wavelength-conversion pattern. If the wavelength conversion from w_1 to w_2 is available at node s for connections along direction d, then $^d V_{ij}^{cw \to w_2}$ exists. The channel-capacity constraint holds to implement that no connection longer than a full circle is permitted.

Strategy 4 differs from strategy 1 in that part of the inter-ring traffic is switched without electronic multiplexing and de-multiplexing. In comparison to strategies 2 and 3, not all the inter-ring traffic has to be carried by a logical ring in strategy 4. These differences make the traffic decomposition more complex. To incorporate all traffic-decomposition possibilities, the additional traffic-load constraint for strategy 2 is amended. In the formulation, $\overline{T} = \{\overline{t_{ij}}\}$ represents the traffic on the wavelengths that are optically switched across the border of the two rings, and $T' = \{t'_{ij}\}$ represents the traffic carried by the wavelengths that traverse through only one ring.

Figure 1.8 shows the problem of two-point-connected double rings solved by using the above formulations. The grooming ratio is considered to be 1. Each ring has three nodes. Uniform and full-mesh traffic demand is carried by the network. Figure 1.8a–d provides an optimal solution obtained from strategies 1 to 4. Figure 1.8a represents the most expensive solution for a particular traffic demand in network configuration. Fourteen SADMs are used. In Figure 1.8b, to get maximum savings, two directions of some bidirectional connections are routed along different paths. In Figure 1.8c, wavelength conversion is used along with ADMs. This strategy gets a minimum of 10

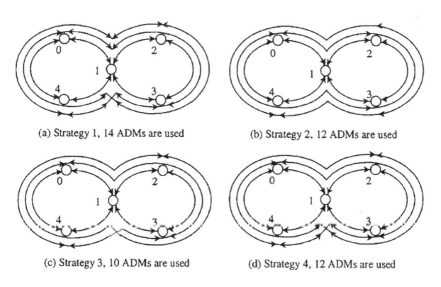

(a) Strategy 1, 14 ADMs are used   (b) Strategy 2, 12 ADMs are used

(c) Strategy 3, 10 ADMs are used   (d) Strategy 4, 12 ADMs are used

FIGURE 1.8 Double-ring optimal solution with (a) strategy 1 (b) strategy 2 (c) strategy 3 (d) strategy 4 [1].

SADMs in this case. In Figure 1.8d, we consider wavelength-continuous OXC with TDM-switching capability, and no wavelength conversion is used.

Heuristic Design

Computer simulation using mathematical formulations performs a detailed simulation to find an optimal solution. The traffic-grooming problem in both single-ring network and multi-ring network is NP-complete [12]. The standard ILP solver, "CPLEX," does not provide a solution to the problem of formulation in a single seven-node bidirectional network with uniform traffic within a reasonable period of time. Hence, it requires heuristics to resolve this problem.

A greedy heuristic in [8] provides the solution. In the greedy approach, there are two steps. In the first step, all connections are allotted to sub-channels (or TDM time slots). The connections are allotted on the basis of their length where longer connections are assigned before shorter ones and shorter connections have been filled in the gaps left behind by longer connections. Normally, routing uses the shortest-path approach. In the second step, sub-channels are chosen, one at a time, and are packed on to high-speed wavelengths. For each wavelength, an unassigned sub-channel with the largest number of adds/drops is chosen as its first branch. This approach was improved by incorporating a simulated-annealing algorithm.

Inter-Ring Traffic using Interconnection Strategies

We consider interconnecting two rings – one ring having N nodes and another having M nodes, carrying traffic uniformly distributed in the network. In this network, the internal traffic on each ring is proportional to N^2 and M^2, respectively. The inter-ring traffic is approximately proportional to $2 \times N \times M$. If N and M are of the same order, total inter-ring traffic is ~ twice the intra-ring traffic on each ring [1].

In strategy 1 (with DXC at intersection node), each inter-ring connection has two parts mixed with intra-ring traffic on the local rings. The greedy heuristic needs to be re-optimized because all the inter-ring traffic (most of the time the dominant component of all traffic on a ring) has to pass through the intersection node. This kind of traffic is normally considered as the egress/ingress traffic. If the egress/ingress traffic alone is taken, then each sub-channel has two bidirectional connections since the intersection node has only two interfaces. Since no connection takes a route longer than half the ring, most SADMs in the network cannot be used to their maximum capacity. An algorithm for computing the solution on a unidirectional egress ring has been discussed in [12].

There is no traffic exchanged between the logical ring and local rings in strategies 2 and 3. This requires the decomposition of the unified traffic representation, so that all inter-ring traffic and part of the intra-ring traffic are transported through the logical ring. All the inter-ring traffic (usually the dominant component on the logical ring) has to be transmitted via the intersection nodes. Global shortest-path routing is likely to provide the path of traffic in most nodes using one side of their interface (closer to the intersection node) rather than using both sides. As a result, most SADMs are used for half of their capacity, and the other half is left idle [1].

Strategy 4 is used in an OXC with TDM-switching function to link the two rings, providing maximum flexibility in traffic grooming. If none of the TDM-switching ports is used, then this strategy is similar to strategy 2 or 3, depending on whether wavelength conversion is needed or not. If all the traffic is sent to TDM switches, then it is identical to strategy 1 [1].

In reality, there are several rings inter-connected for reducing the number of SADMs to carry all the traffic. If DXC-based switching is used at all intersection nodes, a long connection may travel multiple hops (rings) to reach its destination. This not only increases the multiplexing and demultiplexing cost at each intermediate intersection node, but also increases the switching complexity. An alternative approach is to use OXCs at intersection nodes in building one or multiple large rings as hyper-rings, which can cover all local rings. Hyper-rings act as a "highway" to transmit inter-ring traffic. Hence, a hyper-ring connects several local rings. This approach can be repeatedly used to build a hierarchical network to increase geographical coverage. In the first hop, traffic moves from its source to an intersection node, where it is digitally cross-connected to the hyper-ring. In the second hop, traffic traverses along the hyper-ring. In the last hop, traffic moves to the local ring on which the destination node resides. For multiple rings connected by the layered strategies, the traffic-grooming heuristic for each local ring remains the same as the one developed for strategy 1. The traffic on a hyper-ring has the same characteristics as those of a regular ring. This is easy to see when every local ring has only one intersection with the hyper-ring [1].

Traffic-Grooming Algorithms

In order to groom the egress/ingress traffic on the rings, interconnected with strategy 1, we require circular sub-channels, so that the number of wavelengths is reduced. Here, we use a single-point-connected network. Since there is only one intersection node in each ring, a sub-channel should transmit bidirectional traffic. If there is a pair of traffic from one node to the intersection node (or vice versa), they use a common bidirectional sub-channel-1. We assign connections for each node sequentially with the following procedure.

We route the egress/ingress traffic for each node one by one. For the first node, all outgoing traffic from it is transmitted alternately in two different directions, and at the same time all the incoming traffic is also routed alternately from two different directions. Since there may be a mismatch in the number of outgoing and incoming connections, there are some free spots left in the sub-channels which are used by the first node. For the second node, we begin with some new sub-channels to form full circles, and come back to check if there are free spots to accommodate some newly generated unbalanced (more outgoing or incoming) traffic. The procedure for node 2 must be repeated for all other nodes. After we finish the assignment of all the traffic into sub-channels, we use the greedy heuristic [21] to transmit them onto as few wavelengths as possible.

When strategy 1 is used to interconnect two rings, a local ring transport includes not only the egress/ingress type of traffic, but also its local traffic. After the use of the above heuristic to groom the inter-ring traffic, there are some free spots available in the used sub-channels, which are available to transmit local traffic. We have the following two approaches of grooming: The first approach classifies the traffic into two types (inter-ring and intra-ring), and the local traffic is assigned from an unused sub-channel, while the second approach inserts as much local traffic as possible onto the existing egress/ingress sub-channels before grooming the remaining traffic. The first approach is usually economical than the second because of the uniform distribution of traffic within the distorted local ring [1].

The logical rings are to be formed when strategies 2 and 3 are used. There are two points in logical double rings where the two bottleneck nodes are used at the two intersections. In these bottlenecks, all the inter-ring traffic is to be cross-connected at one intersection node restricting a sub-channel to transport two (bidirectional) inter-ring connections. The heuristic for strategy 2 is similar to that in strategy 1, where it forms a full circle on each sub-channel. When connections are assigned to sub-channels, we first take bidirectional connections and route each of the two directions along different half circles of a unidirectional sub-channel. Then, the greedy heuristic [21] is used for these circles and unidirectional connections. In Figure 1.8c, wavelength conversion reduces the number of SADMs in interconnected ring networks. For strategy 3, the hyper-ring is considered to be wavelength-continuous (with no wavelength-conversion capability at intersection node), and traffic-grooming algorithm designed for strategy 2 is applied. If there are any unused wavelength segments, adjacent to the common interconnection node, they can be connected by utilizing the wavelength-conversion capability of OXCs.

Results of Interconnected Double Rings

For conducting simulation experiments, two bidirectional rings are considered to be interconnected at two points. The first ring has seven nodes, and the second ring has eight nodes as shown in Figure 1.9. These are interconnected at two consecutive nodes 0 and 6. The experiment is performed under three grooming ratio (C) assumptions of 4, 16, and 64. The traffic pattern is designated by a random traffic matrix. Each element in the traffic matrix is a uniformly distributed random integer between 0 and 15. The traffic considered between nodes 0 and 6 is distributed uniformly between the two local rings. For strategies 1 and 4, inter-ring traffic (or part of it) is taken to be switched at one of the intersection nodes decided by the shortest-path routing algorithm. If there are two equal-cost paths, traffic is bifurcated. For strategies 2 and 3, the logical ring follows the node-number sequence [21].

When grooming ratio is low ($C = 4$), the optically switched strategies (2 and 3) yield low-cost solutions. In this low grooming ratio case, the TDM-level switching at the intersection nodes plays an active role in reducing the number of SADMs at the non-intersection nodes [1]. However, this benefit is not enough to compensate for the extra cost that has to be spent at the switching node. With increase of the grooming ratio ($C = 16$ and $C = 64$), the advantage of using TDM-level switching can be gradually seen. For any grooming ratio, the OXC-connected network always uses fewer wavelengths. For $C = 4$, strategies 2 and 3 provide ~11% savings on SADM

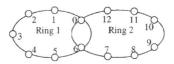

FIGURE 1.9 Interconnected double ring with two interconnection points [1].

cost and 9.5% savings on wavelength cost in comparison to strategy 1. But for $C = 16$, strategy 1 provides about 4% savings on SADM cost with the addition of 10% more wavelengths when compared with strategies 2 and 3 [1].

Strategy 4 gets more ADM savings to accommodate the traffic. There are several ways to decompose the inter-ring traffic for strategy 4 and find the best operating point. The cost of the logical ring is very low if the grooming ratio is small or large, provided all the inter-ring traffic is switched optically. For networks with medium grooming ratio, TDM-level switching is considered.

1.5 PACKET COMMUNICATION USING TUNABLE WAVELENGTH ADMs

ADMs are one of the key devices in fiber-based high-speed ring networks. We have already discussed ADMs using cascaded Mach Zehnder (MZ) devices in Chapter 3 of volume 1. In [23], the access protocols are mentioned for MAC in all-optical packet networks based on WDM multi-channel ring topologies where nodes are equipped with one fixed-wavelength receiver and one wavelength-tunable transmitter. Access delays and throughputs were taken as performance indices for a simulation-based comparison of the proposed protocols, in the case of a 16-node multi-ring network with either balanced or unbalanced traffic [23]. Simulation results showed that the throughput limitations and the fairness problems inherent in the network topology could be overcome with relatively simple protocols. In multi-wavelength ring networks, each node comprises one dynamic-wavelength ADM (DWADM) and one SADM. Rings provide slot synchronization at high data rates and also allow the efficient use of the available optical bandwidth for packet communications. The ADM is used for cost-effective traffic grooming in WDM ring networks. Thus, a ring network designed with a DWADM, ADM, and WDM provides good performance. Since all source nodes need to transmit the signal to a destination by sharing the channel, MAC protocols are used to transmit packets to adjudge the access to the shared channels [21].

DWADM

A DWADM (named as ROADM) adds and drops signals on a selected wavelength in a main fiber bus using a tunable transmitter and a tunable receiver. It operates over a range of speeds, with high selectivity and low loss. Figure 1.10 represents the operation of DWADM in ring topology and its difficulty in developing a protocol by comparing with the architecture proposed in [23]. In general WDM networks, each node having one or more tunable (or fixed) transmitters and receivers can independently accept the packets on some wavelengths while transmitting some other packets on some other wavelengths. If the wavelength received at some nodes is λ_1, the wavelength transmitted must be the same wavelength λ_1; and, furthermore, if the node has a packet to send to another node d, then node d should also use the wavelength λ_1 to receive it. Figure 1.10b shows an architecture with DWADM where a virtual loop is created, and such loops can change dynamically over time, as per new packet arrivals at the various nodes. Figure 1.10d represents an architecture with DWADM which is one of the best methods to enhance channel utilization and prohibit channel collision. For a ring network with 16 nodes, maximum eight wavelengths are

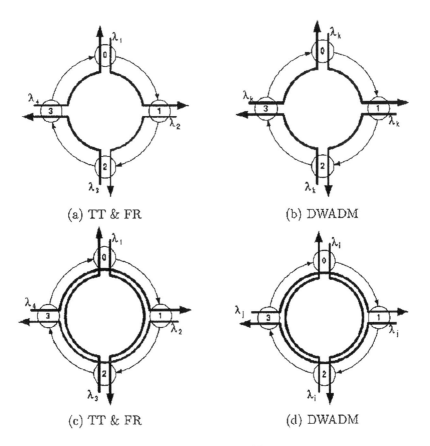

(a) TT & FR   (b) DWADM

(c) TT & FR   (d) DWADM

FIGURE 1.10 Operation of DADM in ring topology [1].

required for transmission. Naturally, by this limitation in wavelength usage, the performance in the ring network having DWADMs is limited in comparison to that in a WDM ring network without DWADMs [1].

1.5.1 Protocol

System Architecture

The challenge in an efficient protocol is its implementation. We consider the following for designing the same.

- There are $N \leq 16$ nodes and $W + 1$ wavelengths of which W wavelengths (λ_1 $\lambda_2 \ldots \lambda_w$) are for data channels and one wavelength (λ_0) is for a control channel. Each node is equipped with two ADMs – one static and one dynamic – as shown in Figure 1.12.
- The static (SONET) ADM is permanently tuned to a specific channel, called as control channel, with λ_0 at all nodes. Nodes are communicated over λ_0 in TDM fashion to exchange information on a node having the packets to

send to another node, and acknowledgments if the node uses the channel to transmit, as well as the wavelength channel number.

• The DWADM at each node is considered to obtain actual data transfer over the data channels ($\lambda_1, \lambda_2, \lambda_w$), and the information transfer has just a single hop over a data channel.

• A slot is the time duration to transmit a packet from node i to the next node $i + 1$. Each slot has the same time slot.

• A cycle is a sequence of N consecutive slots. The duration of a cycle is equal to or more than the propagation delay around the ring.

• All packets are of same fixed length, and the slot size is such that one packet exactly fits into one slot.

• The ring network is synchronized throughout.

• Nodes have one separate logical queue for every destination. Thus, $N-1$ packet queues are necessary at each node. Every queue handles packets in a first-in-first-out (FIFO) order. The queues are to be assigned a packet to be transmitted in a cyclic fashion.

• A transmitted packet has no priority in the next cycle and is queued in the destination queue, with newly arrived packets, for later selection by a packet-selection protocol.

Notations considered here are given below:

D_i is the destination node of a packet to be sent from node i.
C_i is the ith channel number for a packet to be transmitted from node i.
P_{ij} is a packet with destination node j generated at source node i.
p_{ij} is the probability that a packet originating at node i is directed to node j.

$$i \ominus j = (i - j) \bmod N.$$

Traffic Type

We consider that the traffic at node i originates with an exponential distribution having mean α_i. The traffic characteristic are different depending on the network operation, and the destination node is chosen with the same probability. For a client–server-based network in Figure 1.11, the server node is selected with a larger probability by client nodes. There are two types of traffic that are analyzed:

Type-I: Balanced traffic: All source nodes generate the same amount of traffic, and $N-1$ destinations are selected with equal probability randomly.
Type-II: Unbalanced traffic: A pre-fixed server node selects the client destination node out of $N-1$ client nodes with the same probability. The client node's destination can be the server node with a higher probability and the other nodes with a lower probability. This is unbalanced traffic.

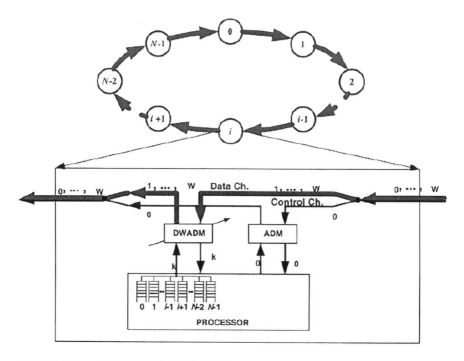

FIGURE 1.11   Architecture of a DWADM ring network with structural design of a node.

1.5.2 Algorithm of Virtual Path Creation and Assigning Wavelengths

At any given time, the channels to transmit a packet and to drop a packet at a node are taken to be of the same wavelength in a DWADM. If node i is scheduled to send a packet to node j via wavelength λ_1, in cycle t, and if node j is taken to be the destination node of node i's packet transmission, one should consider how and when node j can send its packets to node m to get the best performance.

Both balanced and unbalanced traffic can be considered for analysis. We consider that node j transmits the packets to node m using the wavelength λ_k in cycle t in which node i also transmits the packets to node j. Node j is easily able to send or receive a packet using the same wavelength λ_k simultaneously, like an ADM. This method results in efficient channel utilization, and in the other case, node j has to transmit its packets to node m using some other wavelength in the next cycle $t + 1$. Since we use a DWADM, after node i transmits a packet to node j via wavelength λ_k in cycle t, node j has to transmit the packet to node m using wavelength λ_k or some other wavelength in cycle $t + 1$. This method misuses the channel and may give a longer packet delay. Finally, if node j receives a packet from node i and node j also sends a packet to node i and has another packet to send to node i, these packet-transmission paths can be combined into a circle [1].

So for a traffic matrix, this method generates virtual circles (or paths) as much as possible by connecting as many paths as possible between source nodes and destination nodes, and then assigning the wavelengths to achieve low packet delay and high channel utilization. For a fixed node, the following operations have to be done: sending a control packet, gathering information about traffic (packet backlog) between nodes, executing the algorithm to create virtual circle paths and assigning wavelengths, and sending acknowledgments with the results.

1.5.3   PRIORITY SCHEMES

The selection of priority for a node is a vital problem dealt with by the following priority schemes [1,24–26].

Fixed priority at node i (FiPi):
   Priority is provided to a fixed node say node i. Since the computation makes virtual circle paths started at node i, node i has access chance, and node $i-1$ contains the worst access chance. If node i has a packet for node $i-1$ and node $i-2$ also has a packet for the same node $i-1$, since node i has the best access chance, node $i-2$ transmits a packet to $i-1$ in the same cycle

   Random priority (RanP):
      The priority node is found randomly among the N nodes.
   Round-robin priority (RRP):
      In this case, node i always has a priority, and priority is passed from a node to the next node continuously along a ring path in a round-robin manner, even if the node has no packet to send.
   Special-round-robin priority (SRRP):
      This is the same as round-robin priority except that the priority node is set only when it has a packet to send.

1.5.4   PACKET-SELECTION PROTOCOLS

Since each node has $N-1$ destination queues, we have to select a scheme for sending packets. Any one of the selection procedures mentioned below can be selected [1].

   FIFO(FIFOS):
      The packets are chosen according to the order in which they were placed in the queues. This will decrease the access delay.
   Random (RanS):
      The packets are to be transmitted in random manner among the non-empty queues. This protocol is easy to implement, but each node has a fairness problem when selecting the destination node.
   Round-robin scheme (RRS):
      The packet to be transmitted is chosen in a round-robin manner.
   Counter-clockwise-round-robin scheme (CRRS):
      This protocol considers scanning the packets in the destination queues in a counter-clockwise manner.
   Longest-waiting-queue-first scheme (LWQFS):
      The packets to be transmitted are selected on the basis of front the longest queue.

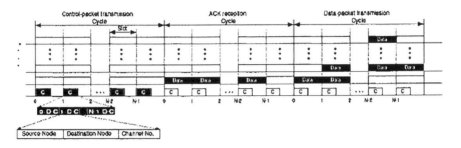

FIGURE 1.12 Packet format with its control.

Farthest-node-queue-first scheme (FNQFS):

   The packets to be transmitted are chosen from the destination queue of the farthest node first. This protocol provides a long virtual path in comparison to others.

Nearest-node-queue-first scheme (NNQFS):

   The packets to be transmitted are selected from the destination queue of the nearest node first. This protocol provides a better throughput by transmitting as many packets as possible by avoiding collisions.

Packet sent out Procedure

Figure 1.12 shows the control packet format and the packet-transmission procedure. The control packet consists of N fields, one for each node, and each field has three mini fields – source node address, destination node address, and channel number. The number of slots in a cycle is equal to the number of nodes. Three cycles are needed to transmit a packet: control-packet transmission cycle, acknowledgment reception cycle, and data-packet transmission cycle [1].

The server node is node 0, and the priority node is node i. The control packet at node 0 uses the prearranged destination node of node 0 by the packet-selection protocol and is sent to the next node 1. Each source node addresses the mini-field packed at the server node when the control packet leaves the server node. Node 1 receives the packet and sends the destination node information to the destination node address, if node 1 has a packet to send. Then, the control packet is transmitted to the next node, and this procedure continues until node $N-1$ gets its control information. Finally, server node 0 will get the control packet with information from all the nodes about their respective destination nodes. After receiving the control packet, by using the algorithm, server node 0 generates virtual circle paths and assigns wavelengths to each of them. Now, the priority node gets the packet-transmission chance in comparison to the other nodes. The algorithm for creating virtual paths and assigning wavelengths for priority assignment is given below.

// Priority-node-selection protocol
   1. $s = k$; or // Fixed-priority at node k.
   2. $s = i$, or // Round-robin priority.
   3. if there is a packet to transmit at node i, $s = i$;
   // Special round-robin priority.
      {

// Packet-selection protocol from the destination queues [1]

Select the packet P_s, D_s

a. first entered; or // FIFO

b. randomly, or // Random

c. in round-robin order; or // Round-robin

d. in counterclockwise round-robin; or // Counterclockwise round-robin

e. at the longest-waiting queue; or // Longest-queue-first

f. at the nearest-destination-node queue; or // Nearest-node-queue first

g. at the farthest-destination-node queue; // Farthest-node-queue first while

((if P_s, D_s exists) &&

(the path from D_s to D_{Ds} was not already assigned a wavelength))

{ Create a path from s to D_{Ds}, by adding the path from D_s to D_{Ds}; $D_s = D_{Ds}$)

Find an available channel and assign the path a channel;

if (there is no available channel to assign), Collision;

if (D_D was already assigned a wavelength), Collision;

$i = i \oplus 1$;

if

(every selected packet is considered

),

end;

}

The control information fields are included in the channel number mini-field in the control packet and are sent again to each node with the ACK reception cycle (Figure 1.13). If a node is unsuccessful, assign a wavelength; its corresponding field is left empty. The appropriate data packets will be transmitted in the next data-packet transmission cycle via the assigned wavelengths by each node simultaneously.

1.5.5 IMPLEMENTATION OF ALGORITHM

In a sample network (Figure 1.13), we consider $N = 5$, and node 0 is the server node having the priority. Using the packet-selection protocol, node 0 pick a packet to transmit to node 2, node 1 choose node 2, node 2 take node 3, node 3 select node 0, and node 4 choose node 1. Then, the server node, node 0, computes a source and destination node matrix. Since we suppose that the server node is node 0, the task of creating virtual circle paths and assigning wavelengths is done at node 0. As seen, a collision occurs between node 0 and node 1 because their packets have the same destination. Since node 0 is the priority node and has priority, node 1 cannot have a chance to transmit a packet in this cycle. After the virtual path from node 0 to node 2 is created, since the destination of node 2 is node 3, this path is added to the virtual circular path previously created from node 0 to node 2. So, the resulting virtual path is from node 0 to node 3. Since node 3 wants to transmit a packet to node 0, one long virtual circular path from node 0 to node 2, and to node 3, and to node 0 is made, and it is assigned to the wavelength λ_1. Since node 4 transmits a packet to node 1, and node 1 transmits or receives any packet in this cycle, node 4 can transmit a packet to node 1 using some other wavelength λ_2 if this system has more than two wavelengths.

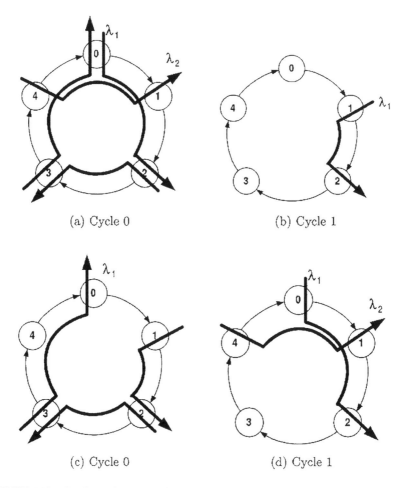

(a) Cycle 0   (b) Cycle 1

(c) Cycle 0   (d) Cycle 1

FIGURE 1.13   Configurations with different priority-node selections [1].

If node 1 has a priority in the next cycle or there are no packets to transmit from other nodes, node 1 can transmit the packet in the next cycle. This arrangement is illustrated in Figure 1.13a,b. In addition to this result, if we change the priority node 0 to node 1, the arrangement is represented in Figure 1.13c,d. After assigning the channels, node 0 transmits the control packet with the computed channel number.

1.6   ONLINE CONNECTION PROVISIONING USING ROADMs

OADMs are used for this provisioning. The (single-wavelength) fiber-optic SONET rings are relatively considered to upgrade them for operation over multiple wavelengths by employing these OADMs and appropriate terminal equipment at the network nodes; i.e., the same fiber cable plant can be reused. In such a W-wavelength ring network (where W becomes as high as 32 or 64), an OADM "drops" information from and "adds" information to the ring on a given wavelength in which it is established to operate. OADMs significantly decrease the network cost by permitting traffic to bypass intermediate nodes without expensive O-E-O conversion [27]. An OADM operated on a fixed wavelength is known as a fixed OADM (FOADM), and OADMs operated dynamically and tuned to different wavelengths in some control mechanism are known as ROADMs [28]. The ROADMs are used to add and drop traffic onto/from different wavelengths at different times and provide desirable flexibility and enable fast provisioning of dynamic traffic. We need tuning constraint for the tuning-based ROADM architecture on online connection provisioning, and several heuristics as well as other constraints are reported [29].

1.6.1 TUNING CONSTRAINT

There are two different ROADM architectures – An architecture shown in Figure 1.14a using switches to achieve add/drop functionality is called as switching-based architecture [1], and another ROADM architecture shown in Figure 1.14b using tunable devices such as fiber Bragg grating (FBG) to selectively add/drop a designated wavelength and called as tuning-based architecture. Since the tuning-based architecture is more inexpensive than the switching-based one, the tuning-based architecture is normally preferred. In the tuning-based architecture, a ROADM adds/drops different wavelengths by tuning the head to the corresponding wavelengths. If a ROADM is not adding/dropping, the tuning head can stay or park itself on any free wavelength or between any two adjacent wavelengths.

ROADMs at the source and destination nodes are adjusted to the wavelength which is allotted to set up a connection. If the tuning process takes place with the allotted wavelengths, it causes a service interruption in the traffic carried by those wavelengths which is unacceptable and should stay away from the connection set up.

In a network, we consider ROADM 1 and ROADM 2 tuned to be on wavelengths λ_1 and λ_2, respectively, and a connection is to be set up between these two ROADMs. In Figure 1.14b, there is traffic on wavelength λ_3 crossing these two ROADMs. If without crossing wavelength λ_3 the ROADM cannot be tuned to be on the same wavelength, the connection has to be blocked. This imposes another constraint on

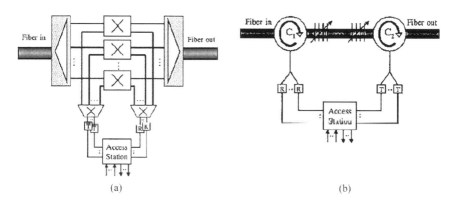

FIGURE 1.14 ROADM architectures: (a) switch based (b) FPG based [1].

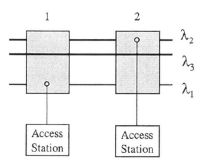

FIGURE 1.15   Tuning procedure: an example.

setting up a connection, and it is considered to be the tuning constraint. Since the tuning head of a ROADM is not allowed to cross working wavelengths, it can only tune within a certain tuning range. Figure 1.15 shows the tuning procedure with an example

1.6.2   PROBLEM STATEMENT

A network node has several ROADMs to add/drop multiple wavelengths simultaneously. Two or more ROADMs at a node are connected in sequence. The adding/dropping at one ROADM depends on the tuning range of the other ROADMs at the same node. While setting up a connection, it is required to select one of the ROADMs used at the source and destination nodes, as well as the wavelength and direction, if the ring is bidirectional. The assigned wavelengths are in the tuning range of both the ROADMs. This is called ROADMs assignment subproblem [1]. It needs to also fix the tuning ranges of the ROADMs and hence has a significant impact on the establishment of future connections. This is called the tuning-head positioning (TP) subproblem.

The online provisioning problem is formulated as follows.

1. Given: a WDM ring network using tuning-based node architecture with tuning constraint, and dynamic traffic, which arrives one by one.
2. Determine: direction, wavelength, and ROADM assignment (DWRA) for the connection and TP for the affected ROADMs.

Objective: minimize the overall connection blocking probability.

1.6.3   HEURISTICS

The tuning constraint is needed for setting up connections. A connection can be made only if the ROADMs at the source and destination nodes are assigned with the same wavelength, no matter how many ROADMs are there at both end nodes and how many wavelengths are available along the path [1]. In the DWRA problem of heuristics, the following parameters are to be considered.

Hop distance (bidirectional rings): Hop distance is the number of intermediate nodes If the hop distance is shorter, fewer wavelength-links are used.

Utilization of ROADM: A ROADM has two ports – add-port and drop-port – where a ROADM uses one port without using the other. However, the use of only one port does not fully utilize the capability of the ROADM. If the connection can be set up by using the unused port of a working ROADM, i.e., two connections share one ROADM, then a ROADM can be saved.

Number of selected ROADMs: Different solutions of the DWRA sub problem depend on different numbers of ROADMs along the path. The connection will classify the tuning range of the affected ROADMs, causing the tuning range to reduce in size. The flexibility of the ROADMs is decreased.

For prioritization of wavelength assignment, four heuristics for the DWRA sub-problem are reported:

Least-Hop-First-Fit (LHFF) chooses lowest hop count and first-fit wavelength assignment for tie-breaking.

Least-Affected-ROADMs-First-Fit (LAFF) chooses the least number of ROADMs and first-fit wavelength assignment for tie-breaking.

Least-Affected-ROADMs-Maximum-Sharing (LAMS) considers the least number of ROADMs. If there are multiple solutions, the one that can share more ROADMs with existing connections is chosen.

Least-Hop-Least-Affected-ROADMs (LHLA) considers the lowest hop count, and if there is a tie, the one that influences the least number of ROADMs is chosen.

There are three heuristics for the TP sub-problem.

Tune-to-Large (TL): When the tuning range is divided into two parts, this heuristic always keeps the tuning head in the part with a larger range.

Proportional-Tuning (PT): PT utilizes a probabilistic method to estimate the tuning with the probability that the tuning head is proportional to the size of the tuning range of that part.

Max-Cover-Tuning (MCT): In the two heuristics mentioned above, the tuning head of a ROADM is located independently. MCT seeks to place the tuning heads of the ROADMs at the same node in a more coordinated way. The tuning heads are such that the number of wavelengths tuned to, at least, one ROADM at the node after setting up the connection is maximum. If there are multiple solutions, it chooses the one with the maximum number of wavelengths that are tuned to by at least two ROADMs at the node. It repeats this procedure for tie-breaking till it finds the solution [1].

1.6.4   COMPARISON OF HEURISTICS SCHEMES USING NUMERICAL EXAMPLES

For comparisons, we consider a bidirectional WDM ring with eight nodes in which each node consists of four ROADMs for each direction and each unidirectional link having eight wavelengths. Traffic (consisting of unidirectional connection requests) is uniformly distributed among all the node pairs, and each connection needs the entire bandwidth of a wavelength. The traffic arrival is a Poisson process, and the connection-holding time obeys exponential distribution (whose average value is normalized to unity in our studies reported here). The 100,000 connection requests are considered to obtain the network performance using different heuristics [1].

Using the heuristic MCT for TP, heuristics LHLA and LAMS perform well in comparison to heuristics LHFF and LAFF, and among LHLA and LAMS, LHLA performs slightly better than LAMS. Reducing the number of wavelength links, the number of affected ROADMs is reduced. We then compare the performance of different heuristics for the TP subproblem [1]. TL performs poorly in comparison to PT and MCT. MCT provides lowest blocking probability performance among the three heuristics. This is due to the fact that MCT manages the tuning heads of the ROADMs at a node to get a larger tuning range for the node. The tuning constraint will significantly enhance the traffic blocking probability for the same network configuration.

SUMMARY

WDM optical ring provides many benefits in the metro arena, including scalable capacity, transparency, and survivability. Moreover, many techno-economic studies also show the cost-effectiveness of WDM for bit rates beyond OC-12, boosts further by falling component prices. As a result, WDM has grown as a metro-core solution and has various architectures. Meanwhile, metro-edge networks represent an integration of the optical and electronic domains, combined many user protocols onto large metro-core wavelength tributaries. Several metro-edge solutions have been developed, ranging from next-generation SONET/SDH and multi-service provisioning platforms to "IP-based" packet rings. The choice of edge solution will clearly depend on an operator's needs, but those including WDM technology are useful.

Section 1.2 gives an overview of research on traffic grooming in SONET/WDM ring networks, including node architecture, single-hop grooming, multi-hop grooming, dynamic grooming, and traffic grooming in interconnected rings.

Section 1.3 discusses formal mathematical specifications of the traffic-grooming problem in several ring networks, i.e., single-hop and multi-hop (with a single hub) cases of unidirectional and bidirectional rings. Then, a simulated-annealing-based traffic-grooming algorithm was used for the single-hop case and a greedy heuristic for the multi-hop case. The simulated-annealing-based heuristic provides a solution to the sub-optimal problem caused by the two-step strategy in the previous greedy heuristic. The multi-hop approach provides more ADM savings when the grooming ratio is neither too small nor too large, but it usually results in more wavelength usage due to the extended connection length.

Section 1.4 compares four interconnection strategies ranging from using traditional digital cross-connects to various optical cross-connects. Both mathematical formulations and heuristic approaches are mentioned for solving the traffic-grooming problem in interconnected rings. Full-circle-forming strategies to improve the performance of the interconnected-ring environment are also included. The TDM-switching capability will be useful in reducing the total network cost. The optically cross-connected strategies provide more SADM cost savings when the grooming ratio is not too large and also reduce the number of wavelengths and switch port count. So the OXCs with TDM-switching function provides the advantages of either side. A true mixed-switching strategy is difficult to implement, so a more practical solution is to either use pure optical switching or TDM-level switching depending on what benefits you are interested in.

Section 6.5 describes a new ring architecture and introduced protocols that use an SADM and a DWADM at each node which have the advantages of WDM and ADM. The use of DWADM in multi-wavelength ring networks improves dynamic packet transport. But the DWADM has a restriction that a node must transmit and receive on the same wavelength at the same time. If both the priority-node-selection protocol and the packet-selection protocol are selected properly, then shorter packet delays can be achieved.

DWADM/ROADM provides flexibility to the network, enables online connection provisioning in the network, and reduces the cost of the network. A tuning-based ROADM in the network tunes from one wavelength to another wavelength and is not allowed to cross a working wavelength because it will interfere with the traffic carried by the wavelength. The tuning constraint of tuning-based ROADMs plays an important role in the online connection provisioning environment. Section 6.6 has investigated the online provisioning problem which has two sub-problems – the problem in which DWRA is considered and the problem of TP.

EXERCISES

1.1. Consider the network in Figure exercise 1.1. The network is an unidirectional ring with two wavelengths (λ_1 and λ_2) and two time slots on each wavelength channel (C_1 and C_2). Assume that the traffic matrix $T = [t_{ij}]$ is 4×4.
   (a) Write the objective function for node 1.
   (b) Write the traffic-load constraints for the connection between node $L_{12}^{C_1\lambda_1}$ 1 and node 2 (assume $t_{12} = 3$).
   (c) Write the channel-capacity constraints for the physical link.
   (d) Write the transmitter and receiver constraints for node 1 and node 2. What is the grooming ratio?

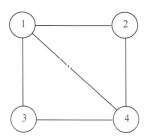

1.2. For interconnected double rings, what switching strategy should we consider for small, medium, and large traffic-grooming ratios?

1.3. The ring network in Figure exercise 1.2 contains uniform traffic request and grooming ratio 4. Consider that each request requires one unit of sub-channel capacity. Find a solution to put all the connection requests on two wavelengths such that the minimum number of ADMs are used. Validate the minimum number of ADMs used with the tabular result given by the ILP solver using the simulated-annealing-based traffic-grooming algorithm.

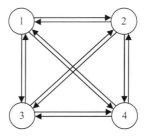

1.4. Consider the bidirectional single-point-connected double-ring network shown in Figure exercise 1.3. Given the following traffic matrix, show

$$\begin{pmatrix} 0 & 2 & 3 & 2 & 4 & 2 & 0 & 6 \\ 2 & 0 & 3 & 4 & 2 & 1 & 2 & 5 \\ 3 & 4 & 0 & 2 & 2 & 2 & 4 & 5 \\ 2 & 7 & 8 & 0 & 3 & 2 & 1 & 3 \\ 5 & 3 & 2 & 5 & 8 & 0 & 2 & 4 \\ 2 & 5 & 7 & 9 & 3 & 2 & 0 & 2 \\ 4 & 3 & 2 & 9 & 3 & 2 & 4 & 0 \end{pmatrix}$$

(a) the decomposed traffic matrix on each ring for strategy 1 (DXC)
(b) one reasonable decomposition of the traffic matrix on each ring for strategy 2 (OXC).

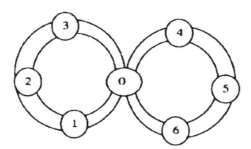

1.5. Discuss the performance of strategy 1 (DXC) compared to the strategies 2 and 3 (OXC) under uniform traffic using ADM and traffic grooming.

1.6.   Consider the following traffic matrix for a unidirectional ring.

$$
\begin{pmatrix}
0 & 23 & 24 & 20 \\
23 & 0 & 21 & 23 \\
24 & 22 & 0 & 24 \\
25 & 23 & 21 & 0
\end{pmatrix}
$$

Assume $c = 1$ and transmission in the clockwise direction only. For a virtual connection assignment, estimate the number of ADMs and the number of wavelengths required.

1.7.   Derive the mathematical problem formulation for traffic grooming in single-hub multi-hop networks, extend it to the bidirectional formulation for uniform traffic, and analyze its complexity in terms of big-0 notation.

1.8.   Derive the mathematical problem formulation for traffic grooming in a single-hop unidirectional ring, extend it to the bidirectional formulation for uniform traffic, and use an algorithm and analyze its time complexity.

1.9.   Derive the tuning constraint for online traffic grooming in a single-hop unidirectional ring and extend it to the bidirectional formulation.

REFERENCES

1. B. Mukherjee, *Optical WDM Networks*, Springer-Verlag, 2006.
2. T. Hills, "Next-gen SONET," Lightreading Report, May 2002.
3. G. Keiser, *Optical Fiber Communication*, McGraw Hill, 2002.
4. Fiber Channel Association, *Fiber Channel: Connection to the Future*, Austin TX: Fiber Channel Association, 1994.
5. A. Banerjee, G. Kramer, and B. Mukherjee, "Fair queuing using service level agreements (SLAs) for open access in an ethernet passive optical network (EPON)," *Proceedings, IEEE International Conference on Communications (ICC)'05*, Seoul, Korea, May 2005.
6. J. Wang and B. Mukherjee, "Interconnected WDM ring networks: Strategies for interconnection and traffic grooming," *SPIE Optical Networks Magazine*, vol. 3, no. 5, pp. 10–20, 2002.
7. N. Ghani, J.-Y. Pan, and X. Cheng, "Metropolitan Optical Networks," *Optical Fiber Telecommunications*, Vol. 4, Elsevier Press, March 2002, pp. 329–403.
8. X. Zhang and C. Qiao, "An effective and comprehensive approach for traffic grooming and wavelength assignment in SONET/WDM rings," *IEEE/ACM Transactions on Networking*, vol. 8, no. 5, pp. 608–617, 2000.
9. J. Wang, W. Cho, V. R. Vemuri, and B. Mukherjee, "Improved approaches for cost-effective traffic grooming in WDM ring networks: ILP formulations and single-hop and multihop connections," *IEEE/OSA Journal of Lightwave Technology*, vol. 19, pp. 1645–1653, November 2001.
10. ANSI T1X1.5, 2001_062, "Synchronous Optical Networks (SONET)," January 2001.
11. B. N. Desai, "An optical implementation of a packet-based (Ethernet) MAC in a WDM passive optical network overlay," *Proceedings, OFC'01*, Anaheim, CA, pp. WN5, March 2001.
12. A. L. Chiu and E. H. Modiano, "Traffic grooming algorithms for reducing electronic multiplexing costs in WDM ring networks," *IEEE/OSA J Lightwave Technology*, vol. 18, no. 1, pp. 2–12, 2000.

13. C. Lee and S. Chang, "Balancing loads on SONET rings with integer demand splitting," *Computer Ops Res,* vol. 24, no. 3, pp. 221–229, 1997.
14. K. Zhu, H. Zhu and B. Mukherjee, *Traffic Grooming in Optical WDM Mesh Networks,* Optical Network Series, Springer, 2005.
15. P. J. Wan, G. Calinescu, L. Liu, and O. Frieder, "Grooming of arbitrary traffic in SONET WDM BLSRs," *IEEE Journal on Selected Areas in Communications,* vol. 18, no. 10, pp. 1995–2003, 2000.
16. J. Simmons, E. Goldstein, and A. Saleh, "Quantifying the benefit of wavelength add-drop in WDM rings with independent and dependent traffic," *Proceedings, OFC'98,* San Jose, CA, pp. 361–362, February 1998.
17. J. Simmons, E. Goldstein, and A. Saleh, "Quantifying the benefit of wavelength add-drop in WDM rings with independent and dependent traffic," *IEEE/OSA Journal of Lightwave Technology,* vol. 17, no. 1, pp. 48–57, 1999.
18. O. Gerstel, R. Ramaswami, and G. H. Sasaki, "Cost-effective traffic grooming in WDM rings," *IEEE/ACM Transactions on Networking,* vol. 8, no. 5, pp. 618–630, 2000.
19. R. Berry and E. Modiano, "Reducing electronic multiplexing costs in SONET/WDM rings with dynamically changing traffic," *IEEE Journal on Selected Areas in Communications,* vol. 18, no. 10, pp. 1961–1971, 2000.
20. O. Gerstel, P. Lin, and G. Sasaki, "Wavelength assignment in a WDM ring to minimize cost of embedded SONET rings," *Proceedings, IEEE INFOCOM'98,* San Francisco, CA, pp. 94–101, March 1998.
21. A. Fumagalli, I. Cerutti, M. Tacca, F. Masetti, R. Jagannathan, and S. Alagar, "Survivable networks based on optimal routing and WDM Self-Healing Rings," in *Proceeding of IEEE INFOCOM '99,* New York, USA, March 21–25 1999.
22. W. Cho, J. Wang, and B. Mukherjee, "Improved approaches or cost-effective traffic grooming in WDM ring networks: Uniform-traffic case," *Photonic Network Communications,* vol. 3, no. 3, pp. 245–254, 2001.
23. M. A. Marsan, A. Bianco, E. Leonardi, M. Meo, and F. Neri, "MAC protocols and fairness control in WDM multirings with tunable transmitters and fixed receivers," *IEEE/OSA Journal of Lightwave Technology,* vol. 14, no. 6, pp. 1230–1244, 1996.
24. B. C. Chatterjee, N. Sarma, and P. P. Sahu, "Priority based dispersion-reduced wavelength assignment for optical networks", *IEEE/OSA Journal of Lightwave Technology,* vol. 31, no. 2, pp. 257–263, 2013.
25. B. C. Chatterjee, N. Sarma, and P. P. Sahu, "Priority based routing and wavelength assignment with traffic grooming for optical networks", *IEEE/OSA Journal of Optical Communication and Networking,* vol. 4, no. 6, pp. 480–489, 2012.
26. B. C. Chatterjee, N. Sarma, and P. P. Sahu, "A heuristic priority based wavelength assignment scheme for optical networks", *Optik – International Journal for Light and Electron Optics,* vol. 123, no. 17, pp. 1505–1510, 2012.
27. C. R. Giles and M. Spector, "The wavelength add/drop multiplexer for lightwave communication networks," *Bell Labs Technical Journal,* vol. 4, no. 1, pp. 207–229, 1999.
28. H. Zhu and B. Mukherjee, "Online connection provisioning in metro optical WDM networks using reconfigurable OADMs (ROADMs)," *IEEE/OSA Journal of Lightwave Technology,* vol. 23, pp. 2893–2901, 2005.
29. K. Zhu and B. Mukherjee, "Traffic grooming in an optical WDM mesh network," *IEEE Journal on Selected Areas in Communications,* vol. 20, no. 1, pp. 122–133, 2002.

2 Queuing System and Its Interconnection with Other Networks

In Chapter 1 of Volume-1 we already discussed the physical and data-link layers of the open system interconnection and TCP/IP model in which the theoretical basis is mentioned for successive and reliable transmission of information. The node-to-node transmission in the network depends on characteristics of various media through which this information can be sent. At the same time, it is also required to access the network efficiently. This can be made either in a deterministic fashion or by means of a contention method. In this direction, most analytical models of computer systems and communication networks need queuing theory which is based on waiting-line phenomena [1–3]. Queuing depends on service resources and customer populace. The customers arrive at a system, form a queue and wait for their turn to get the service, and then quit after serviced by the system.

The system has terminals, buffers, communication links, controllers, and computer-system elements such as protocol-handling devices, memories, processors and disks. The individual queues are represented in terms of the types of resources, the number of distinct resources, the queuing (scheduling) disciplines, and the probability distributions for the service times of jobs at the queues.

2.1 QUEUING MODELS

The performance of a network is a function of the interaction of the system workload and the resources [1]. A queuing model is shown in Figure 2.1 where a node is analyzed by a single-server queue. The node consists of a buffer where messages arrive and queue up and wait for service from a single processing element (a single server).

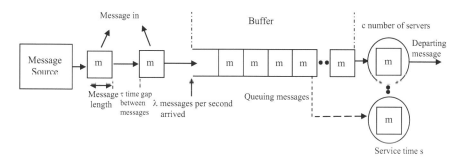

FIGURE 2.1 A simple queuing model having message a source, buffer and server.

The messages arrive at the buffer from a group of message sources connected to the node. In a finite-source system, the arrival rates depend on the number of message sources of the system where larger the number of sources, the higher the message arrival rate. In an infinite-source system, the arrival rate is influenced by the number of message sources in the system. The following assumptions are considered:

- The arrival rate of messages at the buffer is denoted as λ messages per second and the length of data is represented in terms of bits, bytes, characters, etc.
- All messages have the same length L.
- Operations at a node depend on the service rate and queuing. The service rate is expressed in terms of the number of jobs departing the node per unit service time. This service rate is defined as the node processes jobs are influenced by the length of the associated queue order/load of the system.

The queuing order is used to find the arrangement in which the queued-up jobs get the service. There are different ways to order the queue – a first-come-first-served (FCFS) discipline, first-come-last-service (FCLS) discipline, and priority service [4].

2.1.1   FCFS System

Here we consider that the messages are served on the basis of FCFS [4] at a rate of C data units per second, C = the capacity of the link. If a message arrives and there are n messages in front of it in the buffer, then the total time T taken to serve this message is the sum of time w required to wait in the queue and the message-processing time s.

$$T = \text{waiting time in queue}(w) + \text{service time}(s)$$

$$= \frac{L}{C} + \frac{nL}{C} = \frac{1+n}{\mu} \tag{2.1}$$

Where $\mu = C/L$ = the service rate in messages per seconds. A simple model of a single-server queue is shown in Figure 2.2. The total delay time T is a random variable

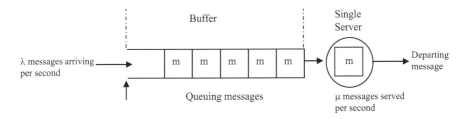

FIGURE 2.2   A simple queuing model having message a source, buffer, and single server.

Poisson Process

In queuing systems, both the message arrival rate and the service time are random. Since these two parameters are represented in statistical terms, a probability distribution relates them. The message arrival process is based on Poisson statistics which is widely followed by model telephone traffic, and it also follows with measurements on message statistics taken in actual computer in communication networks. Poisson statistics is a discrete distribution of events. For a system with a large number of independent customers, Poisson statistics gives the probability $P(n(t))$ of exactly n customers arriving during a time interval of length t, which is written as [4]

$$P(n(t)) = \frac{(\lambda t).e^{n-\lambda t}}{n!} \tag{2.2}$$

Where λ is the mean arrival rate and $n = 0, 1, 2, \ldots$

If $N = \{n = 0,1,2,\ldots\}$ is the set of the possible number of messages arriving at a node, where the probability of n arrivals is represented by equation (2.2), then the expected number of message arrivals in a time interval T is written as [5]

$$E(N) = \sum_{n=0}^{\infty} n.P(n(t)) = \lambda T \tag{2.3}$$

Where $E[x]$ indicates the expectation value of the quantity x and the expected number of arrivals in the time interval $(0, T)$ is λT. Using the Poisson arrival process, the messages enter the queuing system at times $t_0, t_1, t_2 \ldots t_n$ where $t_0 < t_1 < t_2 \ldots < t_n$. The random variables $\tau_n = t_n - t_{n-1}$ (where $n \geq 1$) represent the inter arrival times, which form a sequence of independent and identically distributed random variables. τ represents an arbitrary inter arrival time having a probability density function $n(t)$; the Poisson arrivals result in an exponential inter arrival probability density. In an infinitesimal time interval Δt, the probability $a(t)\Delta t$ represents the probability that the next arrival occurs after at least t seconds but not more than $(t + \Delta t)$ seconds from the time of the last arrival; i.e., the probability $p_o(t)$ of no arrivals for a time t multiplied by the probability $p_1(\Delta t)$ of one arrival in the infinitesimal time interval Δt is

$$a(t)\Delta t = p_o(t).p_1(t) \tag{2.4}$$

Where, from equation (2.2), we have $p_o(t) = e^{-\lambda t}$ and $p_1(\Delta t) = \lambda \Delta t.e^{-\lambda \Delta t}$. If $\Delta t \to 0$, the exponential term in $p_1(\Delta t)$ approaches unity and

$$a(t)dt = \lambda e^{-\lambda t} dt \tag{2.5}$$

In Figure 2.1, the time τ between arrivals is a continuously distributed exponential random variable indicating the Poisson arrival process as having exponential arrival times.

The service time required to complete the message transmission is directly related to how long the message is. These messages that are in a queue are served by a server in a fixed-capacity outgoing line that processes messages at a rate of data units per second. The service time of these messages also follows statistical mode as the messages vary in length in a random way. We consider the queuing-theory assumption that the messages are exponentially distributed in length with average length L. For servers having a capacity of C data units per second, a message of L data units long is transmitted or serviced in L/C seconds. The service-time distribution $S(t)$ is considered to be

$$S(t) = \frac{C}{L} e^{-ct/L} \tag{2.6}$$

Which has an expectation value (that is, average message-service or transmission time) of

$$E[S(t)] = \frac{L}{C} = \frac{1}{\mu} \tag{2.7}$$

The service-time distribution also obeys Poisson statistics.

2.1.2 REPRESENTATION OF QUEUE MODELS

Queuing systems are represented by the symbols A/B/c/K/N/Z [6], that represent the following:

- A – inter arrival time distribution
- B – service discipline
- c – number of servers
- K – system capacity (maximum queue length or buffer size)
- N – number of potential customers in the given source population
- Z – queuing discipline.

Since there is no restriction on the length of the waiting line (the queue), the number of sources is infinite, and the jobs are accepted on an FCFS basis, the shortened notation A/B/c is generally used. Some of the symbols traditionally used for A and B are [5]

- GI indicates general independent inter arrival times.
- G indicates general service time distribution.
- M indicates exponential (Markov) inter arrival or service time distribution.
- D indicates deterministic (constant) inter arrival or service time distribution.

An M/G/1 queue has Poisson arrivals and a general service distribution, and similarly in an M/D/1 queue, D indicates fixed or constant service time [5].

Figure 2.1 shows the simplest model in queuing theory which is an M/M/c queue, whereas Figure 2.2 shows a queuing model having a single server represented as M/M/1. In these models, messages arrive with a Poisson distribution, enter an infinite buffer with an exponential service time distribution, and are handled by a server on an FCFS basis. The M/M/1 queuing system has been widely used since the probability distributions describing the input process and the service process use a realistic approach of Poisson process to a queuing system since customer arrival patterns follow many real-world systems such as a Poisson probability distribution.

2.1.3 RANDOM VARIABLES AND PARAMETERS

Before analyzing different queues, we have to describe the basic relationships between the random variables indicating the number of messages in various parts of the system and the random variables describing time (e.g., queuing time, service time, etc.).

We consider the following random variables [5]:

$N(t)$ = the number of messages in the system at time t
$N_q(t)$ = the number of messages in the queue at time t
$N_s(t)$ = the number of messages being processed at time t.

The number of messages at any time t

$$N(t) = N_q(t) + N_s(t) \tag{2.8}$$

When a system is in operation, the above equation is difficult to estimate as it is in a transient state, and initially after the system has been turned on, the number of messages in the queue and the number of messages being serviced both depend on how long the system has been in operation and on initial conditions such as the number of messages waiting in the queue at time $t = 0$. However, once the system starts operating, the numbers of messages in the system and queue are not dependent on time. When the system is in the steady state, then equation (2.8) is written as

$$\tilde{N} = \tilde{N}_q + \tilde{N}_s \tag{2.9}$$

where although the steady state quantities $\tilde{N}$, $\tilde{N}_q$, and $\tilde{N}_s$ are independent of time, they are still random variables; that is, they are describable by probability distributions. Thus, we can write

$$E\left[\tilde{N}\right] = E\left[\tilde{N}_q\right] + E\left[\tilde{N}_s\right] \tag{2.10}$$

which we express as

$$N = N_q + N_s \tag{2.11}$$

where, in equilibrium, N = mean (average) total number of messages in the system (those waiting in line plus those being serviced). N_q = mean number of messages in the queue alone. N_s = mean number of messages being serviced. Similarly, the total waiting time can be modeled as follows.

If w represents the total time a message spends in the system (queuing time plus service time), q describes the waiting time in the queue, and s is the service time for a message, then clearly [5],

$$W = q + s \qquad (2.12)$$

Similarly

$$E[w] = E[q] + E[s] \qquad (2.13)$$

In the steady state, equation (2.13) becomes

$$T = T_q + T_s \qquad (2.14)$$

where T = average time a message needs to spend in the system (includes time spent in the queue plus service time), T_q = average time spent waiting in the queue, and T_s = mean (average) time required to process a message.

2.2   QUEUES

2.2.1   M/M/1 QUEUES

Here, we analyze a simple M/M/1 queuing system [5,7] which is represented by the state transition diagram shown in Figure 2.3. In equilibrium, the probability of finding the system in a given state n is independent of time. If a customer arrives, the state changes from n to $n + 1$, whereas when a customer has been served and leaves, the state changes from n to $n - 1$. In equilibrium, these two processes must take place at the same rate. This principle is known as detailed balancing which is used here to analyze the probability states of an M/M/1 queuing system.

Following are the assumptions in analyzing this system:

*   The transitions are only between adjacent states.
*   For a system in state n, there are n customers or jobs in the system consisting of queue plus server, and when a new customer arrives, the system moves to state $n + 1$.

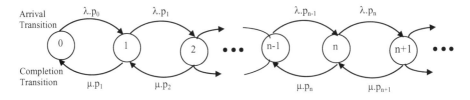

FIGURE 2.3   State transition diagram of M/M/1 queuing system following Markov's chain [5].

- When a customer has been served and departs, the system moves down one state to state $n-1$. Queuing systems in which the only transitions are to adjacent states are known as birth-death systems, the arrival of a customer is the birth, and the departure is the death.

In Figure 2.3, the circles indicate the state of the system, the top arrows show transitions from a lower state to a higher one, and the left-going arrows represent transitions from a high to a low state. For the mean arrival rate of λ customers per second, the mean number of transitions per second from state n to state $n+1$ is $\lambda.p_n$, where p_n is the equilibrium probability having exactly n customers in the system (queue plus server). If the server process μ customers per second, the transition rate from state $n+1$ to state n is $\mu.p_{n+1}$. In either case, the transition rate is the arrival rate or service rate times the probability of the system being in the initial state, not the final state.

For the M/M/1 queue, the probability p_n that n messages are present in the buffer at time t needs to be determined. The following assumptions are considered:

- For an equilibrium condition, the rate at which messages arrive must be equal to the rate at which the messages depart. So, in equilibrium, the message flow into a state n must be equal to the message flow out of state n.
- To analyze the M/M/1 queue, the arrival rate for any state n is equal to a constant λ, and the service rate for states $n \geq 1$ is equal to a constant μ.
- For the state $n = 0$, $\mu_0 = 0$. Since we are dealing with Poisson arrivals and service rates, a simple way to analyze the state of the buffer is to consider the time interval starting at t and ending at $t + \Delta t$, where Δt is an infinitesimal time increment ($\Delta t \rightarrow 0$).

During the time interval $(t, t + \Delta t)$, the following possible transitions could occur:

1. A message reaches state n from state $n-1$ with a flow rate λp_{n-1}.
2. A message reaches state n from state $n+1$ with a flow rate μp_{n+1}.
3. A message departs state n and enters state $n-1$ with a flow rate μp_n.
4. A message departs state n and enters state $n+1$ with a flow rate λp_n.

We have

$$\text{Flow into state } n = \lambda p_{n-1} + \mu p_{n+1}$$
$$\text{Flow out of state } n = \lambda p_n + \mu p_n$$

(2.15)

At equilibrium, for $n \geq 1$,

$$\lambda p_{n-1} + \mu p_{n+1} = (\lambda + \mu) p_n \tag{2.16}$$

When a message arrives while the queue is in state 0 (the empty state), the queue moves to state 1. The transition rate for this condition is λp_0. Alternatively, the departure of a message from state 1 moves the queue to state 0. This occurs at a flow rate of μp_1. In equilibrium, these two flows must be equal, i.e.,

$$\lambda p_0 = \mu p_1 \tag{2.17}$$

$$P_1 = \frac{\lambda}{\mu} P_0 = \rho p_0 \tag{2.18}$$

where $\rho = \lambda/\mu$ is known as the traffic intensity. The parameter ρ determines the minimum number of servers required to keep up with the arrival of messages. The unit associated with the traffic intensity ratio is called the Erlang after the Danish mathematician A. K. Erlang. If for an example of this ratio, in a queuing system, the average inter arrival rate $1/\lambda$ is 10 s and the average service rate $1/\mu$ is 5 s, then the traffic intensity ratio is 0.5 Erlangs. If the average service rate were 15 s, then the traffic intensity ratio would be greater than 1 (1.5 Erlangs) which indicates that messages arrive faster than they can be processed. In that case, two servers are needed to handle the rate of arriving messages.

A recursion expression can be represented by equation (2.18). First from equation (5.16) with $n = 1$, we have, using equation (2.18),

$$P_2 = (\rho + 1) p_1 - \rho p_0 = \rho^2 p_0 \tag{2.19}$$

Since this process is repeated recursively, the general expression for p_n is written as

$$P_n = \rho^n p \tag{2.20}$$

From the probability conservation, we can express the summation of all probabilities with the following equation:

$$\sum_{n=0}^{\infty} p_n = 1 \tag{2.21}$$

As $n \to \infty$, p_n must approach zero. Since the state probabilities must decrease with n, it is apparent from equation (2.21) that we must have $\rho = \lambda/\mu < 1$. It shows that the average number of arrivals per unit time λ must be less than the system capacity μ. There is a situation where the queue in the buffer would come indefinitely; otherwise a steady state cannot be achieved.

Equation (2.22) shows the formula for p_0:

$$\sum_{n=0}^{\infty} p_n = 1 = p_0 (1 + \rho + \rho^2 + \ldots + \rho^n + \ldots) = \frac{p_0}{1 - \rho}$$

$$p_0 = 1 - \rho \tag{2.22}$$

For $n \geq 0$, the probability of occupancy of the various states of the M/M/1 queue is written as

$$p_n = (1-\rho)\rho^n \qquad (2.23)$$

From equation (2.23), we derive that $\rho < 1$ and that, as the traffic intensity ρ increases, the higher states become relatively more probable.

Little's Theorem

There are two important parameters in queuing systems – average number of waiting customers/messages and the mean time that these messages spend in the system waiting for service. Four different quantities N, N_q, T and T_q (in 2.1.4) represent the system. A simple formula known as Little's theorem determines any of these quantities [4–6]. A simple heuristic derivation of Little's theorem is represented by the graph shown in Figure 2.4.

Here we consider that in a system, the messages are processed in the order of arrival on an FCFS basis in which $\alpha(\tau)$ denotes the number of message arrivals in the interval $[0, \tau]$ and $\delta(\tau)$ the number of messages departing in this interval. These parameters are represented as the top and bottom lines, respectively, of the shaded area in Figure 2.4 where a message i arrives at time t_{ia} and departs at $t_{id}(i = 1, 2, \ldots$ messages). If the system is vacant at time $\tau = 0$, the number of messages $N(t)$ in the system at time t is

$$N(t) = \alpha(t) - \delta(t)$$

The total time $\gamma(t)$ all messages have spent in the system at time t is cumulative in the interval $[0, t]$ and represented as

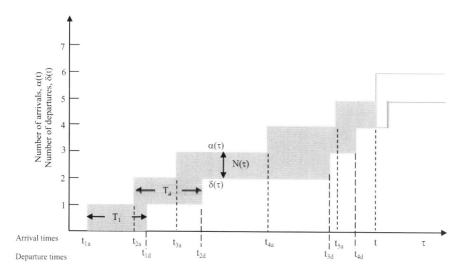

FIGURE 2.4   Graphical representation of heuristically Little's theorem [5].

$$\gamma(t) = \int_0^t N(\tau) d\tau \qquad (2.24)$$

which is also equal to

$$\sum_{i=1}^{\delta(t)} T_i + \sum_{i=\delta(t)+1}^{\alpha(t)} (t - t_i)$$

where T_i is the time message i spends in the system.

$$N_t = \lambda_t T_t \qquad (2.25)$$

The average number of messages in the interval [0, t] is written as

$$N_t = \frac{1}{t} \int_0^t N(\tau) d\tau$$

and the average time spent by a message in the system is equal to

$$T_t = \frac{1}{\alpha(t)} \left[\sum_{i=1}^{\delta(t)} T_i + \sum_{i=\delta(t)+1}^{\alpha(t)} (t - t_i) \right]$$

$$\lambda(t) = \frac{\alpha(t)}{t}$$

$$= \text{average message arrival rate}$$

Equation (2.25) is Little's theorem [5] derived for FCFS service in the finite interval [0, t]. In equilibrium, t keeps increasing (that is, $t \to \infty$) and N_t, λ_t, and T_t become finite and approach their equilibrium values N, λ and T, respectively.

So Little's theorem becomes

$$N = \lambda T \qquad (2.26)$$

which indicates that the mean number of customers in a queuing system is equal to the product of average arrival rate of customers to that system and mean time spent in the system.

Similarly we have the following relationship for a queuing system:

$$N_q = \lambda T_q \qquad (2.27)$$

Equation (2.27) formulates the average number of customers in the queue which is the average arrival rate of customers multiplied by the average amount of time that a customer spends waiting in the queue.

Total Queue Time of M/M/1 Queue

To find the formulas of T and T_q for an M/M/1 queue, equation (2.16) is used for finding all statistical interest for the queue. The mean number of messages N in the system is written as

$$N = E\left[\tilde{N}\right] = \sum_{n=0}^{\infty} np_n = \frac{\rho}{1-\rho} \qquad (2.28)$$

It is seen that for ρ less than 0.5, the average number of messages N is less than 1. For large values of ρ, the number of waiting messages increases rapidly. Using Little's theorem represented by equation (2.26), we then find the average waiting time in the system to be

$$T = \frac{N}{\lambda} = \frac{\rho}{\lambda(1-\rho)} = \frac{1}{\mu(1-\rho)} = \frac{1}{\mu}(1+N) \qquad (2.29)$$

The average time that a message spends in the system is a function of the traffic intensity, ρ. For $\rho = 0$, T is the average service time of a message; the message does not wait in the system and is serviced in $1/\mu$ seconds on the average. As the traffic intensity ρ tends to unity, both the average number of messages and the average time spent in the system increase abruptly. For the M/M/1 queue, an extreme penalty is paid if the system runs near its capacity. The average service time per message is given by $1/\mu$, and the average time T spent in the system is the sum of the average service time and the average queue time T_q:

$$T = \frac{1}{\mu} + T_q \qquad (2.30)$$

So queue time is formulated as

$$T_q = T - 1/\mu = \frac{\rho}{\mu(1-\rho)} = \frac{\lambda}{\mu - \lambda} \qquad (2.31)$$

Applying Little's formula as given by equation (2.27), the mean number of messages in the queue is

$$N_q = \lambda T_q = \frac{\rho^2}{1-\rho} \qquad (2.32)$$

We can also determine the probability of finding at least n messages in the system. We need to determine the buffer size needed for a specified probability of buffer overflow. The probability that the number of messages waiting in an M/M/1 queue is greater than N is given by

$$P(n > N) = \sum_{n=N+1}^{\infty} p_n = (1 - \rho) \sum_{n=N+1}^{\infty} \rho^n = \rho^{n+1} \qquad (2.33)$$

Since $\rho < 1$, the probability of the number of messages in the system exceeding some number N is a geometrically decreasing function of that number.

2.2.2 M/M/1/K QUEUES

In reality, there is no system having an infinite amount of buffer space. So we have a queuing system in which a maximum number of messages are reserved. The M/M/1/K queue is a model of a system which reserves a limit of k messages including the one being serviced [4,5]. Once the system is constrained for storing k messages, any messages arriving further will have no entry into the system. Figure 2.5 is a state-transition diagram for this model. Assuming that messages arrive and are processed at constant rates, the transition-rate coefficients are [5]

$$\lambda_n = \begin{cases} \lambda & \text{for } n = 0,1,2,\ldots,k-1 \\ 0 & \text{for } n \geq k \end{cases}$$

and

$$\mu_n = \begin{cases} \mu & \text{for } n = 0,1,2,\ldots,k \\ 0 & \text{for } n > k \end{cases} \qquad (2.34)$$

In this model, new messages also follow Poisson process, like the M/M/1 case, but only those that find the system with strictly less than k messages in it will be allowed to enter.

The steady state probabilities are given by

$$P_n = p_0 \prod_{i=0}^{n-1} \frac{\lambda}{\mu} = \rho^n p_0 \quad \text{for } n \leq k \qquad (2.35)$$

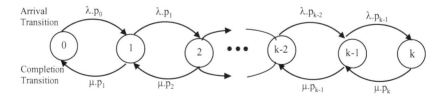

FIGURE 2.5 State transition diagram of M/M/1 queuing system following Markov's chain [5].

$$p_n = 0 \quad \text{for } n > k \tag{2.36}$$

To solve for p_0, we note that the probabilities of the finite set of states must sum to one. Thus we have

$$\sum_{n=0}^{k} p_n = 1 = p_0 \sum_{n=0}^{k} p^n = p_0 \frac{1 - p^{k+1}}{1 - p} \tag{2.37}$$

We can write

$$P_0 = \frac{1 - p}{1 + p^{k+1}} \tag{2.38}$$

Substituting equation (2.38) into equation (2.35) then yields

$$P_n = \begin{cases} \dfrac{(1-p)p^n}{1-p^{k+1}} & \text{for } 0 \le n \le k \\ 0 & \text{otherwise} \end{cases} \tag{2.39}$$

Since there can never be more than k messages in the system, a steady state can be achieved for all λ and μ values. For $\lambda < \mu$, the system is in equilibrium. If $\lambda = \mu$, then $p = 1$ and we can write

$$\lim_{p \to 1} P_n = \lim_{p \to 1} \frac{(1-p)p^n}{1-p^{k+1}} = \frac{1}{k+1} \tag{2.40}$$

For $0 \le n \le k$.

For $p^k \ll 1$, equation (2.39) reduces to the infinite-buffer result. The probability that the buffer is occupied completely and that messages are blocked from entering the system and the probability that there are k messages in the buffer is written as [4].

$$P_k = \left\{ \frac{(1-p)p^k}{1-p^{k+1}} \right\} \tag{2.41}$$

Using Little's formula, the parameters N, T, N_q, and T_q are formulated. If $\lambda < \mu$, then the expected number of messages N is written as

$$N = E\left[\tilde{N}\right] = \sum_{N=0}^{K} np_n$$

$$= \frac{p}{1-p} - \frac{(k+1)p^{K+1}}{1-p^{K+1}} \tag{2.42}$$

The average number of messages waiting in the queue N_q is given by equation (2.43)

$$N_q = N - N_s \tag{2.43}$$

where $N_s = E\left[\tilde{N}\right] = \left(1 - p_0\right) \times 1 = 1 - p_0 \tag{2.44}$

Since the average number of messages being serviced is merely the probability that the system is nonempty $(1 - p_0)$ times the average number of messages that are serviced under this condition (which is 1). Once there are k messages in an M/M/1/K queuing system [5], any additional messages are not permitted into the system until another existing message has been served. This occurs with a probability pk as given by equation (2.41). The probability that a message can enter the system is then $1 - p_k$, and the average rate λ_a of messages entering the system is given by

$$\lambda_a = \lambda\left(1 - p_k\right) \tag{2.45}$$

where λ is the message arrival rate. Using Little's formula, the average time for a message to pass through the system is written as

$$T = \frac{N}{\lambda_a} \tag{2.46}$$

and the average time a message waits in the queue is

$$T_q = \frac{N_q}{\lambda_a} = \frac{N - N_s}{\lambda(1 - p_k)} \tag{2.47}$$

2.2.3   M/M/m QUEUES

The M/M/1 queuing system was already mathematically modeled by a Poisson input process [7] and an exponentially distributed service process [7]. The M/M/m queuing model is similarly analyzed, but in this model, the number of servers is greater can reach a maximum of m as shown in Figure 2.6. We consider that all m servers

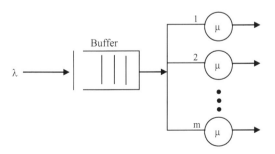

FIGURE 2.6   M/M/m queue system [5].

have the same destination. A network node using M/M/m has several outgoing transmission channels, any of which could be selected to send out a message to a neighboring node. We also consider

- a system having an infinite buffer
- the arrival rate λ_n for any state n is equal to a constant λ
- all m servers are taken to be identical so that each has same processing capacity C for the service rates

$$\mu_n = \text{minimum}\left[n\mu, m\mu\right]$$

$$= \begin{cases} n\mu & \text{for } 0 \le n \le m \\ m\mu & \text{for } m \le n \end{cases} \qquad (2.48)$$

The server utilization factor U for a multiple-server system is given by $U = \rho/m$ where m is the number of servers required to handle the incoming traffic rate. The factor U is also indicated as the expected fraction of busty servers considering each server has the same distribution of service time. U can be the expected fraction of the system capacity that is in use.

The state-transition architecture for the M/M/m queue is represented in Figure 2.7. To find state-transition probability p_n, we separate the solution into two parts since the transition rates when $n \ge m$ and $n \le m$ are different. We follow the same argument made in the derivation of equation (2.16) to get the detailed-balance equation (since we have again assumed an equilibrium condition):

$$(\lambda + n\mu)\, p_n = \lambda\, p_{n-1} + (n+1)\mu\, p_{n+1} \quad \text{for } n \ge 1 \qquad (2.49)$$

$$(\rho + n) p_n = \rho . p_{n-1} + (n+1) p_{n+1} \qquad (2.50)$$

Analogous to the derivation of equation (2.20), equation (2.50) is written recursively to obtain

$$p_n = \frac{\rho^n}{n!} p_0 \quad \text{for } n \le m \qquad (2.51)$$

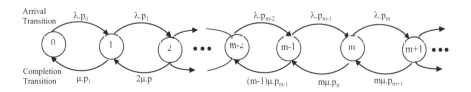

FIGURE 2.7 State transition diagram of M/M/m queuing system following Markov's chain [5].

Similarly, for $n \geq m$, we have the following detailed-balance equation in the equilibrium case:

$$(\lambda + m\mu) p_n = \lambda p_{n-1} + m\mu p_{n+1} \tag{2.52}$$

Equation (2.52) can be written in terms of ρ as

$$(\rho + m) p_n = \rho p_{n-1} + m p_{n-1} \quad \text{for} \quad n \geq m \tag{2.53}$$

Again, solving this recursively, we have

$$P_n = \frac{\rho^n p_0}{m! m^{n-m}} \quad \text{for} \quad n \geq m \tag{2.54}$$

To find the expression for p_0, we use the probability conservation relation given in equation (2.21), together with equations (2.51) and (2.54), which then gives

$$p_0 = \left[\sum_{n=0}^{m-1} \frac{\rho^n}{n!} + \sum_{n=m}^{\infty} \frac{\rho^n}{m! m^{n-m}} \right]^{-1}$$

$$= \left[\sum_{n=0}^{m-1} \frac{\rho^n}{n!} + \frac{pm}{m!} \frac{1}{\left(1 - \rho/m\right)} \right]^{-1} \tag{2.55}$$

Now we have to find four main measures of system performance N_q, T_q, $N1$, and T. The queue length is found by considering $n \geq m$ since a queue will be formed only if all m servers are busy. So we can write

$$N_q = E\left[\tilde{N}_q^{'} \right] = \sum_{n=m}^{\infty} (n - m) p_n$$

$$= \sum_{k=0}^{\infty} k p_{m+k}$$

$$= \sum_{k=0}^{\infty} k \frac{\rho^{k+m} p_0}{m! m^k}$$

$$= \frac{(\rho)\left(\rho/m\right)}{m!\left(1 - \rho/m\right)^2} p_0 \tag{2.57}$$

By using Little's formula [5], we find

$$T_q = \frac{N_q}{\lambda}$$

$$N = T\lambda$$

(2.58)

where $T = T_q + T_s = T_q + E[s] = T_q + \frac{1}{\mu}$ and mean message processing time $E[s] = 1/\mu$.

For the M/M/m queue, the probability that all m servers are busy is obtained, considering an arriving message joins the queue, using equations (2.54) and (2.55):

$$P[\text{queuing}] = \sum_{n=m}^{\infty} p_n$$

$$= \sum_{n=m}^{\infty} p_o \frac{\rho^n}{m! \, m^{n-m}}$$

$$= \frac{\dfrac{\rho^m}{m!} \dfrac{1}{(1 - \rho/m)}}{\displaystyle\sum_{n=0}^{m-1} \frac{\rho^n}{n!} + \frac{\rho m}{m!} \frac{1}{1 - \rho/m}}$$

(2.59)

This expression is known as Erlang's C formula or Erlang's delay formula used in telephony. It provides the probability that no trunk (i.e., server) is available for an arriving call in a system of m trunk lines.

2.2.4 M/M/∞ Queue System

Figure 2.8 shows the state transition diagram of M/M/m for the limiting case where $m = \infty$ (the number of servers is infinity). The recursive relation and probability from the figure are obtained as,

$$\lambda p_{n-1} = n\mu p_n, \quad n = 1, 2, \ldots$$

$$p_n = p_0 (\rho)^n \frac{1}{n!}, \quad n - 1, 2, \ldots$$

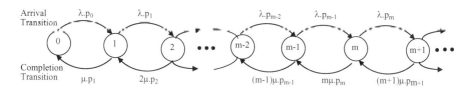

FIGURE 2.8 State transition diagram of M/M/∞ queuing system following Markov's chain [5].

$$\text{where} \sum_{n=1}^{\infty} p_n = 1 \text{ and } p_0 = \left[1 + \sum_{n=1}^{\infty} (\rho)^n \frac{1}{n!}\right]^{-1} = e^{-\rho}$$

$$\text{So we can write } p_n = (\rho)^n \frac{e^{-\rho}}{n!} \quad n = 1, 2, \ldots \qquad (2.60)$$

In steady state, the number of messages in the system follows Poisson distribution with parameter ρ. The average number of messages in the system is written as

$$N = \rho \qquad (2.61)$$

and using Little's theorem, the average delay is written as

$$T = 1/\mu$$

i.e., time delay is equal to average service time as there is no need for a message to wait in the queue because the system has an infinite number of servers.

2.2.5   M/M/M/M QUEUE SYSTEM

Figure 2.9 shows state transition structure of M/M/m/m model in the case of telephone system. If an arriving message finds that all m servers busy, it avoids coming to the system. The last m in M/M/m/m system indicates the maximum number of customers in the system is limited up to m[8]. In this model, arrivals are represented as connection requests for virtual circuit connections between two nodes, and the number of virtual circuits is m. The average service time $1/\mu$ is the average duration of virtual circuit conservation.

$$\lambda p_{n-1} = n\mu p_n, \quad n = 1, 2, \ldots m$$

$$p_n = p_0 (\rho)^n \frac{1}{n!}, \quad n = 1, 2, \ldots m$$

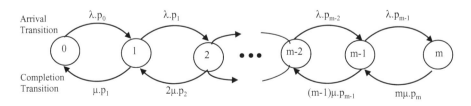

FIGURE 2.9   State transition diagram of M/M/m/m queuing system following Markov's chain [5].

$$\text{where } \sum_{n=1}^{m} p_n = 1 \text{ and } p_0 = \left[1 + \sum_{n=1}^{m} (\rho)^n \frac{1}{n!} \right]^{-1}$$

So, the probability that an arrival can find all m servers busy and become lost is written as [5]

$$P_m - \frac{\rho^m / m!}{\sum_{n=0}^{m} \rho^n / n!} \qquad (2.62)$$

In steady state, the number of messages in the system obey Poisson distribution with parameter ρ. The average number in queue is obtained as

$$N_q = P_m \frac{\rho}{1 - \rho} \qquad (2.63)$$

and using Little's theorem, the average queue delay is obtained as

$$T_q = \frac{N_q}{\lambda} = P_m \frac{\rho}{\lambda(1 - \rho)} \qquad (2.64)$$

The average total time delay is obtained as

$$T = T_q + 1/\mu$$

2.2.6 M/G/1 QUEUES

As discussed earlier, we consider that arriving messages have exponentially distributed lengths [5,9]. In many data transmission systems, the message-length distributions are more accurately represented by non-exponentially distributed lengths. Most messages in a response network or in a packet-switched network have a fixed length. The system having non-exponentially distributed messages can be handled analytically in an M/G/1-queuing representation having a general service-time distribution. The basic assumptions for the M/G/1 model are the same as those in the previous sections. We assume the following:

- The system is characterized by a Poisson arrival process with an average rate of λ arrivals per second. Only nearest-neighbor state transitions are allowed.
- The messages are processed on an FCFS basis.
- The service-time distribution has an arbitrary or general form $B(t)$ with average service time of $1/\mu$.

The formulations of M/G/1 system are made using the Pollaczek–Khinchin (P–K) mean-value formulas for the average number of messages in a queue and the average time delay. The customers are serviced in the order in which they arrive. The service time of ith arrival is X_i. In this case, the service times are distributed randomly and identically with X_1, X_2, which are mutually independent and interdependent of the inter arrival times:

$$\overline{X} = E(X) = 1/\mu = \text{average service time}$$

$$\overline{X^2} = E(X^2) = \text{second moment of service time.}$$

P–K Formula

It is based on the concept of average residual service time where the service time has two parts – time of sending the packets (i.e., serving customers) and time of sending control information or making reservations for sending the packets. The parameters are taken as [10]

W_i = waiting time of ith customer in queue, where $i = 1,2,\ldots$
R_i = residual service time of ith customer. If jth customer is already being served when ith customer arrives, R_i is the time required for customer j to be completely serviced. If no customer is in service (i.e., the system is empty when customer i arrives), then R_i is zero.
X_j = service time of jth customer.
N_i = number of customers found waiting in queue by ith customer upon arrival.

The W_i is obtained as

$$W_i = R_i + \sum_{j=i-N_i}^{i-1} X_j \tag{2.65}$$

From the result of equation (2.65), we obtain

$$E\{W_i\} = E\{R_i\} + E\left\{\sum_{j=i-N_i}^{i-1} E\{X_j|N_i\}\right\} = E\{R_i\} + \overline{X}E\{N_i\}$$

Considering the limits $i\to\infty$, we get

$$W = R + \frac{1}{\mu}N_q \tag{2.66}$$

where R = mean residual time = $\text{Limit}_{i\to\infty}E\{R_i\}$ and average waiting time W and average number of customers in queue N_q are obtained by taking the limits of customer index $i\to\infty$. These limits exist provided $\lambda < \mu$. Equation (2.46) is obtained by considering these limits. Using Little's theorem we obtain

$$N_q = \lambda W$$

Using equation (2.66), we obtain $W = R + \rho W$ and finally we get

$$W = \frac{R}{1-\rho} \tag{2.67}$$

The residual service time depends on time statistically and randomly. The residual service time $r(\tau)$ depends on time τ. The average residual service time over time interval $[0, t]$ is derived as

$$\frac{1}{t}\int_0^t r(\tau)d\tau = \frac{1}{t}\sum_{i=1}^{M(t)}\frac{1}{2}X_i^2 \tag{2.68}$$

where $M(t)$ is the number of service completions within $[0,t]$, and we can write equation (2.68) as

$$\frac{1}{t}\int_0^t r(\tau)d\tau = \frac{1}{2}\frac{M(t)}{t}\frac{\sum_{i=1}^{M(t)} X_i^2}{M(t)}$$

Over the time interval $[0, \infty]$, we obtain

$$\text{Limit}_{t\to\infty}\frac{1}{t}\int_0^t r(\tau)d\tau = \frac{1}{2}\text{Limit}_{t\to\infty}\frac{M(t)}{t}\text{Limit}_{t\to\infty}\frac{\sum_{i=1}^{M(t)} X_i^2}{M(t)} \tag{2.69}$$

The two limits on right side are the time averages of the departure rate and second moment of the service time, respectively, whereas the limit on left side represents average residual time. Taking time averages, we obtain

$$R = \frac{1}{2}\lambda\overline{X^2} \tag{2.70}$$

Using equations (2.67) and (2.70), the expected customer waiting time is obtained as

$$W = \frac{\lambda\overline{X^2}}{2(1-\rho)} \tag{2.71}$$

Equation (2.71) represents P–K formula. The above derivation shows that customers are serviced in the order of their arrival. This formula depends on the order of servicing customers till the order is estimated independently. The P–K formula is also suitable even if any service order is a sequence of reversals in queue position, i.e., the queue positions of customers is exchanged. This formula is not considered if the service order depends on service time.

The P–K formula provides an expression for different parameters in queue model. The total time delay consisting of queue time and service time is written as

$$T = \overline{X} + \frac{\lambda \overline{X^2}}{2(1-\rho)} \tag{2.72}$$

Using Little's formula for W and T [5], the expected number of customers in queue N_q and expected number in system are obtained as

$$N_q = \frac{\lambda^2 \overline{X^2}}{2(1-\rho)} \tag{2.73}$$

$$N = \rho + \frac{\lambda^2 \overline{X^2}}{2(1-\rho)} \tag{2.74}$$

For M/M/1 system, $\overline{X^2} = 2/\mu^2$, and expected customer waiting time is obtained as

$$W = \frac{\rho}{\mu(1-\rho)}$$

For M/D/1 system (D represents deterministic distribution), $\overline{X^2} = 1/\mu^2$, and expected customer waiting time is obtained as

$$W = \frac{\rho}{2\mu(1-\rho)}$$

The $\sigma^2 b$ is the variance of the service time, and considering a standard definition of the variance, the ratio of $\sigma^2 b$ to the mean service time is

$$C_b^2 = \frac{\sigma_b^2}{\mu^2}$$

The average number of messages in the system $E[\tilde{N}]$ is then given by the P–K mean-value formula:

$$N = E\left[\tilde{N}\right] = \rho + \rho^2 \frac{1 + C_b^2}{2(1-\rho)}$$

which is given in terms of three known quantities, namely, the traffic intensity factor ρ, the message length $1/\mu$ (or, alternatively, the average service time $1/\mu$), and the variance σ_b^2 for the service time.

Using equation (2.11), the average number of messages in the queue N_q and average number of messages being processed N_s are obtained (the average message service time $1/\mu$ multiplied by the arrival rate λ) as

$$N_q = N - N_s = N - \rho$$

$$= \rho^2 \frac{\left(1 - C_b^2\right)}{2(1 - \rho)}$$

Considering the message length to be fixed (or, equivalently, the service time be fixed), $\sigma_b^2 = 0$. Then we can write

$$N = E\left[\tilde{N}\right] = \rho + \frac{\rho^2}{2(1 - \rho)}$$

$$= \frac{\rho}{1 - \rho} - \frac{\rho^2}{2(1 - \rho)}$$

In this case, M/G/1 system has $\rho^2 / \left[2(1 - \rho)\right]$ which is fewer than average messages in M/M/1 system.

The system follows M/D/1 queue demonstrating that the average number of messages in the system increases in direct proportion to the variance of the service-time distribution. The average time delay of messages passing through the system is determined by using Little's formula given by equation (2.26):

$$T = \frac{1}{\lambda} E\left[\tilde{N}\right] = \frac{1}{2\mu(1 - \rho)}\left[2 - \rho\left(1 - \mu^2\sigma^2\right)\right] \qquad (2.75)$$

Queue time delay $= T_q = T - 1/\mu$

For exponential message lengths, $\sigma^2\mu^2 = 1$. Total time delay becomes $T = 1/[\mu(1 - \rho)]$ for an M/M/1 queue. Similarly, for fixed message lengths (fixed service times), the time delay is less than that of an M/M/1 queue.

2.2.7 M/G/1 Queues with Vacations

Here, at the last part of each busy period, the server goes on vacation for a random interval of time. A new arrival to an idle system waits for the end of vacation period to get serviced. If the system is idle at the end of the vacation, a new vacation may begin [9]. During vacations, there are basically transmissions of various kinds of control and record keeping packets when there is lull in data traffic. We consider V_1, V_2, ... as vacation times for the server as shown in Figure 2.10. The assumptions are as follows:

- The vacations are not dependent and distributed randomly.
- The vacation times are not dependent of the customer inter arrival times and service times.

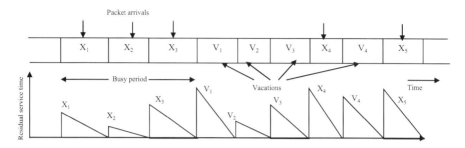

FIGURE 2.10 M/G/1 system with vacations: at the end of busy period, the server goes on vacations and residual time distribution during busy periods and vacations [5].

- The arrivals are based on Poisson distribution, and the service times follow general distribution.
- A new arrival to the system waits in the queue till the completion of the current service or vacation and then till all the customers waiting before it is serviced.

The analysis of the new system is as same as that of P–K formula except for the following: vacations are included in the residual service times $r(\tau)$, $M(t)$ is the number of services completed at time t, and $L(t)$ is the number of vacations completed by time t. The following equation is obtained:

$$\frac{1}{t}\int_0^t r(\tau)d\tau = \frac{1}{t}\sum_{i=1}^{M(t)}\frac{1}{2}X_i^2 + \frac{1}{t}\sum_{i=1}^{L(t)}\frac{1}{2}V_i^2 \qquad (2.76)$$

For a steady state, $M(t)/t$ tends to λ as t increases. So, the first term of equation (2.76) tends to $\lambda\overline{X^2}/2$. As $t\to\infty$, the time spent for serving customers tends to ρ, and time occupied with vacations is $1-\rho$. Time averages of these parameters are written for $t\to\infty$:

$$\text{Limit}_{t\to\infty}\frac{1}{t}\int_0^t r(\tau)d\tau = \frac{1}{2}\text{Limit}_{t\to\infty}\frac{M(t)}{t}\text{Limit}_{t\to\infty}\sum_{i=1}^{M(t)}\frac{1}{2}X_i^2/M(t)$$

$$+\text{Limit}_{t\to\infty}\frac{L(t)}{t}\text{Limit}_{t\to\infty}\sum_{i=1}^{0L(t)}\frac{1}{2}V_i^2/L(t) \qquad (2.77)$$

The second term tends to $(1-\rho)\overline{V^2}/(2\overline{V})$, whereas the first term approaches to $\frac{1}{2}\lambda\overline{X^2}$ (where $\overline{V}$ and $\overline{V^2}$ are the first and second moment vacation intervals). The average residual time R is obtained as

$$R = \frac{1}{2}\lambda\overline{X^2} + (1-\rho)\overline{V^2}/(2\overline{V}) \qquad (2.78)$$

The average queue time is obtained from the above equation as

$$W = \frac{\lambda \overline{X^2}}{2(1-\rho)} + \frac{\overline{V^2}}{2\overline{V}} \qquad (2.79)$$

The lengths of vacation intervals are not independent of already completed service times and arrival times. Since $\overline{V}$ and $\overline{v^2}$ are functions of the underlying M/G/1 process, it is not easy to find these quantities.

Priority Queuing

Priority queuing model is based on M/G/1 system having customers who have arrived at a network and having n different priority classes [9]. Classs-1 has highest priority, classs-2 has second-highest priority, and class-n has nth highest priority. The first and second moments of service time of class k are indicated by λ_k, $\overline{X_k} = 1/\mu_k$ and $\overline{X_k^2}$, respectively (where $k = 1, 2,...,n$). The following assumptions are considered:

- Arrival processes of all classes are not dependent on.
- Arrivals of all classes obey Poisson distribution.
- Arrival processes of all classes are not dependent on service times.

The two types of priority are non-preemptive priority and preemptive priority.

Non-Preemptive Priority

In non-preemptive priority, the service to a customer is permitted to be completed without interruption even if a new customer of higher priority arrives in the network. When the server becomes free, the first customer has the highest non-preemptive priority class in the queue waiting for service. This approach follows this rule.

Here, the equations for the average delay of each priority class are developed in which P–K formula is used [9]. The following are considered:

N_q^k = average number in queue for priority k
W_k = average queue time for priority k
$\rho_k = \lambda_k/\mu_k$ = system utilization for priority k
R = average residual service time for priority k.

The overall system utilization is written as

$$\rho = \rho_1 + \rho_2 + \cdots + \rho_n \qquad (2.80)$$

ρ is assumed to be less than 1($\rho < 1$). There will be some priorities that average time delay of a class less than k will be infinite and that of a class more than k will be finite. Using P–K formula and Little's theorem, the highest priority class is represented as

$$W_1 = R + \frac{1}{\mu_1} N_q^1$$

$$(2.82)$$

$$N_q^1 = \lambda_1 . W_1$$

Using the above equation, we can write

$$W_1 = \frac{R}{1 - \rho_1}$$

For the second-highest priority, the average queue delay W_2 also has additional queue delay due to customers having higher priority arriving in the network and waiting in queue. So we can write

$$W_2 = R + \frac{1}{\mu_1} N_q^1 + \frac{1}{\mu_2} N_q^2 + \frac{1}{\mu_1} \lambda_1 W_2$$

$$(2.83)$$

Using Little's theorem, we get

$$W_2 = R + \rho_1 W_1 + \rho_2 W_2 + \rho_1 W_2$$

So, we can write

$$W_2 = \frac{R + \rho_1 W_1}{1 - \rho_1 - \rho_2}$$

Substituting W_1, we can write

$$W_2 = \frac{R}{(1 - \rho_1)(1 - \rho_1 - \rho_2)}$$

The derivation is the same for other classes $k > 1$. We can get W_k as follows:

$$W_k = \frac{R}{(1 - \rho_1 - \cdots - \rho_{k-1})(1 - \rho_1 - \rho_2 - \cdots - \rho_k)}$$

$$(2.84)$$

The total time delay of a customer of priority class k is written as

$$T_k = \frac{1}{\mu_k} + W_k$$

$$(2.85)$$

The average time delay is also affected by the introduction of priority in a customer's service. The average time delay is reduced if a customer having short service time is given higher priority and a customer having longer service time is given lower priority.

Preemptive Priority

In the case of non-preemptive priority, the average delay of a priority class is a function of the arrival rate of lower-priority classes, and a high-priority customer should wait for a low-priority customer to be serviced. In the case of preemptive priority, a customer of lower priority already in service is interrupted if a customer of higher priority arrives and is serviced immediately after the cancellation of service to the lower-priority customer. Its average queue time delay follows P–K formula, but in the presence of k customers, the $k + 1$ customer's arrival may not affect this calculation. Under this condition, W_k is obtained as follows:

$$W_k = \frac{R_k}{(1 - \rho_1 - \rho_2 - \ldots - \rho_k)} \tag{2.86}$$

where $R_k = \dfrac{1}{2} \displaystyle\sum_{i=1}^{k} \lambda_i X_i^2$. The unfinished services of customers in queue considering sum of remaining service times of all customers in the system are independent of the priority discipline of the system. In all queuing systems, we should consider that the servers are busy when the system is not empty and customers depart the system after getting their required service. The total time delay of a customer of priority class k is obtained as

$$T_k = \frac{1}{\mu_k} + W_k + \left(\sum_{i=1}^{k-1} \rho_i \right) T_k \tag{2.87}$$

where the third term in the equation (2.87) provides the average waiting time for customers of priorities 1 to $k-1$ who arrive while customers of class k are in the system.

$$\sum \frac{1}{\mu_i} \lambda_i T_k = \left(\sum_{i=1}^{k-1} \rho_i \right) T_k$$

$$\text{For } k = 1,\ T_1 = \frac{(1/\mu_1)(1 - \rho_1) + R_1}{1 - \rho_1} \tag{2.88}$$

$$\text{For } k > 1,\ T_k = \frac{(1/\mu_k)(1 - \rho_1 - \cdots - \rho_k) + R_k}{(1 - \rho_1 - \cdots - \rho_{k-1})(1 - \rho_1 - \cdots - \rho_k)} \tag{2.89}$$

The average time delay is also affected by introducing priority in customer service.

2.3   NETWORKS OF QUEUES

There are many storage devices at a node where traffic information can be stored with queues [9]. These queues can interact with the messages departing from a node permitting one or more other queues and correspondingly simultaneously accepting

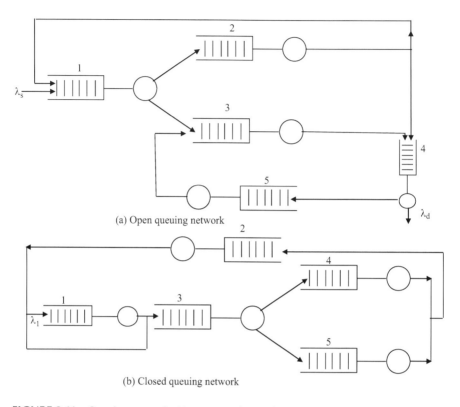

FIGURE 2.11   Queuing network: (a) open queuing and (b) closed queuing [5].

traffic from other queues. Two generic classes of queuing networks are the "open" and "closed" queuing systems shown in Figure 2.11. Open systems are characterized by the fact that the number of messages in the system is not fixed; i.e., traffic can enter and depart the system. In contrast, closed systems have a fixed number of the messages in the network, and arrivals or departures are not allowed. The network should be relieved of queuing systems which are widely used for modeling multi-access networks [4].

For determination of the analytical model, some assumptions and restrictions are made. The difficulty in establishing a model is correlating message inter arrival times with message lengths, when it serves the first queue at its entry point in the network.

- In a virtual-circuit system, a message follows a path consisting of a sequence of links through the network.
- For two transmission links, the message lengths follow exponential distribution and are independent of each other.
- The message lengths are independent of the inter arrival times at the first queue.
- The length of a message is fixed after it is chosen from a particular exponential distribution.

This assumption means that the next node cannot be considered as an M/M/1 queue since the service times are not independent [9]. For transmission, several messages are in a row at the first queue, and at the second queue between two of these messages, the first message's inter arrival time is equal to the transmission time of the second message. So at the second queue, there is a strong correlation between message length and inter arrival time, since long messages have shorter waiting times than short messages; i.e., it takes more time to transmit long messages at the first queue, providing the second queue more time to empty out.

To avoid these difficulties, a simple model is considered in which each stage of the network is represented as a separate M/M/1 queue. This model follows Kleinrock's independence assumptions:

- If a new choice of message length is made independently at each queue (that is, at each outgoing link), then separate M/M/1 queuing models can be used for each communication link between nodes. So, for virtual-circuit systems having Poisson message arrivals at the entry points, message lengths are exponentially distributed.
- For a densely connected network, it moderate to heavy traffic loads. It also holds if messages join a queue from several different input lines so that there is sufficient mixing of message types at each node.

The modeling of a sequence of links independent M/M/1 queues provides Jackson's theorem which is not invalid for a network with FCFS approach. The M/M/1 approximation is also not valid for all networks where message packets follow different routes between a pair of source and destination nodes. Each arriving packet is allotted to the queue having the least number of bits. In this case, the network needs to be modeled as a queue system having more than one server. So M/M/1 approximation is not accurate.

Jackson's Theorem

To derive Jackson's theorem [9], an open network of M queues is considered. A typical queue system (queue i) is shown in Figure 2.12. Here the parameters q_{ij} are the routing or branching probabilities that a message completing service at queue i is

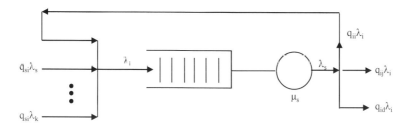

FIGURE 2.12 A queue in an open queuing network [5].

routed to queue j. The subscripts d and s indicate the destination and source, respectively. The arrival rates λ_i obey Poisson distribution, and the queue service rate is μ_i. The vector n represents the global state $(n_1, n_2, \ldots, n_M)$ where n_i = the state of queue i.

Every queue requires continuity of flow at the input and the output. We consider λ_i to be the total arrival rate at queue i from the source and from all the queues. We can write

$$\lambda_i = q_{si}\lambda_s + \sum_{k=1}^{M} q_{ki}\lambda_k \tag{2.90}$$

Jackson's theorem states that the equilibrium probability $p(n)$ that the network is in state n is represented as

$$p(n) = p(n_1, n_2, \ldots, n_m)$$
$$= p_1(n_1)p_2(n_2)\ldots, p_m(n_m) \tag{2.91}$$

where $p_i(n_i)$ is the equilibrium probability that queue i is in state n_i. In this case, a global balance equation is satisfied by a product-from solution as given in equation (2.91). To establish a global balance equation for a state n, it is required to have the total rate of departure from state n equal to the total rate of arrival into n. So we can write

$$\left[\lambda_s + \sum_{i=1}^{M} \mu_i \right] p(n) = \lambda_s \sum_{i=1}^{M} q_{si}p(n_1, n_2, \ldots, n_i - 1, \ldots, n_M)$$

$$+ \sum_{i=1}^{M} q_i \mu_i p(n_1, n_2, \ldots, n_i + 1, \ldots, n_M)$$

$$+ \sum_{i=1}^{M} \sum_{j=1}^{M} q_{ij} \mu_j p(n_1, n_2, \ldots, n_i - 1, \ldots, n_j + 1, \ldots, n_M) \tag{2.92}$$

In the equation, the left-side term represents the total departure rate from state n, since there is an arrival with rate λ or a departure at a rate μ_i from any one of the M queues. The total arrival rate into state n represented by the right-hand side of equation (2.92) has three terms as given below [9,10]:

- The first term shows arrivals from the external source at queue i which is in state $n_i - 1$. The arrival rates are $\lambda_s q_{si}$.
- The second term shows departures at queue i directly to the destination d. The departure rates from queue i are $q_{id}\mu_i$.
- The third term describes transitions from queue j, which is in state $nj + 1$, to queue i, which is in state $n_i - 1$.

The equation (2.92) reduces to the following by using equation (2.91) to eliminate the parameter qsi:

$$\lambda_i p(n_1, n_2, \ldots, n_i - 1, \ldots, n_m) = \mu_i p(n_1, n_2, \ldots, n_i, \ldots, n_m) \tag{2.93}$$

and its variations

$$\lambda_j p(n_1, \ldots, n_i - 1, \ldots, n_m) = \mu_j p(n_1, \ldots, n_i - 1, \ldots, n_j + 1, \ldots, n_m) \tag{2.94}$$

$$\lambda_i p(n_1, \ldots, n_i, \ldots, n_m) = \mu_i p(n_1, \ldots, n_i + 1, \ldots, n_m) \tag{2.95}$$

Equation (2.93) is based on the concept of reversibility introduced by Reich. In equilibrium, it represents the transitions from one state to another in reverse time occurring with rates same as those of the same transitions in forward time. Thus, the arrival stream is equal in all respects to the departure stream in reverse time.

With these substitutions, equation (2.92) then reduces to the expression

$$\lambda_s = \sum_{i=1}^{M} q_{id} \lambda_i \tag{2.96}$$

which is the condition for flow conservation from source to destination. Thus, the remaining condition in equation (2.92) is satisfied by equation (2.93).

We derive the following expression for the joint probability $p(n)$ from equation (2.93) for ith queue:

$$p(n) = \left(\frac{\lambda_i}{\mu_i} \right) p(n_1, n_2, \ldots, n_i - 1, \ldots, n_m) \tag{2.97}$$

Repeating this process n_i times, we obtain

$$p(n) = \left(\frac{\lambda_i}{\mu_i} \right)^{n_i} p(n_1, n_2, \ldots, n_i - 1, \ldots, n_m) \tag{2.98}$$

Following the same steps for all the other queues, we get

$$p(n) = \prod_{i=1}^{M} \left(\frac{\lambda_i}{\mu_i} \right)^{n_i} p(0) \tag{2.99}$$

where $p(0)$ is the probability that all M queues are vacant. To find $p(0)$, we take $\rho_i = \lambda_i \mu_i$. Considering all possible states, the total probability is equal to 1:

$$\sum_n p(n) = p(0) \left[\sum_n \left(\prod \rho^n i \right) \right] = 1 \tag{2.100}$$

For solutions to exist, the term in brackets must be finite. Thus, the sums and products are interchanged to get

$$\sum_{n} \left(\prod_{i=1}^{M} \rho^{n} i \right) = \prod_{i=1}^{M} \sum_{i}^{\infty} \rho^{n} i = \prod_{i=1}^{M} (1 - \rho i)^{-1} \qquad (2.101)$$

Thus taking equations (2.99), (2.100), and (2.101) into consideration, $p(n)$ is written as

$$p(n) = \prod_{i=1}^{M} (1 - \rho_i) \rho^{n_i} i = \prod_{i=1}^{M} p_i (n_i) \qquad (2.102)$$

which is Jackson's theorem.

Jackson's theorem in equation (2.91) is also extended to the networks where each queue i has m_i servers rather than a single server. In that case, the formula applied to $p_i(n_i)$ is used for the M/M/m_i case.

Application to an Open Queuing Network

The mean network-wide time delay averaged over all M links is considered in an open queuing network. The open queuing model in equilibrium represented in Figure 2.13 consists of a collection of M queues. If γ is the arrival rate in messages per second connected with path s, then the total arrival rate in the network is $\gamma = \Sigma \gamma_s$. From Little's formula, the network time delay T averaged over the entire network is

$$\gamma T = E[n] \qquad (2.103)$$

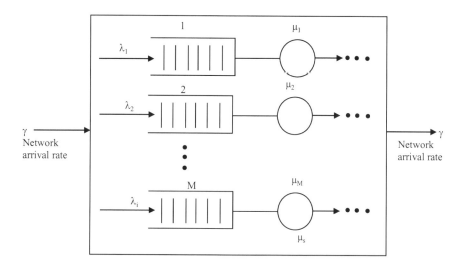

FIGURE 2.13   Average time delay in open queuing model [5].

Here $E[n]$ is written as

$$E[n] = \sum_{i=1}^{M} E[n_i] \tag{2.104}$$

where $E[n_i]$ is the average number of messages in queue or in service at node i. For an M/M/1 queue [5], we have

$$E[n_i] = \lambda_i T_i$$

$$= \frac{\lambda_i}{\mu_i - \lambda_i} \tag{2.105}$$

Thus the average time delay is

$$T = \frac{1}{\gamma} \sum_{i=1}^{M} \frac{\lambda_i}{\mu_i - \lambda_i} \tag{2.106}$$

Equation (2.106) neglects propagation delay over the links. If this is included, then for an average propagation delay Td_i over link i, equation (2.106) become

$$T = \frac{1}{\gamma} \sum_{i=1}^{M} \left(\frac{\lambda_i}{\mu_i - \lambda_i} + \lambda_i T_{d_i} \right) \tag{2.107}$$

The average delay per packet of a traffic stream transmitting through a path p is given by

$$T_p = \sum_{\substack{all(i,j) \\ on\ path,p}}^{M} \left(\frac{\lambda_{ij}}{\mu_{ij}(\mu_{ij} - \lambda_{ij})} + \frac{1}{\mu_{ij}} + T_{d_{ii}} \right)$$

where subscript (ij) represents link (i,j) for the path p and service time for link (i,j) is $1/\mu_{ij}$.

2.4 TIME REVERSIBILITY – BURKE'S THEOREM

Time reversibility is the property of the detailed balanced equation used for M/M/1, M/M/m, M/M/∞, and M/M/m/m systems which are represented using Markov chains [5,9] considering birth–death process. These equations depend on the relation that for any state j, the steady state probability of j times the transition probability from j to $j + 1$ is equal to the steady state probability of state $j + 1$ times the transition probability from $j + 1$ to j. An irreducible periodic and discrete time Markov chain X_n, X_{n+1}, having transition probabilities P_{ij} and stationary distribution $\{p_{ji} \ j \geq 0\}$ is considered. The sequence in states is taken as backwards in time chain X_n, X_{n-1}, and probability is written as

$$P\{X_m = j | X_{m+1} = i, X_{m=2} = i_2, \ldots X_{m+k} = i_k\} = \frac{p_j.P_{ji}}{p_i}$$

The state condition at time $m + 1$ and at time m does not depends on that at times $m + 2$, $m + 3$,... The backward transition probabilities are written as

$$P_{ij}^* = P\{X_m = j | X_{m+1} = i\} = \frac{p_j.P_{ji}}{p_i}, \quad i,j \geq 0.$$

If $P_{ij}^* = P_{ij}$ for i,j i.e. the forward and backward transition probabilities are identical. The following are the characteristics of the reversed chain:

- The reversed chain is irreducible, periodic and has a stationary distribution same as that of the forward one. It is written as

$$p_j = \sum_{i=0}^{\infty} p_i P_{ij}^*$$

  The reversed equality is also true.
- The non-negative numbers p_i is found for $i \geq 0$ summation are unity, and the transition probability matrix $P^* = \{P_{ij}^*\}$ such that

$$p_i P_{ij}^* = p_j P_{ji} \quad i,j \geq 0$$

  Then the stationary distributions $\{p_i | i \geq 0\}$ and P_{ij}^* are transition probabilities of the reversed chain. For overall j, we write

$$\sum_{j=0}^{\infty} p_j P_{ji} = p_i \sum_{j=0}^{\infty} P_{ij}^* = p_i$$

  These characteristics show time reversibility of the chain.
- A chain is time reversible if and only if the detailed balanced condition holds:
  This time reversibility concept extends to continuous Markov chains. This analysis is carried out by time discretization in the intervals of length δ. For the continuous case, $\delta \to 0$. For a continuous time chain, transition rates are q_{ij}, and the following assumptions are considered:
- The reversed chain is irreducible, and a continuous Markov chain has the same stationary distribution as the forward chain with transition rates

$$q_{ij}^* = \frac{p_j q_{ij}}{p_i}, \quad i,j \geq 0$$

- If a probability distribution $\{p_j | j \geq 0\}$ non-negative numbers q_{ij}^*, $i, j \geq 0$ is found such that

$$p_i q_{ij}^* = p_j q_{ji} \quad i, j \geq 0$$

then the stationary distribution and for all $i \geq 0$

$$\sum_{j=0}^{\infty} q_{ij} = \sum_{j=0}^{\infty} q_{ij}^*$$

$\{p_j | j \geq 0\}$ is the stationary distribution for both forward and reversed chains, and q_{ij}^* are transition rates of the reversed chain.

For overall j, we obtain

$$\sum_{j=0}^{\infty} p_j P_{ji} = p_i \sum_{j=0}^{\infty} P_{ij}^* = p_i$$

This property proves the chain is time reversible.
- Forward chain is time reversible if and only if the detailed balanced condition holds [9]

$$p_i q_{ij} = p_j q_{ji} \quad i, j \geq 0$$

Due to the time reversibility property, both forward and reversed systems are statistically not distinguishable in steady state. The initial state is selected according to the stationary distribution so that the queuing systems are in steady state in all times.

Burke's Theorem

For a steady arrival rate λ, the theorem states the following:

- The departure process is a Poisson process with rate λ.
- At each time t, the number of customers in the system does not depend on the sequence of the departure times prior to t.
- If the customers are served in the order in which they reach the system and a customer leaves at time t, the arrival time of that customer does not depend on the departure process prior to t.

Proof:

Since the system is time reversible, both forward and reversed systems are statistically not distinguishable, so the departure rate in the forward system is basically arrival rate in reversed system.

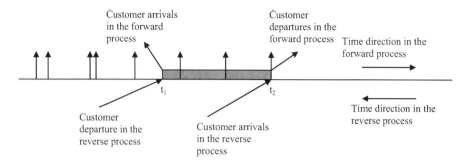

FIGURE 2.14   Forward and reverse processes [5].

The first statement holds:

- For a fixed time t, the departures prior to t in forward system are arrivals after t in the reversed system. So the arrivals in the reversed case are not dependent on Poisson process. Also the future departure process does not depend on the current number of messages in the system.
- We take a customer arriving at time t_1 and departing at time t_2. In the reversed system, the arrival process is not dependent on Poisson process, and the arrival process before time t_2 is not dependent on times spent in the system by customers who arrived after time t_2. So the time interval t_2-t_1 is not dependent on the reversed system before time t_2, and in forward system, the same interval is not dependent on the departure process before time t_2 (Figure 2.14).

2.5   INTERCONNECTION WITH OTHER NETWORKS

The concept of getting access to mainframes, services, and users in the network is internetworking which effectively forms a single large loosely coupled network of many different local- and wide-area networks. The large network is the Internet also called as an extended network. Three interconnections of networks in the Internet are connections between homogeneous networks, connections between heterogeneous networks, and connections of a network to a long-haul public network.

Interconnection Issues
The basic task involved in setting up an interconnection between the networks is to connect the hardware and software permitting stations on the two networks with each other conveniently and efficiently. Devices for interconnecting different networks are relays operating at one of several layers in the TCP/IP protocol model as [5]

1. repeaters operating at the physical link layer
2. bridges interconnecting LANs at the transport layer
3. routers functioning at the network access layer
4. gateways handling higher-level Internet protocols – mainly transport layer and application layer.

2.5.1 GATEWAYS

The gateways are referred to as bridges [11] with certain additional rules which are described later in this chapter. A gateway simply operates in the host node of each network in which the gateway receives the messages from the senders including intermediate nodes in the network, transmits from one network-protocol hierarchy to another, and then sends these newly formatted messages to the receivers including the next intermediate node in the network. Each network often operates a half-gateway, with the two halves connected to the networks A and B by a transmission line as shown in Figure 2.15. The fundamental issue concerning network interconnection is to find out a protocol for transmitting and forwarding messages such that the gateways are functionally simple, efficient, flexible, and do not need changes in the architecture of any the network segments being connected [11]. The following are to be considered:

1. Datagrams versus virtual circuits: The packet switched networks may provide either a datagram (connectionless) service or a virtual-circuit (connection-oriented) service. These types of interfaces are related since a virtual circuit interface is realized within a network by using datagrams. If there is a datagram-to-virtual-circuit interface, the task of the Internet is to transport independent, individually addressed datagram signals.

2. Naming, addressing, and routing: Names, addresses, and routes are three operations/functions in an Internet environment. A name indicates some resources (such as a process, device, or service), an address indicates the location of the resource, and a route specifies a path from the sender to the receiver.

   To implement routing, a decision-making algorithm is used to determine the next node that a packet is sent to on its way through the network. This can be achieved by means of a routing table in each station and gateway that gives, for each possible destination network, the next gateway to which the Internet packets are transmitted. There are static (fixed) and adaptive (dynamic) routing strategies that exist. In a static method, the same single route is always used to send messages from a source to a destination, whereas in an adaptive strategy, the route is not fixed but changes dynamically as a function of prevailing network conditions. These routing strategies are discussed in Chapter 3.

3. Fragmentation and reassembly: Fragmentation is needed when a gateway links two networks that have different maximum and/or minimum packet sizes or when it connects a network that transmits packets with one that transmits messages.

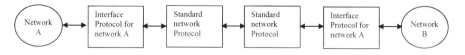

FIGURE 2.15 A gateway represented as two parts having half-gateways.

4. Flow control and congestion control: The flow control and congestion control are often required for successful transmission. Flow control is concerned with regulating the flow of messages between an s–d pair, so that the source does not send data at a rate greater than what the receiver can process. In the Internet, the responsibility of flow control belongs to the end-to-end protocols. Internet congestion control, on the other hand, is network-wide mechanism.
5. Security: Since a great deal of information passes through a network, we may be required to provide security for the information being transmitted. This would generally be done at the end; otherwise the Internet would become rather complex and, in addition, users would become sensitive about the content of their gateway.

2.5.2 BRIDGES

A bridge normally links two or more LANs (either similar or dissimilar) with the media-access level of the transport layer of TCP/IP protocol through a media-access-control (MAC)-level bridge [11]. The interconnected individual networks (known as segments) are bridged local networks. After filtering, the bridge controls traffic between two electrically independent cable systems attached to it. There are two standards in IEEE 802.1 making that LANs are extended by bridges to obtain consistent characteristics- the first standard is part D of IEEE Standard 802.1 describing a bridge for connecting LAN segments based on the IEEE-802 MAC specifications. The extended version of IEEE standard 802.5 is used specifically to link to token-ring LAN bridges connected to 802-type LANs (including Ethernet) called local bridges. Bridges having one or more port interfaces to long-haul backbone networks are remote bridges. A bridge is introduced between two existing LANs so that they work correctly for current stations. Two routing algorithms are used for a bridged LAN environment for dynamically creating a spanning tree topology [12,13] as a transparent bridge and the others taken a source routing approach.

2.5.2.1 Spanning Bridges

There are two steps – bridge forwarding and bridge learning – in bridge routing process in the transparent bridge concept which is considered in IEEE 802.1 as a standard for interconnecting 802-type LANs.

Bridge Forwarding. When a frame arrives to a bridge port, the destination address in the frame header is checked in a forwarding database as shown in Figure 2.16. This database has a list of group and individual station addresses together with information that relates these addresses to one or more of the bridge ports. This information has both source and destination addresses. If the destination address is not in the forwarding database, it is sent out on all ports of the bridge on which the frame was transmitted by using the process of flooding [12]. When the destination address remains in the forwarding database, the port identifier of the stored address is checked with the identifier of the port on which the address was obtained at port x. If the two identifiers are same, the frame is addressed to a station on the same LAN in which it originated.

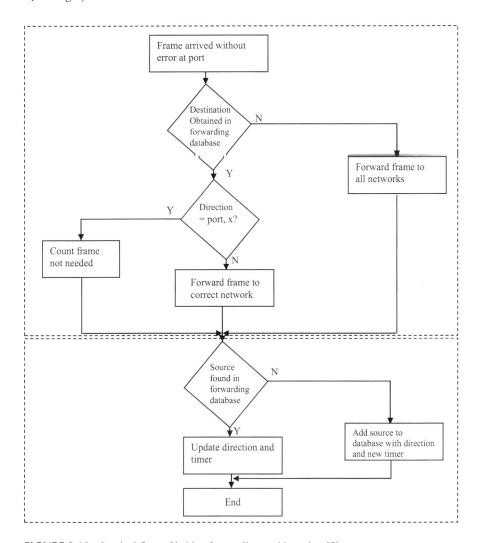

FIGURE 2.16   Logical flow of bridge forwarding and learning [5].

Bridge Learning. In bridge learning in Figure 2.16, when a frame reaches
a bridge, its source address is checked against the information residing in
the forwarding database [12]. If the source address is not there, the bridge
makes a new entry (to the database) having the address, the identifier of
the port on which the address was obtained, and a timer value representing
the timing of receipt of the address. If the address already remains in the
database and a different port identifier is enclosed with it, then the entry
is revised to report the new information. If the information is same as the
database information, then no modification is required. The timer for a particular entry length is established by network management.

Spanning Tree Algorithm. For these forwarding and learning processes to operate properly, the path of bridges and LANs between any two segments are in the entire bridged LAN. The operation is made using spanning tree algorithm [5].

2.5.2.2   Source Routing Bridges

In source routing, the route is specified in the frame, and the frame is transmitted through it. Source routing operates in LANs including token rings, token buses, and CSMA/CD buses. The source routing algorithm is used dynamically to locate a route to a given destination [14,16]. The frames transmit through all possible paths between the source and destination stations. Upon reaching the destination, all the routes of discovery frames should return to the source along the recorded path. The source chooses the path in this method. The transparent bridges do not change frame, a source-routing bridge adds a routing information field to the frame. This field is added immediately after the source address in every frame is sent to the destination station. The routing information field size is maximum up to 18 octets in length and has a two-octet routing control (RC) field and a variable number of route designator fields, and each is of two octets long and has a LAN segment number and a bridge number.

2.5.2.3   Quality of Bridge Services

The quality of service offered by a bridge depends on availability, frame mishaps, transit delay, frame lifetime, undetected bit errors, frame size, and priority.

Availability. The MAC layer provides assurance of the delivery of frames. Higher-level protocols guarantee end-to-end integrity and take care of high frame-loss rates [14].

Frame Mishaps. New bridges should be introduced in the topology so that frames' transmission guarantees no duplicity.

Transit Time. If frames' arrival time at bridges is faster than the processing rate, the congestion of links and bridges arises, and there is frame loss.

Frame Lifetime. The higher-level protocols are employed for retransmission with some degree of confidence that guarantees no duplicate frame. Thus, for bridge system management, some knowledge about maximum frame lifetime is needed.

Error Rate. The MAC layer should have a very low bit error rate through frame check sequence (FCS) verification for a frame. Thus, a bridge rejects the frames having an incorrect FCS.

Priority. The MAC layer may consider priority with respect to other frames sent from the same source address. A bridge considers this priority to find the order in which frames are forwarded.

2.5.3   ROUTERS

The router is not transparent to user protocols [15] used at the network layer where the Internet protocol (IP), DNA, XNS, and ISO 8473 are used [5]. A router ends at the MAC and LLC layers of each connected LAN permits translation between different address domains. Thus, routers are used to interface 802-type LANs to other networks from IBM and WANs. It has more efficient routing and flow control than a bridge, since it operates at the network level and exploits the traffic management procedures which are part of that level. It depends on parameters such as transit delay, congestion at other routers, or the number of routers between the source and destination stations [5].

Availability. The router handles the failures in links, stations, and other routers.
Frame Mishaps. Routers can tolerate disordering or duplicity of the packets.
Frame Lifetime. No a priori knowledge of packet lifetime is needed by the router, as this information remains in each packet.
Error Rate. The error checks are performed on the entire packet or the network-layer header only. Error packets are discarded. In addition to the use of FCS, it reduces the probability of undetected bit errors.
Priority. The networks may require a high-priority packet delivery service. The packets of emergency messages are kept before lower-priority traffic on transmission queues. Routing control information is made high-priority traffic for the urgent messages.

2.5.4 REPEATERS

Repeaters operate at the physical link layers. They not only use the same type of networks but also amplify the signal and reshape the signal waveform for the reduction of signal errors. They are used for long-haul networks having large coverage areas. They are basically interconnected devices used in the networks and were already discussed in Chapters 2 and 3 of volume 1.

SUMMARY

Queuing theory analyzes the performance of computer and communication networks. Since 1970, network systems have been employing queuing theory to analyze throughputs, queue lengths, and mean response times. These queuing models consistently estimate throughputs and time delay with accuracy of 5%. The basic assumptions are based on the fact that the queuing network has time-invariant parameters and have exponential distributions of service times at all FCFS devices, are often seriously violated in practice. The basic queuing models for computer and communication networks include the M/M/1, M/M/m, and M/G/1 queues which are discussed in detail in this chapter. Little's formula allows us to quickly calculate any one of the four important measures of network performance (total average number of messages N in the system, average number of messages N_q in the queue alone, average time T that a message spends in the system, and average time T_q spent waiting in the queue) once one of these quantities is known.

The interconnection between two independent networks is achieved by means of a special device which is generically called a gateway. The gateway accepts messages from a transmitting network, translates them from this network's protocol hierarchy to that of the connecting network, and then forwards these newly formatted

78   Advances in Optical Networks and Components

messages to the next network. Devices used for interconnecting networks at the data-link level of the transport layer are called bridges, whereas at the network level, routers are used.

In designing a network and its associated routing methodology, a one-dimensional performance criterion is usually defined, and an attempt is made to minimize this quantity to optimize network performance. The simplest criterion is to choose the least-cost route. A common method is to base the cost on the queuing delay in which case the least cost minimizes the message delay.

EXERCISES

2.1. Consider a computer-disk distribution center with a single server at which engineering students arrive according to a Poisson process. Let the mean arrival rate be one student per minute and assume the serving time is exponentially distributed with an average of 30 s per student. (a) What is the average queue length? (b) How long does a student have to wait in the queue? (c) Find the average queue length and waiting time in the queue if the service time increases to 45 s per student.

2.2. Consider a network node that has a buffer which is modeled as an infinite M/M/1 queue. Assume a terminal connected to this node generates a 40-bit message every 0.4 s. (a) If the capacity of the outgoing link is 1000 b/s, what is the average buffer occupancy in units of message? (b) What is the average waiting time? (c) If the line capacity is only 500 b/s, find the average buffer occupancy and the total waiting time. (d) What is the time a message spends in the queue in each case?

2.3. A buffer at a network node is modeled as an infinite M/M/1 queue. Suppose 10 terminals are attached to the node and their outputs are time-multiplexed into the buffer. (a) If each terminal transmits 100-bit messages at 300 b/s, what is the required line capacity if the total time delay is to be 50 ms? (b) Now suppose only one such terminal is attached to the node. What line capacity is required in this case?

2.4. Suppose an M/M/1 queue has a state-dependent Poisson arrival rate λn and a state-dependent departure rate μn. (a) Draw a state-transition diagram for this case.

(b) Show that the equation for state after equilibrium has set in is given by

$$(\lambda_n + \mu_n) p_n = \mu_{n+1} p_{n+1} + \lambda_{n-1} p_{n-1} \quad \text{for} \quad n \geq 1$$

(c) Show that the solution to this equation is given by

$$p_n = \frac{\lambda_0 \lambda_1 \lambda_2 \ldots \lambda_{n-1}}{\mu_1 \mu_2 \ldots \mu_n} p_0$$

2.5. Consider a system where arriving messages tend to get discouraged from joining the queue when more and more messages are present in the system.

Suppose the arrival and departure coefficients in this case are modeled as follows:

$$\lambda_n = \frac{\alpha}{n+1} \quad \text{for } n = 0,1,2,\ldots$$

$$\mu_n = \mu \quad \text{for } n = 1,2,3,\ldots$$

where α and μ are constants. (a) Draw a state-transition diagram for this case (b) Find p_n using the equation given in Prob. 5.10 in volume 1. (c) Show that the expected number of messages in the system is given by α/μ. (d) Using Little's formula, find an expression for the average time T spent in the system.

2.6. Consider a buffer in an M/M/1/K queue. Compare the probabilities of messages being blocked in each of the following cases: (a) $K = 2$, $\rho = 0.1$; (b) $K = 4$, $\rho + 0.1$; (c) $K = 4$, $\rho = 0.8$; (d) $K = 10$, $\rho = 0.8$. What is the probability in each case that the buffer is empty?

2.7. Given that a continuous random variable x has a density function $f(x)$ over the interval a to b, its variance is defined by

$$\text{var}[x] = \int_a^b \left(x - E[x]\right)^2 f(x)\,dx$$

where $E[x] = \int_a^b xf(x)\,dx$ is the average value of x. Thus, given a service time which is modeled by the gamma distribution

$$f(x) = \frac{\beta(\beta x)^{k-1} e^{-\beta x}}{\Gamma(k)} \quad \text{for } x > 0 \text{ and } k > 0$$

where β is a constant and $\Gamma(k)$ is the gamma function, show that $E[x] = k/\beta$ and $\sigma^2 b = k/\beta^2$.

2.8. A constraint on a CSMA/CD-based Ethernet system running at 10 Mb/s is that the farthest station-to-station span is limited to 2.5 km. Show how an Ethernet-to-Ethernet bridge can eliminate this limitation. What are some operational requirements of the bridge?

2.9. A message consisting of 1600 bits of data and 160 bits of header is to be transmitted across two networks to a third network. The two intermediate networks have a maximum packet size of 1024 bits which includes a 24-bit header on each packet. If the final network has a maximum packet size of 800 bits, how many bits (including headers) are delivered at the destination? Into how many packets is the original message fragmented?

2.10. The simplest form of flow control is based on a stop-and-wait protocol. Here a receiver agrees to accept data either by sending a poll or by responding to a select.

After the reception of transmitted data, the receiver must return an acknowledgment to the source before more data can be sent. Suppose a sequence of data frames $f_1, f_2, ..., fn$ is to be sent using the following polling procedure.

Station S_1 sends a poll to station S_2

Station S_2 replies with f_1

S_1 sends an acknowledgment

S_2 responds with f_2

S_1 acknowledges

$\vdots$

.

S_2 sends out f_2

S_2 acknowledges

(a) Letting t_{prop} = propagation time between S_2 to S

t_{proc} = message processing time

t_f = frame transmission time

t_{poll} = station polling time

t_{ack} = frame acknowledgment time

find expressions for the time to initiate a sequence (t_l), the time to send one frame, and the total time needed to send the sequence of n frames.

(b) Assuming that for a long sequence of frames t_l, t_{prop}, and t_{ack} are small and can be ignored, show that the utilization or efficiency of the line is

$$U = \frac{1}{1 + 2a}$$

where $a = t_{prop}/t_f$

2.11. (a) Using the following definition for the parameter a

$$a = \frac{\text{Propagation time}}{\text{Transmission time}}$$

derive an expression for an in terms of the distance d of the link, the propagation velocity v, the frame length L (in bits), and the data rate B.

(b) Using this expression, determine for what range of frame sizes a stop-and-wait protocol gives an efficiency of at least 50% for a 4-Kb/s channel giving a 20-ms propagation delay.

(c) Repeat part b for a 10-Mb/s channel having 5 μs propagation delay.

REFERENCES

1. L. Kleinrock, *Queuing Systems: Theory*, New York: Wiley, 1975.
2. A. G. Alien, "Queuing models of computer systems," *IEEE Communications Magazine*, vol. 13, pp. 13–24, 1980.
3. L. Kleinrock, *Queuing Systems: Computer Applications*, New York: Wiley, 1976.
4. H. Kobayashi and A. G. Kosheim, "Queuing models for computer communication system," *IEEE Transactions on Communications*, vol. 25, pp. 2–29, 1977.
5. G. E, Keiser, *Local Area Networks*, New York: McGraw Hill, 1999.

6. D.G. Kendall, "Some problems on theories of queue," *The Journal of the Royal Statistical Society*, vol. 13, pp. 151–185, 1951.
7. E. Gelenbe and I. Mitrani, *Analysis and Synthesis of Computer Systems*, New York: Academic Press, 1980.
8. E. Reich, "Waiting times when queues are in tandem," *The Annals of Mathematical Statistics*, vol. 28, pp. 768–772, 1957.
9. D. Bertsekas and R. Gallager, *Data Networks*, Englewood Cliffs, NJ: Prentice Hall, 1992.
10. J. P. Buzen and P. J. Denning, "Measuring and calculating Queuing length distribution," *Computer*, vol. 13, pp. 33–44, 1980.
11. A. L. Chaplin, "Standards for bridges and gateways," *IEEE Networks*, vol. 2, pp. 90–91, 1988.
12. W. M. Seifert, "Bridges and routers," *IEEE Networks*, vol. 2, pp. 57–64, 1988.
13. L. Zhang, "Comparison of two bridges routing approaches," *IEEE Networks*, vol. 2, 37–43, 1988.
14. R. C. Dixon and D. A. Pitt, "Addressing and bridging and source routing," *IEEE Networks*, vol. 2, pp. 25–32, 1988.
15. A. S. Tanenbaum, *Computer Networks*, Englewood Cliffs, NJ: Prentice Hall, 1981.
16. M.C. Hammer and G. R. Samsen, "Source routing bridge implementation," *IEEE Networks*, vol. 2, pp. 33–36, 1988.

3 Routing and Wavelength Assignment

This chapter discusses long-haul wide coverage optical networks based on arbitrary (mesh) topology in which nodes use wavelength-routing switches/optical cross-connects (OXCs) for setting up wavelength-division multiplexer (WDM) channels between node pairs [1,2]. It finds a practical approach to resolve the routing and wavelength assignment (RWA) problem of light paths in networks. A large RWA problem has two smaller sub-problem routings for selection of path and assignment of wavelength on routing path, each of which may be resolved independently and efficiently using different efficient techniques. A multi-commodity flow formulation, along with randomized rounding and different constraints, is required to estimate the routes for light paths. Wavelength assignments (WAs) for light paths are carried out by different techniques such as graph-coloring techniques. Wavelength-routed optical networks are the backbone of networks for large regions nationwide or globally. Access stations/users are attached to the network via wavelength-sensitive nodes for switching/routing.

3.1   LIGHT PATHS

Access stations transmit signals to each other via wavelength channels, called light paths [2]. In order to set up a "connection" between a source–destination (SD) pair, we need to have a "light path" between them. A light path is made using multiple fiber links to provide a "circuit-switched" interconnection between two nodes. Each intermediate node in the light path essentially provides a circuit-switched optical bypass facility to support the light path. In an N-node network, if each node has $(N-1)$ transceivers and if there are enough wavelengths on all fiber links, then every node connection request should be set up by an optical light path, and a networking scheme is required to resolve this problem. The network size (N) should be scalable, and transceivers are expensive, so each node can have only a few of them, and technological constraints dictate that the number of WDM channels that can be supported in a fiber be limited to W (64 maximum). Thus, only a limited number of light paths may be set up on the network.

A virtual topology is developed based on a complete set of light paths in network over which virtual topology connection requests are established and routed. The design of virtual topology is discussed in the next chapter. In this chapter, the traffic/connection discussed for wide area network (WAN) is circuit-oriented; i.e., the discussed traffic consists of a set of connections such that each connection requires the full bandwidth of a light path in order for it to be routed between its corresponding sd pair. The RWA problem can be stated as follows:

- A set of light paths need to be set up for connection requests in the network.
- For a constraint on the number of wavelengths, estimate the routes over which these light paths should be set up and also estimate the wavelengths assigned to these light paths so that the maximum number of light paths are set up.

While the shortest-path routes may often be preferable, this choice allows more light paths to be set up. Thus, RWA algorithms generally estimate several alternate routes for each light path that needs to be established. Light paths are not set up due to constraints on routes and wavelengths and are said to be blocked, so the corresponding network optimization problem minimizes this blocking probability (BP).

Several constraints are required to implement routing and WAs in the network. In this regard, a light path operates on the same wavelength across all fiber links that it traverses, in which case the light path is said to satisfy the wavelength-continuity constraint. Thus, two light paths that share a common fiber link should not be used by the same wavelength. For a switching/routing node with a wavelength converter facility, the wavelength-continuity constraints are not satisfied. For such a network having a set of connection demands to be established, an integer linear programming (ILP) formulation of the RWA problem is taken as a multi-commodity flow problem. There are lower and upper bounds in any generic RWA algorithm's performance – an upper bound on the carried traffic (number of light paths established) and a lower bound on the light path. The approach uses wavelength conversion. The WA problem is NP-complete where a number of heuristics are required to obtain the solutions of the RWA problem [3,4].

In this chapter, we mention well-known algorithms for the RWA [1,5]. The RWA approach consists of algorithms to solve the RWA problem [5] for large network. In RWA, the main objective is to minimize the number of wavelengths for accommodating a certain number of connections in the network based on a certain physical topology. The RWA problem has four different sub-problems – each sub-problem is resolved independently with the results of one stage providing the input to the next stage. Firstly, a formulation is made by using a linear program (LP) relaxation based on the physical topology having the set of connections. One can use a general-purpose LP solver to find out solutions to this problem. In this case, simple and specialized techniques are developed to drastically handle the size of the LP in terms of the number of variables and the number of equations that it needs to handle.

3.2   LP FORMULATION OF RWA AND ITS REDUCTION

Combinatorial formulations are made using mixed-integer linear programs (MILPs) for solving the RWA problem [6–9]. These formulations are used in resolving large problems by using sophisticated techniques such as branch-and-bound methods.

The RWA problem, without the wavelength-continuity constraint, is a straightforward multi-commodity flow problem with integer flows in each link [1]. The ILP with the objective function minimizing the flow in each link corresponds to minimizing the number of lightpaths passing through a particular link.

$\lambda_{sd} = 1$, if there is a light path from s to d, where $\lambda_{sd} =$ traffic (in terms of a light path) from any source s to any destination d.

    $= 0$, otherwise

$F_{ij}^{sd} =$ traffic (in terms of the number of light paths) that is flowing from source s to destination d on link ij. The LP formulation is written as follows [1]:

$$\text{Minimize} \quad F_{max}$$

$$\text{such that} \quad F_{max} \geq \sum_{s,d} F_{ij}^{sd} \ \forall ij \tag{3.1}$$

$$\sum_{i} F_{ij}^{sd} - \sum_{k} F_{jk}^{sd} = \lambda_{sd} \quad \text{if } s = j$$

$$= -\lambda_{sd} \quad \text{if } d = j \tag{3.2}$$

$$= 0 \quad \text{otherwise}$$

where
$$\lambda_{sd} = 0,1$$
$$F_{ij}^{sd} = 0,1$$

It is NP-complete and can be approximately written by using randomized rounding [1].

3.2.1 REDUCTION OF SIZE OF LP FORMULATION

A multi-commodity formulation is made using a number of equations having a number of variables in the formulation. Figure 3.1 shows a network consisting of 10 nodes, 15 links (ij-pairs), and an average of 4 connections per node; i.e., 40 connections (SD pairs) need to be set up in the network.

In the simplest and most general formulation, the number of λ_{sd} variables for Figure 3.1 is estimated as $10 \times 9 = 90$. The number of F_{ij}^{sd} variables are estimated as 90 SD pairs $\times$ 15 ij pairs $= 1350$. The number of variables and the number of equations grow proportionally with the square of the number of nodes. A solution

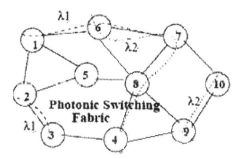

FIGURE 3.1   Network having 10 nodes and 15 physical links in which links are bidirectional.

is obtained only by considering the variables $\lambda_{sd} = 1$ and reducing the number of λ_{sd} variables from 90 to 40. Also, this approach also decreases the number of F_{ij}^{sd} variables to be $40 \times 15 = 600$. For further reduction of the number of variables, the light path needs to pass through few ij links. The links make the path through which a light path passes and we can only consider those links as the F_{ij}^{sd} variables for that particular SD pair. For a light path SD that passes through four links on average, there will be approximately 160 F_{ij}^{sd} variables.

Hence, for a particular light path, the size of the LP problem formulation can be decreased, and hence the number of equations and number of variables can be reduced for large networks.

3.2.2 RANDOMIZED ROUNDING

The randomized rounding technique is highly probabilistic providing an integer solution in which the objective function finds a value close to the optimum of the rational relaxation [10]. This is a sufficient condition to show the near-optimality of our 0–1 solution since the optimal value of the objective function in the relaxed version is better than the optimal value of the objective function in the original 0–1 integer program. This technique has been effectively used in multi-commodity flow problems.

In randomized rounding, there are two types of ILP relation – showing existence outcomes of feasible solutions to an ILP in terms of its fractional relaxation and using the information obtained from the solution of the relaxed problem in order to obtain a good solution to the original ILP.

The multi-commodity flow problem is an undirected graph $G(V,E)$ having k commodities required to find the path. Here, various vertices are different sources and sinks for a particular commodity. Each edge $e \in E$ where E has a capacity $c(e)$ as an upper limit on the total amount of flow in E. The flow of each commodity in each edge is either 0 or 1. The objective function is to reduce the common capacity in each link for unit flows for all commodities. The problem is known to be NP-complete [2] although the non-integral version can be solved using linear programming methods in polynomial time. The algorithm has the following three major steps:

- resolving a non-integral multi-commodity flow problem
- path stripping
- randomized path choice.

Non-Integral Multi-Commodity Flow

The requirement of the 0–1 flows is relaxed to allow fractional flows in the interval [0,1]. The relaxed capacity-minimization problem is resolved by a suitable linear programming method. If the flow for each commodity i on edge $e \in E$ is indicated by $f_i(e)$, a capacity constraint of the form [10]

$$\sum_{i=1}^{k} f_i(e) \leq C \tag{3.3}$$

is then satisfied for each edge in the network, where C is the optimal solution to the non-integral, edge-capacity optimization problem. The value of C is also a lower bound on the best possible integral solution.

Path Stripping

The concept in this step is to translate the edge flows for each commodity i into a set of possible paths. For each commodity i, the following steps are required to be performed:

a. Find a loop-free, depth-first, directed path $e_1, e_2, \ldots, e_p$ from the source to the destination.
b. Let $f_m = \min f_i(e_j)$, where $1 \le j \le p$. For $1 \le j \le p$, substitute $f_i(e_j)$, by $f_i(e_j), -f_m$. Include path $e_1, e_2, \ldots, e_p$ to τ_i along with its weight f_m.
c. Remove any edge with zero flow from the set of edges that carry any flow for commodity i. If there is non-zero flow leaving s_i, repeat step b. Otherwise, continue to the next commodity i.

The weights of all the paths in τ_i are considered to be 1. Path stripping gives a set of paths τ_i that send an optimum flow of commodity i.

Randomization

For each i, obtain a $|\tau_i|$ with face probabilities equal to the weights of the paths in τ_i. Assign to commodity i the path whose face comes up. The integer capacity of the solution produced by the above procedure does not exceed [10]

$$C + \sqrt{3C \ln \frac{|E|}{\varepsilon}}$$

where $0 < E < 1$ with a probability of $1 - \varepsilon$ at least.

In this problem, the F_{ij}^{sd} variables are permitted to take fractional values. These values are required to find the fractional flow through each of a set of alternate paths. A coin-tossing experiment chooses the path over which to transmit the light path λ_{sd} as per the probability of the individual paths. This technique is very good to resolve the problems of routing in the networks having a large number of nodes.

3.2.3 GRAPH COLORING

Once the path is chosen for all connections, the number of light paths passing through any physical fiber link is called as congestion on that particular link. Now, the wavelengths are assigned to each light path in such a way that any two light paths passing through the same physical link are assigned different wavelengths. If the intermediate nodes cannot perform wavelength conversion, a light path having the same wavelength throughout is allotted. This is basically a wavelength-continuity constraint reducing the utilization of the wavelengths in the network, since a light path finds a free wavelength of the same color in all of the physical fiber links of the path. This wavelength utilization is enhanced by the use of wavelength converters in a switching node. In the graph coloring model, the objective is to minimize the number of wavelengths (colors) under the wavelength-continuity constraint, which is achieved by the following steps [10]:

1. drawing the graph of $G(V, E)$, which is a function of number of nodes V and number of edges connected between different nodes. There is an undirected edge between two nodes in graph G if the corresponding light paths pass through a common physical fiber link.
2. coloring the nodes of the graph G in a light path such that no two adjacent nodes have the same color.

This problem is NP-complete, as it is not easy to minimize the number of colors (chromatic number $x(G)$ of the graph G) needed to color. The efficient sequential graph coloring algorithms estimate the number of colors. In a sequential graph coloring approach, vertices are sequentially incorporated to the portion of the graph already colored, and new colorings are determined to take in each newly adjoined vertex. At each step, the total number of colors is needed to be kept as minimum as possible. A particular sequential node coloring gives an $x(G)$ coloring. A_i is the set of the nodes colored i by a $x(G)$ coloring of G. For any ordering of the vertices $V(G)$, i.e., for all members of A_i before any member of A_j for $1 \leq i \leq j \leq x(G)$, the corresponding sequential coloring is an $x(G)$ coloring.

If $A(G)$ is the maximum degree in a graph, then $x(G) < A(G) + 1$. If a graph contains only a few nodes of very large degree, then coloring of these nodes stays away from the requirement of using a very large set of colors. The following theorems are required for coloring.

Theorem:

If G is a graph with $V(G) = v_1, v_2, ..., v_n$, where $\deg(v_i) \geq \deg(v_{i+1})$ for $i = 1,2,..., n - 1$, then, $x(G) \leq \max_{1 \leq l \leq n} \min \{i, 1 + \deg(vi)\}$. Estimation of a sequential coloring procedure corresponding to such an ordering will be called as largest-first algorithm [1].

The sequential coloring approach demonstrates that for a given ordering $v_1, v_2, ..., v_n$ of the vertices of a graph G, the corresponding sequential coloring algorithm does not need more than k colors where $k = \max_{1 \leq i \leq n} \{1 + \deg_{<v_1,v_2...v_n>}(v_i)\}$ and $\deg_{<v_1, v_2, ..., vn>}(v_i)$ indicates the degree of node vi in the vertex-induced subgraph represented by $<v_1, v_2, ..., v_n>$. A vertex ordering that minimizes k was found and can be determined using the following procedure [1,11,12]:

1. For $n = |V(G)|$, v_n should be chosen such that it has minimum degree in G.
2. For $i = n - 1, n - 2, ..., 2,1$, v_i should be chosen such that it has have minimum degree in $<V(G)- v_n, v_{n-1}, ..., v_{i+1}>$.

For any vertex ordering $v_1, v_2, ..., v_n$ determined, we have

$$\deg_{<v_1,v_2...v_n>}(v_i) = \min_{1 \leq j \leq i} \deg_{<v_1,v_2...v_n>}(v_j)$$

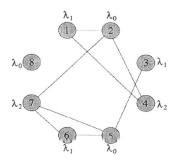

FIGURE 3.2   Auxiliary graph, $G(V,E)$, for the light paths in the network [1].

for $1 \leq i \leq n$, so that such an ordering is considered as a smallest-last (SL) vertex ordering. Any SL vertex ordering minimizes k over the $n!$ possible orderings estimated [12]. The SL coloring algorithm determines the color of the light paths for wavelength allotment (Figure 3.2).

3.2.4   ANALYSIS OF ILP

For static and dynamic traffic analyses, a randomly generated physical topology consists of N nodes, with each node having a physical nodal degree uniformly distributed between 2 and 5. All links are made to be unidirectional, and there are V number of directed links in our simulation network. In the RWA problem, mainly three types of traffic are reported in the literature [1,13] – (a) static traffic, (b) incremental traffic, and (c) dynamic traffic.

a. In case of static traffic [6], all traffic/connection requests are identified in advance, and the light paths are set up to satisfy the maximum number of traffic requests.
b. In case of incremental traffic [1], traffic/connection requests arrive in the system sequentially. The light path is set up for each traffic/connection request in indefinite time.
c. In case of dynamic traffic [13], traffic/connection requests arrive in the system randomly based on a statistical distribution [11], on the basis of mainly Poisson process, and a light path is set up for each traffic/connection request released after some finite amount of time.

The offered traffic follows the followings:

1. A set of light paths are required to be established between randomly chosen SD pairs, and all SD pairs are considered equally.
2. For each (source) node, d' is an average number of light path connections provided by a source node. In an N-node network, the probability that a node has a light path with each of the remaining $(N-1)$ nodes equals $d'/(N-1)$, and d' is taken as the average "logical degree" of a node.

There are enough transceivers at the destination nodes/source nodes to accommodate all of the light path requests that need to be established, so no light path request will be blocked due to the lack of transceivers at these nodes.

Static Light path Establishment (SLE)

For static light path establishment (SLE) of the physical topology having the set of connections to be routed, an LP formulation is made to find a lower bound on the number of wavelengths for the RWA problem [14]. To reduce the size of the LP formulation, K alternate shortest paths between a given SD pair are taken. Only the links constituting these alternate paths are represented as the F_{ij}^{sd} variables, and a standard K-alternate-path algorithm derives K number of alternate shortest paths of an SD pair. This LP problem is solved by a standard LP solver package [15], and the resultant flow values for the F_{ij}^{sd} variables are input to the randomized rounding algorithm. The value of the objective function indicates the lower bound on congestion that is achieved by any RWA algorithm. Considering the individual variables F_{ij}^{sd}, the path-stripping technique and the randomization technique (mentioned in the previous section) are used to assign physical routes for different light paths. Once the light paths have been allotted on physical routes, a wavelength is allotted on each individual light path. This is carried out by coloring the nodes of graph G such that adjacent nodes get different colors, and this corresponds to allotting wavelengths correctly to the light paths with the SL vertex coloring as mentioned in Section 3.2.3.

For larger values of K, a number of wavelengths are required to have coloring of all light paths, and this number is a little higher but very close to the maximum congestion in the network. The maximum network congestion gives the number of wavelengths required in the network for the intermediate switching nodes having wavelength converters in which no wavelength-continuity constraint is followed for a light path. To resolve the large time complexity required for solving larger LP problems having a large number of alternate paths, two or three alternate paths may be taken.

Dynamic Light path Establishment (DLE)

In static problems, light path requests are known in advance. LP formulation and subsequent algorithms are needed to determine the number of wavelengths used to assign all of the light path requests. When the light path requests are dynamic and are not known in advance, the optimal algorithms used in SLE are not advisable. A dynamic algorithm is required for dynamic light path establishment (DLE) where it is adaptive to find a route for the incoming traffic/connections over different paths based on congestion on the different links. A heuristic based on least-congested-path (LCP) algorithm [1] is used for DLE. In the LCP algorithm, a light path is allotted on the LCP selected from a set of alternate paths between an SD pair. The wavelength assigned on this path is the first wavelength available among all wavelength channels where wavelengths are numbered randomly.

The congestion and number of wavelengths for assignment of connection requests mainly depend on the order in which the connection requests reach. In the DLE case, existing connections are not rerouted to assign a new connection blocked due to lack of free wavelengths. The set of connections are the same as those in the static case. The congestion using LCP routing is optimal after verification using the static optimization case based on randomized rounding algorithm for the same set of connection requests in the network. So the LCP algorithm reduces the congestion as each connection arrives. As the congestion increases in the network, the number of wavelengths required in the dynamic case is more than that needed in the static case. In the static case, the coloring algorithm assigned wavelengths to the light paths in a specific order and minimized the number of wavelengths required. This approach of graph coloring for assigning a wavelength to a light path cannot be used in the dynamic case in which light paths arrive in a random order. Due to the wavelength-continuity constraint, the optimal allocation of wavelengths cannot be easily achieved in the dynamic case. There are some approaches to reassign existing connections to unused wavelengths in order to optimally assign the resources for new connections. This problem is resolved by an algorithm that uses a fundamental operation called color interchange in which two light paths interchange their wavelengths to create spare capacity in the network.

3.3 ROUTING

In an RWA, there are two sub-problems – routing and WA – which together make a hard problem. The RWA problem becomes simple when it is divided into the routing subproblem and the WA subproblem. In this section, we focus on various approaches to routing connection requests by following the ILP for SLE and DLE. The WA subproblem is considered separately in Section 3.4.

3.3.1 ROUTING ALGORITHMS

There are many routing algorithms used for static and dynamic light path establishments such as Dijkstra algorithm, Bellman–Ford algorithm, genetic algorithm, and stimulated algorithm. The route/path has to be determined by the routing function which requires correctness, fairness, simplicity, optimality, robustness, efficiency, and stability. The performance of the routing is estimated based on the following parameters:

- number of hops
- cost
- time delay
- propagation time delay.

For analysis of the above-mentioned routing algorithms, a simple cost factor is considered for each physical link of the network. On the basis of minimum cost, the route of each sd pair is estimated using routing algorithm.

3.3.1.1 Dijkstra's Algorithm

One of the most commonly used algorithms for routing is Dijkstra's algorithm [16–18]. Using this algorithm, the shortest path from a given source to all other nodes as destinations is estimated, and the paths are determined in order of increasing path length. Here, for easy representation and explanation, we consider the least-cost path.

At the kth stage, the least cost paths to k nodes closest source node is estimated, and these nodes are in set M. At the $(k + 1)$th stage, the node that has least-cost path from source node is added to set M. As each node is added to set M, its path from source node is distinct. Before describing the procedure of the algorithm, the following definition and notation are to be noted:

$N =$ set of nodes in the network

$s =$ source node

$M =$ set of nodes incorporated in the inclusion matrix

$d_{ij} =$ link cost from node i to node j where $d_{ij} = 0$ and ∞ denotes that two nodes are not directly connected and $d_{ij} \geq 0$ denotes that two nodes are connected physically

$D_n =$ total cost of least-cost path from source node s to node n that is currently known to be in the network.

The algorithm has three steps: step-2 and step-3 are repeated until $M = N$; i.e., step-2 and step-3 are repeated till final paths have been assigned to all the nodes except node s in the network. The algorithm procedure is given below.

Algorithm 1 Dijkstra's Algorithm

Step-1: Initialize
   $M = \{s\}$, (i.e., the set of nodes so far incorporated in inclusion matrix M consists of the source node)
   $D_n = d_{sn}$ for $n \neq s$ (i.e., the initial path costs to neighboring nodes are simply link costs)

Step-2: Find the neighboring node not in M that has the least-cost path from source node s and incorporate that node into M. This can be expressed as Find $w \notin M$ such that $D_w = \min_{j \neq M} D_j$

Step-3: Update least-cost paths:
   $D_n = \min [D_n, D_w + d_{wn}]$ for all $n \notin M$
   If the term is the minimum, the path from s to n is now the path from s to w, concatenated with the link from w to n.

Step-4: Repeat step-2 and step-3 till $M = N$.

One iteration of step-2 and step-3 adds one new node to M and defines least-cost path from s to that node. That path passes only through nodes that are in M. After k iterations, there are k nodes in M, and the least-cost paths from s to each of the k nodes are derived. Among these paths, there is one of least cost that passes exclusively through nodes in M, ending with a direct link from a node in M to a node not in M. This node is added to M, and the associated path is defined as the least-cost path for that node. Table 3.1 shows the routing table of least-cost routing paths for source node $s = 1$ to all other nodes in the network (Figure 3.3) derived by using Dijkstra's algorithm. At each step, the path to each node and the cost of that path are estimated. After the final iteration, the least-cost path to each node from source node $s = 1$ is derived.

TABLE 3.1

Least-Cost Routing Paths from Source Node $s = 1$ to All Destinations Using Dijkstra's Algorithm in the Network Mentioned in Figure 3.3

Iteration	M	D_2 Path$_2$	D_3 Path$_3$	D_4 Path$_4$	D_5 Path$_5$	D_6 Path$_6$
1	[1]	1 1-2	4 1-3	3 1-4	–	–
2	[1, 2]	1 1-2	2 1-2-3	3 1-4	4 1-4-5	6 1-2-6
3	[1, 2, 3]	1 1-2	2 1-2-3	3 1-4	4 1-4-5	5 1-4-5-6
4	[1, 2,3,4]	1 1-2	2 1-2-3	3 1-4	4 1-4-5	5 1-4-5-6
5	[1, 2,3,4,5]	1 1-2	2 1-2-3	3 1-4	4 1-4-5	5 1-4-5-6
6	[1, 2,3,4,5,6]	1 1-2	2 1-2-3	3 1-4	4 1-4-5	5 1-4-5-6

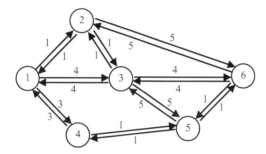

FIGURE 3.3 Sample network with link cost.

Figure 3.4 shows routing paths at iteration 1 (fig(a)), iteration 2 (fig(b)), and iteration 3 (fig(c)) using Dijkstra's algorithm [17].

The following may be a disadvantage of Dijkstra's algorithm: there may be more number of iterations or steps required to achieve routing of all destination nodes for a large network having more number of nodes as the number of iterations is equal to the number of destination nodes.

3.3.1.2 Bellman–Ford Algorithm

Another of the most commonly used routing algorithms is Bellman–Ford algorithm [17–19]. Using this algorithm, one can find the least-cost path from a given source subject to the constraint that the path contains at most one link; find the least-cost path with the constraint that the path contains at the most two links; and compare the costs; and then find the least cost between those of the paths of two links, one link, and so on. Before describing the procedure of the algorithm, the following definitions and notation are mentioned.

N = set of nodes in the network
s = source node
M = set of nodes incorporated in inclusion matrix

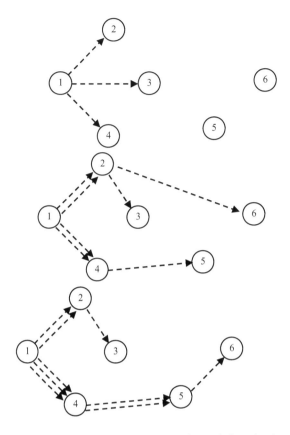

FIGURE 3.4   Routing by Dijkstra's algorithm: (a) iteration 1, (b) iteration 2, and (c) iteration 3.

d_{ij} = link cost from node i to node j where $d_{ij} = 0$ or ∞ denotes that two nodes are not directly connected and $d_{ij} \geq 0$ denotes that two nodes are connected physically

h = maximum number of links in a path at current stage of algorithm = hop size

$D_n^{(h)}$ = total cost of least-cost path from source node s to node n that is currently known to be in the network.

The algorithm has two steps: step-2 is repeated until none of the costs change. The algorithm's procedure is given below.

Algorithm 2 Bellman–Ford Algorithm

Step-1: Initialize
$$D_n^{(0)} = \infty, \text{ for all } n \neq s$$
$$D_s^{(h)} = 0, \text{ for all } h$$

TABLE 3.2

Least-Cost Routing Paths from Source Node $s = 1$ to All Destinations Using Bellman–Ford's Algorithm in the Network Mentioned in Figure 3.3

h	D_2 Path$_2$	D_3 Path$_3$	D_4 Path$_4$	D_5 Path$_5$	D_6 Path$_6$
0	∞ –	∞ –	∞ –	∞ –	∞ –
1	1 1-2	4 1-3	3 1-4	∞ –	∞ –
2	1 1-2	2 1-2-3	3 1-4	4 1-4-5	5 1-2-6
3	1 1-2	2 1-2-3	3 1-4	4 1-4-5	5 1-4-5-6
4	1 1-2	2 1-2-3	3 1-4	4 1-4-5	5 1-4-5-6

Step-2: For each successive $h \geq 0$

$$D_n^{(h+1)} = \min_j [D_j^{(h)} + d_{jn}]$$

The path from s to i node terminates with the link from j to i.

Step-3: Repeat step-2 till paths for all destinations are the same as those of the previous iteration.

For repeating step-2 with $h = K$ and each destination node n, the algorithm compares paths from s to n of $K + 1$ length that are less costly with the path that exists at the end of the previous iteration. If the previous path is less costly, then that path is selected; otherwise it is updated with a new path with more length $K + 1$; i.e., a new path consists of a path of K length plus a direct hop from node j to n. Table 3.2 shows the routing table of least-cost routing paths from source node $s = 1$ to all other nodes in the network (Figure 3.3) derived by using Bellman–Ford algorithm. The iterations are continued till the paths of all destination nodes are the same as those of the previous iteration. After the final iteration, the least-cost path to each node from source node $s = 1$ is derived. Figure 3.5 represents routing paths at iteration 1 (fig(a)), iteration 2 (fig(b)), and iteration 3 (fig(c)) using Bellmen–Ford's algorithm.

It is seen from Tables 3.1 and 3.2 that the number of iterations in Bellman–Ford's algorithm is less than that in Dijkstra's algorithm. Still Dijkstra's algorithm is preferred because of its less time complexity [17]. Although these routing paths are determined on the basis of the cost of links, practically, the routing of paths for SD pairs is determined based on the shortest path of the signal or minimum time delay (where time delay consists of propagation time delay and queuing time delay).

3.3.2 ROUTING APPROACHES

In case of routing, many approaches are used for both static and dynamic routing such as fixed routing, fixed-alternate-path routing, and adaptive routing [1].

3.3.2.1 Fixed Routing

The most simple approach of routing a connection request is fixed routing for a given sd pair. The fixed shortest-path routing for each sd pair is made offline using standard

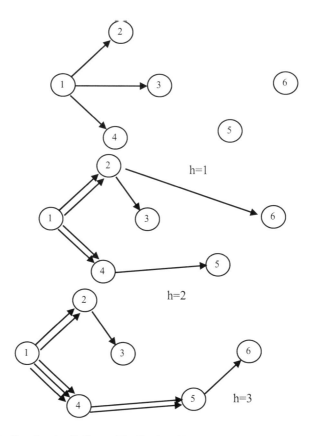

FIGURE 3.5   Routing by Bellman–Ford's algorithm: (a) iteration 1, (b) iteration 2, and (c) iteration 3.

shortest-path algorithms such as either Dijkstra's algorithm or the Bellman–Ford's algorithm discussed in the previous section [17]. Dijkstra's algorithm is preferred because of its less complexity compared to Bellman–Ford's algorithm [17]. The routing table values for a node (source node) are found in that node considering other nodes in the network as destination nodes. As an example, the routing table values for node 1 as a source node of the sample network (Figure 3.3) estimated by using Dijkstra's offline algorithm are given below.

Node-1	
Destination Node	**Path**
2	1-2
3	1-2-3
4	1-4
5	1-4-5
6	1-4-5-6

The routing table values for nodes 2, 3, 4, 5, and 6 estimated by using Dijkstra's offline algorithm are given below.

Node-2		Node-3		Node-4		Node-5		Node-6	
Destination Node	Path	Destination Node	Path	Destination Node	Path	Destination Node	Path	Destination Node	Path
1	2-1	1	3-1	1	4-1	1	5-1	1	6-5-1
3	2-3	2	3-2	2	4-1-2	2	5-4-1-2	2	6-2
4	2-1-4	4	3-2-1-4	3	4-1-2-3	3	5-3	3	6-3
5	2-1-4-5	5	3-5	5	4-5	4	5-4	4	6-5-4
6	2-6	6	3-6	6	4-1-6	6	5-6	5	6-5

The connections of different sd pairs are set up with the pre-determined route. Although this approach to routing connections is simple, there are drawbacks in this approach. WA (discussed later in this chapter) along the path is also fixed; i.e., there is no flexibility of WA if some paths have congestion. In fact, it provides high blocking probabilities in the dynamic case. Secondly fixed routing is unable to handle fault situations in which there are one or more link failures in the network. To handle link faults, the routing scheme must either consider alternate paths to the destination or find the route dynamically.

3.3.2.2 Fixed-Alternate Routing

Fixed-alternate routing is a method to routing that derives multiple paths. In fixed-alternate routing, each node in the network retains a routing table that has an ordered list of a number of fixed routes to each destination node. The order is based on different parameters such as path length, time delay, and number of hops. For example, these routes for sd pair consist of the shortest-path route, the second-shortest-path route, the third-shortest-path route, etc. A primary route for an sd pair is the first route in the list of routes to destination node d in the routing table at node s. An alternate route between s and d is any route that does not share any links (or is link-disjoint) with the first route in the routing table at s. The term "alternate routes" is also used to describe all routes (including the primary route) from a source node to a destination node. All the alternate paths including the primary in the routing table from source node s to other nodes as destinations are determined by using the modified form of offline Dijkstra's algorithm [17].

Routing path-1

Step-1: Initialize

$M = \{s\}$, (i.e., the set of nodes so far incorporated in inclusion matrix M consists of source node)

$D_n = d_{sn}$ for $n \neq s$ (i.e., the initial path costs to neighboring nodes are simply link costs)

Step-2: Find the neighboring node not in M that has the least-cost path from source nodes s and incorporate that node into M. This can be expressed as Find $w \notin M$ such that $D_w = \min_{j \neq M} D_j$

Step-3: Update least-cost paths:

$D_n = \min [D_n, D_w + d_{wn}]$ for all $n \notin M$

If the term is the minimum, the path from s to n is now the path from s to w, concatenated with the link from w to n.

Step-4: Repeat step-2 and step-3 till $M = N$.

Routing path-2

Step-5: All routing paths-1 for all destination nodes are made infinity (very large value).

$M = \{s\}$, (i.e., the set of nodes so far incorporated in inclusion matrix M consists of the source node)

$D_n = d_{sn}$ for $n \neq s$ (i.e., the initial path costs to neighboring nodes are simply link costs)

Step-6: Repeat step-2 and step-3 till $M = N$.

Routing path-3

Step-6: Routing path-1 and routing path-2 for all destination nodes are made infinity (very large value).

$M = \{s\}$, (i.e., the set of nodes so far incorporated in inclusion matrix M consists of the source node)

$D_n = d_{sn}$ or $n \neq s$ (i.e., the initial path costs to neighboring nodes are simply link costs)

Step-7: Repeat step-2 and step-3 till $M = N$.

$$\cdots \qquad \cdots \qquad \cdots \qquad \cdots \qquad \cdots$$
$$\cdots \qquad \cdots \qquad \cdots \qquad \cdots \qquad \cdots$$

Routing path-k

Step-k+3 Routing paths-1, 2 … $k - 1$ for all destination nodes are made infinity (very large value).

$M = \{s\}$ (i.e., the set of nodes so far incorporated in inclusion matrix M consists of the source node)

$D_n = d_{sn}$ for $n \neq s$ (i.e., the initial path costs to neighboring nodes are simply link costs)

Step-k+4: Repeat step-2 and step-3 till $M = N$.

The routing table values of alternate fixed path routing (FR) for the source node are also determined by using the modified form of Bellmen–Ford's algorithm. Consider two alternate routing paths ($k = 2$). The routing table values of fixed alternate path routing for source node $s = 1$ of the network in Figure 3.3 are estimated by using Dijkstra's algorithm and given below.

	Node-1			
Destination Node	Path-1	Cost	Path-2	Cost
2	1-2	1	1-3-2	5
3	1-2-3	2	1-3	4
4	1-4	3	1-3-5-4	10
5	1-4-5	4	1-3-5	9
6	1-4-5-6	5	1-3-6	8

Similarly, routing table values for other source nodes to corresponding destinations are estimated by using the modified form of Dijkstra's algorithm [20]. When a connection request reaches a source node, the source node has to set up a connection on each of the routes in the routing table in sequence, until a route with a valid wavelength is allotted. If no route is obtained from the list of alternate routes, then the connection request is blocked and lost. The routing tables at each node have light paths ordered on the basis of hop numbers/shortest distance/minimum delay. So, the shortest/least-cost path to the destination is the first route in the routing table. When the distance/cost between different routes is the same, any one path may be considered randomly. The advantage of fixed-alternate routing is that it can significantly reduce the connection BP compared to fixed routing [21].

3.3.2.3 Flooding

Flooding is another simple routing technique which does not need any information/parameter values and by which a packet is sent from a source node to every neighboring node [17,22]. At each node, the incoming packet is transmitted on all outgoing links except the link through which it reaches the node. Due to this, multiple copies of the same packet reach the destination node. Figure 3.6 shows the transmission of a packet from node-1 to destination node-5.

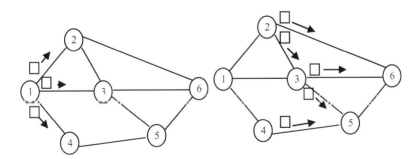

FIGURE 3.6 Flooding in the network from node-1 to node-5.

Flooding approach is used to collect traffic information in dynamic routing for updating routing table based on the collection of traffic information. Flooding is highly robust and may be used to send messages especially in military applications. The principal disadvantage of flooding is that it causes more congestion/traffic overload, even for a small number of connection requests.

3.3.2.4 Adaptive Routing

In adaptive routing, the route of sd pair is found dynamically, depending on the network state. The network state is established by the set of all connections currently in progress. The network state is represented by the following conditions:

- **Node failure:** When a node fails, it can no longer be used as part of a route.
- **Link failure:** When a link fails, it can no longer be used as part of a route.
- **Congestion:** When a particular portion of the network is heavily congested, it is desirable to route the traffic in another direction rather than through the congested area of the network.

For adaptive routing, information about the state of the network is exchanged between the nodes. There is a trade-off between the quality of information regarding network state and the amount of overload.

One approach is an adaptive shortest-cost-path routing which is suitable to wavelength-converted networks. Under this approach, each unused link in the network has a cost of 1 unit, each used link in the network has an infinite cost, and each wavelength-converter link has a cost of c units. If wavelength conversion is not available, then c is infinity. When a connection is established, the shortest-cost path of an sd pair needs to be determined. If there are multiple paths with the same distance, one is selected randomly. By selecting the wavelength-conversion cost c appropriately, paths with wavelength conversion are selected only when paths with wavelength continuity are not found. In shortest-cost adaptive routing, a connection is blocked only when there is no wavelength continuity or wavelength conversion from a source node to a destination node in the network. Adaptive routing path requires control and management protocols to continuously update the routing tables at the nodes.

In the approach based on LCP routing [1,23] for each sd pair, the succession of paths is pre-estimated. Upon the arrival of a connection request, the LCP among the pre-estimated paths is chosen. The congestion on a link is estimated by the number of wavelengths available on the link. Links that have fewer accessible wavelengths are said to be more congested. The congestion on a path is indicated by the congestion on the most congested link in the path. The priority to shortest paths is used in LCP to break any tie that occurs. A disadvantage of LCP is its computational complexity. In choosing the LCP, all links on all candidate paths have to be considered. A variant of LCP is proposed in [24] which only considers the first k links on each path (referred to as the source's neighborhood information), where k is a parameter of the algorithm. Another approach is the distributed adaptive algorithm which estimates the paths as per time delay as performance criteria. Bellman–Ford's algorithm is used for the estimation of routing paths. Each node of the network maintains two vectors:

$$D_i = \begin{pmatrix} d_{i1} \\ \vdots \\ d_{iN} \end{pmatrix} S_i = \begin{pmatrix} s_{i1} \\ \vdots \\ s_{iN} \end{pmatrix} \tag{3.4}$$

where

D_i = delay vector for node i

d_{ij} = current estimate of minimum time delay from node i to node j (d_{ij} – 0)

N = number of nodes in the network

S_i = successor node vector for node i

s_{ij} = next node in the current minimum time delay route from node i to node j.
Periodically (almost every 128 ms), each node exchanges its delay vector
with all of its neighbors. On the basis of all incoming delay vectors, a node
k updates both of its vectors as follows:

$$d_{kj} = \min_{i \in A} [d_{ij} + l_{ki}] \tag{3.5}$$

s_{kj} = i, using i that minimizes the expression above

A = set of neighbor nodes for k and l_{ki} = current estimate of delay from k to i.

This approach of distributed adaptive routing has the following shortcomings [17]:

1. It does not consider line speed but simply queue length.
2. Queue length is in any case an artificial measure of delay, as some variable
   amount of processing time elapses between the arrival of a packet at a node
   and its placement in outbound queue.
3. The algorithm is not very accurate as congestion clears slowly with increase
   in delay.

This distributed approach of adaptive routing can be modified for the improvement of
performance. In this modified approach, in every time interval (10 s), the node computes the average delay on each outgoing link. If there are any significant changes in
delay, the information is sent to all other nodes using flooding. Each node maintains
an estimate of delay on every network link. When new information arrives, it recom-
putes its routing table by using Dijkstra's algorithm. Even after modification, there is
a shortcoming: little correlation between reported values of delay and those already
experienced after routing under heavy traffic load. This is because some links are
loaded with heav traffic and some links are loaded with less traffic. This situation
can be avoided if the threshold value of maximum traffic load is fixed for each link
and the effect of this maximum value is to dictate that the traffic should not be routed
around a heavily utilized line by more additional hops.

3.3.2.5 Fault-Tolerant Routing

While establishing the connections in an optical WDM network, it also provides
some degree of protection against link and node failures in the network by considering a portion of spare capacity [25–27]. Two approaches to fault-tolerant routing are

protection and restoration. The most commonly used approach is protection needed to establish two link-disjoint light paths for the routes sharing any common link used in primary paths for every connection request. One light path, treated as a primary light path, is used for transmitting data, while the other light path is stored as a backup in the event that a link in the primary light path fails. To further protect against node failures, the primary and alternate paths are node-disjoint. Fixed-alternate routing is employed for protection. By selecting the alternate paths on the basis of the fact that their routes are link-disjoint from the primary path, the connection is protected from any single-link failures by allocating one of the alternate paths as a backup path. In adaptive routing, the backup path is also established immediately along with the primary path. The same routing algorithm determines the backup path. The resulting path should be link-disjoint from the primary path.

In restoration, the restoration path is established dynamically as and when the failure occurs. Restoration is successful if sufficient resources are available in the network. When a fault occurs, dynamic discovery and establishment of a backup path under the restoration approach are significantly longer than switching over to the pre-established backup path using the protection approach. The details of protection and restoration are discussed in Chapters 7 and 8, respectively.

The static formulation in Section 3.2 may also provide fault protection in the network. The modified formulation requires additional constraint equations for two light paths to be set up for each connection (one primary light path and one backup light path), and Chapters 7 and 8 may be referred to for protection and restoration.

3.3.2.6 Randomized Routing

Randomized routing is a kind of probabilistic technique which can be used for routing permutation traffic with a reduced number of wavelengths. Apart from these, randomized routing has the advantages of simplicity and suitability for both centralized and distributed routing techniques [17]. The technique is developed for the multistage ShuffleNet for k stages and 2^k nodes per stage. It works in three steps:

- In the first step, the route starts from a source node choosing one of the outgoing links randomly from every node on the route and reaches a random node in the source stage.
- In the second step, the route goes on in the same manner till it arrives at a random node in the destination stage.
- In third step, the route reaches the destination node by using the unique path with length k.

The maximum length of the chosen route is $3k - 1$. Two models are considered and used in their node architecture and wavelength selection.

The model M_1 has wavelength conversion capability. Each link has $3k - 1$ disjoint sets of equal number of wavelength channels. Once the route has been chosen by the above routing algorithm, the following wavelength selection is used. A free wavelength in the set, S_i, $1 \leq i \leq (3k - 1)$, is used by the route in the step i. A routing node can convert wavelengths in S_i to those in S_{i+1} for $1 \le i \le (3k - 1)$ and wavelengths in $S_{k+1}, S_{k+2}, \ldots, S_{3k-1}$ to those in S_{2k}. Using $O(\log^2 N)$ wavelengths, the permutation routing problem can be solved using a high probabilistic (success) guarantee of 1-1/poly(N) for the model, where poly(N) is a polynomial in N.

The model M_2 does not need the routing node having wavelength conversion capability. When a route is found using the above routing algorithm, it chooses randomly a wavelength which is free on all the links of the route. Using $O(\log^2 N)$ wavelengths, the permutation routing problem can be solved using a high probabilistic (success) guarantee of 1-1/poly(N) for the model.

3.4 WA SUBPROBLEM (HEURISTICS)

There are two types of WA approaches – static WA and dynamic WA. Assigning wavelengths to different light paths in a manner that minimizes the number of wavelengths under the wavelength-continuity constraint using the graph-coloring problem was discussed in Section 3.2.3.

By considering the routing approaches in Section 3.3, WA heuristics are discussed in dynamic WA after selecting paths by routing. These heuristics are also considered for the static WA by ordering the light paths, and then the paths are assigned wavelengths as per the order. The heuristic methods allot wavelengths to light paths as soon as the connection request arrive. For the dynamic problem, the number of wavelengths are also minimized, whereas in the static case, the number of wavelengths is fixed (this is the practical situation), and we have to minimize connection blocking.

The following heuristics are reported in the literature [1–5]: (1) random, (2) first-fit (FF), (3) least-used/spread, (4) most-used/pack, (5) min-product, (6) least-loaded, (7) max-sum, (8) relative capacity loss (RCL), (9) distributed relative capacity loss (DRCL), (10) wavelength reservation (WRSV), (11) protecting threshold, (12) priority-based WA, and (11) dispersion reduction WA. These heuristics can all be implemented as online algorithms and can be combined with different routing schemes. Most of the schemes attempt to reduce the overall BP for new connections, while the last two approaches aim to reduce the BP for connections that traverse more than one link. In our discussions, we use the following notation and definitions:

L: number of links.

M_t: number of fibers on link i.

M: number of fibers per link if all links contain the same number of fibers.

W: number of wavelengths per fiber.

$\pi(p)$: set of links comprising path p.

S_p: set of available wavelengths along the selected path p.

D: L-by-W matrix, where D_{ij} indicates the number of assigned fibers on link i and wavelength j. Note that the value of D_{ij} varies between 0 and M_t.

Load: For dynamic traffic, the holding time is exponentially distributed with a normalized mean of one unit, and connection arrivals are Poisson; thus, load is expressed in units of Erlangs.

3.4.1 WAVELENGTH SEARCH ALGORITHM

For assigning wavelengths, available wavelengths are searched in a path. Most of the search algorithms are based on graph coloring. Some of the constraints used in graph-coloring problems were discussed in Section 3.2.3. In this section, different searching algorithms are discussed.

3.4.1.1 Exhaustive Search

The algorithm providing the optimum coloring of a given graph is presented in Figure 3.7. The algorithm splits the possible colorings into two different cases in which each step until the graph is perfect where each node is a neighbor of all the other nodes. In each step, a pair of nodes which are not neighbors are searched [28,29]. Now these nodes can be colored with the same color or with different colors. If the nodes are given the same color, we can clearly merge them into one node inheriting all the neighbors of the merged nodes. Otherwise, if different colors are provided to the nodes, we can make an edge between them (right subtree in the figure). At the end, among all the perfect graphs, we pick the one which has the smallest number of nodes as shown in Figure 3.7.

3.4.1.2 Tabu Search

Tabu search (TS) is a relatively new heuristic method. It is a random local search in which some movements are forbidden (i.e., tabu) [29,30]. Usually a move leading back to the previous point is classified as a tabu move for certain number of rounds.

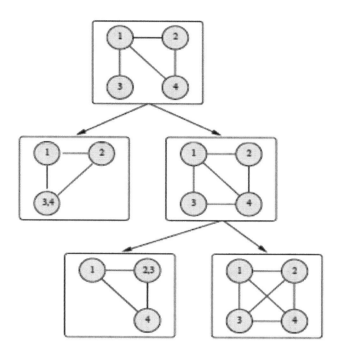

FIGURE 3.7 Coloring of nodes using exhaustive search.

This should make it possible to get away from local minima. The search is ended when the cost function reaches a certain predefined value or a certain number of rounds has elapsed. This algorithm differs from all the previous ones in that it considers finding not the minimum coloring but a legal k-coloring for the given graph, i.e., it tries to select for each node one of the k colors in such a way that no neighboring nodes get the same color.

We consider $s = (V_1, V_2 \ldots V_k)$ be a partition of graph G, where subset V_i of nodes represents those nodes having color i. Define a cost function as $f(s) = \sum_i E(V_i)$, where $E(V_i)$ is the number of edges in subgraph V_i. If there is an edge in some subgraph, it means there are neighbors sharing the same color. So when $f(s) = 0$, we have a legal k-coloring for the graph. We are given a graph G, target number of colors k, length of tabu list $|T|$, number of neighbors rep, and maximum number of iterations nmax [30].

1. Set some initial configuration $s = (V_1, V_2 \ldots V_k)$.
2. Set nbit = 0.
3. Initialize tabu list T.
4. As long as $f(s) > 0$ and nbit < nmax
   a. Find nrep neighbors (where nrep is number of neighbors) s_i for which $s \rightarrow s_i \notin T$ or $f(s_i) \leq Af(s)$.
   b. Choose the best among them (or the first for which $f(s_i) < f(s)$).
   c. Update tabu list T.
   d. Set $s = s_0$ and nbit = nbit + 1.
5. If $f(s) = 0$, we find a legal k-coloring for a given graph; otherwise increase k and repeat again.

In the neighborhood of a partition, we define partitions where one node is moved to another subset. In order to find the minimum k for which the algorithm finds a legal coloring, we must run the algorithm several times with decreasing values of k until the algorithm fails. On the other hand, if we are only interested in getting a feasible k-coloring, the iteration is not required.

3.4.1.3 Simulated Annealing

Simulated annealing (SA) is one of the techniques for wavelength coloring [28,29]. The idea is based on simulated annealing of some object. The objective function represents the energy of the system, and the control variable T represents its temperature. In the algorithm, the higher the temperature is, the greater is the probability of acceptance of a move leading to a higher energy state. The node coloring problem is solved with SA: The energy E of the system is the number of used colors. Following are the steps in the SA algorithm:

1. In the beginning, assign each node a unique color.
2. Set the temperature $T = T_0$ (e.g., $T_0 = 1$).
3. Choose a random node and a random new color for it. Make sure that the new color does not lead to an illegal configuration.
4. Compute the change of energy ΔE.

5. If $\Delta E < 0$ or $e^{\Delta E/KT} >$ random(0, 1), accept the change.
6. If there have been at least M changes or N trials, then set $T = \alpha$. (α is a small constant, e.g., 0.95.)
7. If $T > T_i$, go back to 3.

Also other kinds of formulations have been suggested for the energy function. A drawback with this formulation is that energy can only have discrete values, and this makes it hard for the algorithm to find the right direction to advance.

3.4.1.4   Genetic Algorithms

The genetic algorithm (GA) is another widely used method for wavelength search in graph coloring. In GA, the idea is to simulate evolution. Here vectors represent genotypes, and the node coloring problem, which can be solved by GA, has to be found [28]. In this case, the vectors define the order in which the nodes are colored. So the best ordering to color the nodes has to be found with the greedy algorithm. The choice of crossover operation for permutations is not straightforward, and several different schemes have been proposed. Let A and B be the parents. A is chosen randomly, but those that give good coloring are favored. B is chosen randomly from the whole population. The length of both vectors is N.

The algorithm procedure is as follows:

1. Initialization: Set indices $i_A = i_B = 1$, and set the child C to null.
2. Choose vector A with probability of 0.75 and vector B with probability of 0.25 as parents.
3. Add the next element, pointed to by i_A or i_B, of the chosen vector to the child vector C if it is not already there.
4. Increment the value of index by 1 so that it points to the next element of the parent vector.
5. Repeat this until the child C contains all the values 1, 2, …, N.

So basically the order of both parents is combined to get the child. In each step, the next element of randomly chosen parent is copied to the child if it is not there yet. Index i_A points to the next element of parent A and i_B to that of parent B. As a mutation operator, we simply exchange the places of two random nodes in the vector.

3.4.2   WA HEURISTICS

There are many approaches used for WA/selection: randomized approach, first fit, least used, most used, min-product approach, least loaded, max-sum, relative capacity loss, distributed relative capacity loss, priority-based approach, dispersion reduction assignment, etc.

3.4.2.1   Random WA (R)

In this scheme, the first search is to determine the set of all wavelengths available on the path chosen by routing. Among the available wavelengths, a wavelength (usually with uniform probability) is chosen in the following manner [1,3,5].

All the wavelengths are indexed, and all the indices can be selected when generating the order randomly. The wavelength usage factor is not used and thus does not require any global state of information.

3.4.2.2   First-Fit (FF) Approach

In this approach, we index all wavelengths. For searching the available wavelengths, a lower-indexed wavelength is chosen before a higher-numbered wavelength. Its time complexity is lower than that of random WA since there is no need to search the entire wavelength space for each connection request allotted to the path. The idea of FF is to orient all the in-use wavelengths toward the lower end of the wavelength indices so that the probability of continuous and longer paths being available toward the higher end of the wavelength is high. In FF, BP and fairness, computational overhead, and time complexity are decreased and are also less than those of other schemes [1,3].

3.4.2.3   Least-Used (LU) Approach

LU approach chooses the wavelength which is less used in the network for giving the load balancing of all the wavelengths. This approach considers breaking the long wavelength paths quickly. Here connection requests are set up only on a small number of links serviced in the network. The performance of LU is worse than that of the random approach, because of additional communication overhead needed to compute the LU wavelength). This approach also needs additional storage and has higher computation complexity [1,5].

3.4.2.4   Most-Used (MU) Approach

This approach operates in a way opposite to that of the LU approach. It chooses the MU wavelength in the network. Its performance is better than LU approach [31,32]. The overhead, storage, and computational complexity are the same as those in the LU approach. The MU approach performs slightly better than the FF approach in packing connections into fewer wavelengths and conserving the extra capacity of less-used wavelengths.

3.4.2.5   Min-Product (MP) Approach

This approach is employed in multi-fiber networks [33]. MP approach to a single-fiber network basically operates with a concept similar to that of the FF approach. Here MP approach is based on packing wavelengths into fibers, and it minimizes the number of fibers in the network. MP approach first computes:

$$\prod_{l \in \pi(p)} D_{lj}$$

for each wavelength j, i.e., $1 \leq j \leq W$. If X represents the set of wavelengths that minimize the above value, then MP approach considers the lowest-numbered wavelength in X. This approach has additional computation costs/complexity.

3.4.2.6   Least-Loaded (LL) Approach

Like MP, the LL heuristic is also employed in multi-fiber networks [34]. This heuristic selects the wavelength having the largest residual capacity on the most-loaded link along route p. While using in single-fiber networks, the residual capacity is either 1 or 0, and it chooses the lowest-indexed wavelength with residual capacity 1. Thus, it is also used as an FF approach in single-fiber networks. The LL approach chooses the lowest-indexed wavelength j in S_p.

$$\max_{j \in S_p} \min_{l \in \pi(p)} (M_l - D_{lj}) \qquad (3.6)$$

The LL approach's BP performance is slightly better than those of MU and FF approaches in a multi-fiber network.

3.4.2.7 MAX-SUM (MS) Approach

This approach [35] was employed in multi-fiber networks, but it is also used in single-fiber networks. It uses all possible paths (light paths with their pre-selected routes) in the network for path selection and maximizes the remaining path capacities after light path establishment. The traffic matrix (set of possible connection requests) is known in advance, and the route for each connection is pre-selected. This condition is fulfilled if the traffic matrix is stable for a period of time. To describe the heuristic, the following notations are taken.

φ represents the network state specifying the existing light paths (routes and WAs) in the network. In this approach, the link capacity on link l and wavelength j in state φ, $r(\varphi, l, j)$, is considered as the number of fibers on which wavelength j is unused on link l, i.e.,

$$r(\varphi, l, j) = M_l - D(\varphi)l_j \qquad (3.7)$$

where $D(\varphi) = D$ matrix in state φ. The path capacity $r(\varphi, p, j)$ on wavelength j represents the number of fibers on which wavelength j is available on the most-congested link along the path p, i.e.,

$$R(\varphi, p) = \sum_{j=1}^{\max} \min_{l \in \pi(p)} c(\varphi, l, j) \qquad (3.8)$$

We consider

$\Omega(\varphi, p)$ = no of possible wavelengths that are available for the light path that is routed on path p.

$\varphi'(j)$ represents the next state of the network if wavelength j is allotted to the connection. The MS approach chooses the wavelength j to maximize the quantity expressed as

$$\sum_{p \in P} R(\varphi'(j), p)$$

P is the number of all potential paths for the connection request in the current state. Once the light path is set up for the connection, the network state is changed, and the processing of the next connection request starts.

3.4.2.8   Relative Capacity Loss (RCL) Approach

RCL approach [36] can be viewed as an MS approach that chooses the wavelength j to minimize the capacity loss on all light paths expressed as

$$\sum_{p \in P} \left\{ R(\varphi'(j)) - R(\varphi'(j), p) \right\}$$

where φ is the network state before the light path is set up. The capacity on wavelength j reduces after the light path is set up on wavelength j; MS chooses wavelength j by minimizing the total capacity loss on this wavelength. Total capacity loss is written as

$$\sum_{p \in P} \left\{ r(\varphi'(j)) - r(\varphi'(j), p) \right\}$$

In RCL, wavelength j is chosen to minimize the relative capacity loss written as

$$\sum_{p \in P} \left\{ R(\varphi'(j)) - r(\varphi'(j), p) \right\} / r(\varphi, p, j)$$

It is estimated that RCL approach minimizes capacity loss, but it may not give the best choice of wavelength. Choosing wavelength i may block one light path p_1, whereas choosing wavelength j decreases the capacities of light paths p_2 and p_3 but does not block them. So wavelength j should be chosen over wavelength i, even though the total capacity loss for wavelength j is more than the capacity loss of wavelength i. Thus, the RCL approach calculates the relative capacity loss for each path on each available wavelength and then selects the wavelength that minimizes the sum of the relative capacity losses on all the paths. Both MS and RCL approaches are employed for non-uniform traffic by taking a weighted sum over the capacity losses.

3.4.2.9   Distributed Relative Capacity Loss (DRCL) Approach

There are additional costs in applying all the mentioned algorithms – LU, MU, MP, LL, MS, and RCL – as per global knowledge of the network state in a distributed control network. Information on the network state is exchanged frequently between the nodes to ensure accurate calculations, by using link-state routing protocol. Although the MS and RCL approaches perform better, the implementation of these is not easy but cost effective in a distributed environment, and also MS and RCL both employ fixed routing that does not improve network performance. Two issues have to be dealt with:

- how information of network state is exchanged.
- how the amount of calculation is reduced upon receiving a connection request.

Here, each node in the network keeps record of information on the capacity loss on each wavelength on the basis of table lookup; a small amount of calculation is required upon the arrival of a connection request. To obtain a valid table, the related values are changed after updating the network state. To simplify the computation, the DRCL algorithm is employed. The routing is performed using Bellman–Ford's algorithm [1]. As per Bellman–Ford's algorithm, routing tables are determined in each node with its neighboring nodes as a destination and updates its own routing table as per information of update network state. An RCL table is needed at each node, and the nodes are allowed to exchange their RCL tables as well. The RCL tables are exchanged in a similar manner as the routing tables. Each entry in the RCL table has information of wavelength w, destination d, and rcl(w, d). When a connection request arrives, more than one wavelength is available on the selected path by computation carried out among these wavelengths on the basis of the information acquired at each node. The MS and RCL approaches consider a set of potential paths for future connections. DRCL approach considers all the paths from the source node of the arriving connection request to every other node in the network, excluding the destination node of the arriving connection request. DRCL approach then chooses the wavelength minimizing the sum of rcl(w, d) over all possible destinations d calculated at source node as follows.

– If there is no path from node s to node d on wavelength w, then rcl (w, d) = 0; otherwise

if there is a direct link from node s to node d, and the path from s to d on wavelength w is routed through this, then rcl(w, d) = 1/k, where k is the number of available wavelengths on this link through which s can reach d; otherwise

If the path from source node s to node d on wavelength w starts with node n (n is s node's next node for destination d on wavelength w), and there are k wavelengths available on link of s node to next node n through which s can arrive, then rcl(w, d) at node s is set to be (l/k) and similarly, rcl(w, d) is estimated at next node n.

Most of the WA schemes minimize BP considering that longer light paths have a higher probability of getting blocked than shorter paths and some schemes attempt to protect longer paths. The schemes mentioned in this section need wavelength reservation (WRSV) and wavelength threshold protection (WThr) [37]. Normally, an RWA algorithm gives priority to shorter hop counts in comparison to longer hop counts, blocking longer hop connections. This gives rise to the fairness problem. In order to improve fairness among connections, an appropriate control is employed to regulate the admission of a connection request. There is a BP associated with each sd pair having longer hop connections and the corresponding arrival stream. In the next section, we discuss the same.

3.5   FAIRNESS IMPROVEMENT

A global measure is required to control overreduction of blocking of longer hop connections for fairness improvement. The use of wavelength converters improves blocking performance. A wavelength converter is capable of shifting one wavelength of incoming signal to another wavelength to relax wavelength constraints. Wavelength rerouting is another option to improve BP performance. Although both wavelength converter and wavelength rerouting increase fairness, in both the cases, time complexity and cost of the network increase. A fairness improvement algorithm has the following properties in addition to improving fairness:

a. Wavelength channel utilization must be high.
b. The algorithm must be flexible enough to choose the desired trade off between fairness level and global performance loss.
c. It must be suitable for networks with different degrees of connectivity.
d. The fairness improvement algorithm makes shorter hop connection more penalized than longer hop connections.

For achieving fairness in wavelength routing, the following issues must be addressed [1]:

- WRSV
- WThr
- limited alternate routing (LArout)
- static priority
- dynamic priority.

3.5.1 WAVELENGTH RESERVATION

In wavelength reservation, to lower the BP of traffic stream on a longer route, one or more wavelengths are kept solely for this purpose on every link of the route. The links used for the route are called logical links. When a connection is allowed to try for other wavelengths, others are not allowed to use the reserved logical link. Since an exclusive logical link is reserved for a traffic stream, the chance of being blocked is reduced, and at the same time, setup time is also reduced. Hence, it increases fairness. It is difficult to reserve logical links always. A complex procedure will be required to optimally choose logical links for a topology and a fixed number of wavelengths per fiber. WRSV is either centralized or distributed. There are two ways of WRSV to assign a wavelength for a route: forward reservation and backward reservation. Since these two reservations are used for mesh topology, they are based on a distributed approach because connection requests arrive at the node at which a connection request has to be established and allocated with wavelength.

3.5.1.1 Forward Reservation

The forward reservation method is based on alternate path routing, and it reserves the wavelengths on the links while the control message is passed forward from the source to the destination seeking for a free wavelength on a route. The message carries along with it the set of wavelengths denoted by S_{free} that is free on all the links. When a node is visited by a message through the incoming link of the node, S_{free} is updated by taking the intersection of this set with the set of wavelengths that are free on the outgoing link of the node. Every node maintains a table having the state of each outgoing link with available wavelengths. The light path is the connection_id having the source, destination, and seq_no. The following control messages are used by the forward reservation protocol [1].

- **Res:** This control message is used for reserving wavelengths on the links of a route. It consists of various fields – connection_id, S_{free}, and route. The route contains the route information specifying links. It travels from the source to destination along the route.
- **Res_Succ:** This control message indicates that the reservation of some wavelength has been successful. It consists of various fields – connection_id, S_{free}, and w_{res}, where w_{res} = free wavelength that can be used by the connection request. It travels from the source to destination along the route.
- **Res_Fail:** This control message indicates that the reservation of a wavelength has failed. It consists of various fields – connection_id, S_{free}, and route. This message travels from intermediate nodes to the source node along the route.
- **Rel:** This control message is sent to release the light path on some wavelength. It consists of the connection_id of the light path released. This message travels from the source node to the destination node along the route.

The different parameters used in designing the forward reservation method are mentioned below [1]:

a. **Max_WL:** Max_WL is the maximum number of wavelengths in set S_{free} at the time of starting a new search for a free wavelength on some route. In other words, it is the initial size of the set S_{free}. In this case, the protocol always tries to reserve as many wavelengths as possible on the links of a route. This will reduce setup time and increase the chance of getting a free wavelength. But the reservation conflict increases because a connection request attempts to reserve some wavelengths which are already reserved by other connection requests.

b. **Max_TRIES:** Max_TRIES is the maximum number of times a free wavelength is searched before the source node gives up. The more the tries, the higher the chance of finding a free wavelength.

c. **GAP:** GAP is the retransmission time gap between two successive tries by the source. A small gap time increases control traffic and reservation conflicts. On the other hand, a large gap time makes the state of the network during subsequent tries more uncertain. Moreover, a large gap time increases connection setup time.

d. **BUF_TIME:** BUF_TIME is the buffer time of the control message, i.e., the time till which it is held at the intermediate node. Zero BUF_TIME is simple to implement as there is no need for buffers to queue the pending messages. A non-zero BUF_TIME will increase the chance of finding a free wavelength on the outgoing link.

Protocol Mechanism

There are two situations used for processing requests in forward reservation: successful reservation of a wavelength and reservation failure. The following procedure is used for processing requests arriving from nodes to establish and release light paths [3,5].

Step-1: When a connection request arrives at node s to some other node d, the source node s prepares a Res message with the required information. A new sequence number, seq_no, is obtained. The S_{free} with MAX_WL wavelengths is constructed. The control message is transmitted through the first route on the control channel.

Step-2: When the message is at intermediate node i, it selects the outgoing route from the route information. The set S_{free} is updated by considering set of common available wavelengths with the outgoing link. If S_{free} has available wavelengths, the node forwards the message to the next neighbor node with updated S_{free} along the outgoing link chosen from the route information; otherwise go to the next step.

Step-3: If S_{free} has no available wavelength, the Res message is temporarily queued in the local buffer for BUF_TIME till some required common wavelength is free in the same outgoing link before the expiry of BUF_TIME. After expiry of BUF_TIME, the Res message is removed, and Res_Fail message is generated and sent back to the node (from where Res message comes) along the incoming link and ultimately reaches the source. If Res message is released to the next node of the route, go to the next step.

Step-4: Repeat step-2 and step-3. Res message reaches the destination node d along the links via intermediate nodes of the route selected as primary path. Once the Res reaches the destination d, Res_Succ message is released toward the source with the selection of wavelength W_{res} from available wavelengths in set S_{free}.

The above procedure is repeated for MAX_TRIES number of times leaving a GAP between two successive tries. If everything fails, the connection request is rejected. When data transmission on the allocated light path is complete, the source node s prepares a Rel message carrying the connection_id and route information of the light path. This message is transmitted toward the destination d. When a node receives Rel message, it updates the table with the information that the wavelength used by the light path on the specified link can no longer be used and is now available for another connection request. When the Rel message reaches d, the release operation is completed. Figure 3.8a shows the forward reservation situation, where there is a free wavelength and the reservation is successful. Figure 3.8b shows the forward reservation situation, where there is no free wavelength and the reservation fails.

3.5.1.2   Backward Reservation

In the forward reservation method, wavelengths are reserved on the links of the specified route using control messages during forward transmission from source s to destination d. There may be wavelength conflicts during reservation, since the

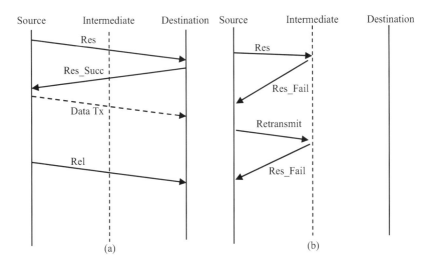

FIGURE 3.8 Forward reservation: (a) free wavelength and successful reservation and (b) unsuccessful reservation.

set of wavelengths denoted by S_{free} participate during reservation in intermediate nodes. This may provide poor wavelength or bandwidth utilization. To remove these shortcomings, backward reservation is used. In this technique, there is no reservation during forward transmission of control messages; rather all the wavelengths are collected during forward transmission. Once the set of wavelengths are known to the destination node, it sends the control message with the set of available wavelengths. While the control message travels backward from destination to source, it reserves the wavelength on the link, and ultimately one wavelength is reserved after the control message reaches the source node. The following control messages are used by backward reservation protocol [1].

- **Collect:** This message is used to collect the set of free wavelengths on the links of a route. It consists of three fields – connection_id, route, and S_{free}. Since the message does not reserve any wavelength, S_{free} contains initially all the wavelengths. It travels along the route from the source to the destination node.
- **Collect_Fail:** This message represents the collection of free wavelengths on a route. It consists of two fields – connection_id and route. It traverses back from intermediate nodes to source node along the route.
- **Res:** This control message is used for reserving wavelengths on the links of a route. It consists of three fields – connection_id, S_{free}, and route. The route contains the route information specifying links. It travels from the source to destination.
- **Res_Succ:** This control message indicates that the reservation of some wavelength has been successful. It consists of three fields: connection_id, route, and w_{res} where w_{res} = free wavelength can be used by the connection request. It travels from the source to destination along the route.

- **Res_Fail:** This control message indicates that the reservation of a wavelength has failed. It consists of two fields – connection_id and route. This message travels from the intermediate node to the source node along the route.
- **Rel:** This control message is sent to release the light path on some wavelength. It carries the connection_id of light path to be released. This message travels from the source node to the destination node along the route.

Protocol Description

When a connection request arrives at the source node, the following steps are to be taken in backward reservation [1].

Step-1: When a request arrives at node s for connection to some other node d, the source node s prepares a Collect message with the required information. A new sequence number seq_no is obtained. S_{free} contains all the wavelengths. The control message is transmitted through the first candidate route on the control channel.

Step-2: When the message is at intermediate node i, it selects the outgoing route from the route information. The set S_{free} is updated by considering a set of common available wavelengths with the outgoing link. If S_{free} has available wavelengths, the node forwards the message to the next neighbor node with updated S_{free} along the outgoing link chosen from the route information; otherwise it goes to the next step.

Step-3: If S_{free} has no available wavelength at node i, the Res message is temporarily queued in the local buffer for BUF_TIME till some required common wavelength is free in the same outgoing link before the expiry of BUF_TIME. After expiry of BUF_TIME, the Res message is removed, Res_Fail message is generated and sent to the destination node along the outgoing link, and Fail_Info is sent to the source node. The Collect_Fail message is also generated so that it can try other routes and go back to the source node along the incoming links which are determined from the route information. When the source node receives Collect_Fail message, it goes to the next step.

Step-4: It searches the next candidate node, when the source node receives Collect_Fail message.

Step-5: Step-2, step-3, and step-4 are repeated with a GAP till the number of retransmissions has reached MAX_TRIES. When the Collect message reaches the destination node d, it generates Res message having a new set of S_{free} consisting of the subset of free wavelengths indicated by the Collect message. The message is transmitted from destination d to source node s.

When the message is at intermediate node i, it selects the next link from route information, and S_{free} is updated by considering common wavelengths on the outgoing link. The above procedure is repeated for MAX_TRIES number of times leaving a GAP between two successive tries. If everything fails, the connection request is rejected. When the data transmission on the allocated light path is complete, the

source node s prepares a Rel message carrying the connection_id and route information of the light path. This message is transmitted toward the destination d. When a
node receives a Rel message, it performs the necessary table update that the wavelength used by the light path on the specified link can no longer be used and is now
available for another connection request. Figure 3.9a shows the backward reservation
situation, where there is a free wavelength and the reservation is successful. Figure
3.9b shows the forward reservation situation, where there is no free wavelength and
the reservation fails. Figure 3.9c represents the situation where the reservation fails
due to non-availability of free wavelengths which the Res control message is trying to reserve. This failure of reservation is due to the fact that during the forward
transmission of the Collect message the wavelength is free but during backward
reservation it is not free.

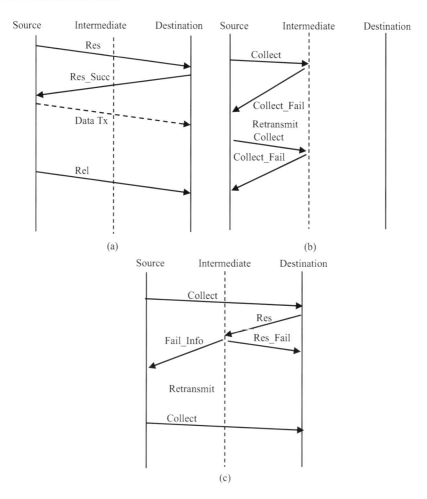

FIGURE 3.9 Backward reservation: (a) successful reservation of wavelength, (b) failure due
to the non-availability of free wavelength, and (c) failure of WRSV.

3.5.1.3 Congestion-Based Routing WRSV Method

Congestion-based routing WRSV is an extension of backward reservation technique [1,38] for a given sd pair; the least-congested route among all the candidate routes is selected. The congestion of a route is measured by the number of light paths that are currently used in the route.

In this case, when a connection request arrives at node s, it starts searching a parallel route. The source node generates Collect messages, one for each candidate route corresponding to the destination node d. All these messages are transmitted to the destination, collecting all the free wavelengths on various links of the candidate routes. When the node d receives Collect messages transmitted from all the candidate routes, it chooses the best route among them on the basis of least congestion. The best route having least congestion is estimated on the basis of the fact that the route has the maximum number of free wavelengths available. The destination node generates a Res message and transmits it to the source along the selected route. S_{free} of the Res message is set to that of the Collect message received on the selected route. When the Res message reaches the source, one wavelength is chosen for assigning the connection request. The situation may arise that the wavelength which is available during forward transmission is not available during backward transmission. This results in reservation conflicts because of attempting reservation for many connection requests. In order to reduce these reservation failures during backward transmission, the Collect message can collect the conflicts for each of the wavelengths on various links. At the same time, each node keeps track of the number of connection requests. On the basis of all the above information, the Collect message chooses appropriate wavelengths on the route for reservation. If there is no free wavelength, then the destination node sends Collect_Fail toward the source node s [1].

3.5.1.4 k-Neighborhood Routing

For LCP routing, all the candidate routes are searched in parallel. In fact, this results in the increase of both control message traffic and connection setup time. The k-neighborhood routing method [1] attempts to overcome the increase of the above difficulties. When a connection request reaches the node, the two steps are followed: route selection and wavelength selection.

In the route selection step, the best route is chosen among all the candidate routes by collecting congestion information about the routes on the first k links. This method ensures that the control traffic and setup time are reduced. The source node s prepares Collect messages, one for each candidate route, and sends them along the candidate routes. Every collect message passes through k links on the route gathering wavelength availability and conflict information. The kth node sends to the source a message with a set of wavelengths available on the first k links. When the source receives a message containing the above information from k neighborhood nodes, the least-congested route is chosen. Instead this information could be collected when the actual demand arises. When another connection request arrives, the route is selected based on the locally available information.

The next phase of k-neighborhood routing is wavelength selection in which a free wavelength is chosen by using forward reservation method and is reserved on the selected route. If no wavelength is available on the chosen route, other routes can be considered in the non-decreasing order of congestion. Backward reservation may also be used for WRSV. In backward reservation, the Collect message is taken for collecting conflict information on the basis of which the wavelength is chosen [1].

3.5.2 WThr Protection

In WThr protection, connections of shorter hops are allocated with a wavelength only if the number of the idle wavelengths on the link is at or above a given threshold. Some free wavelengths can be used by longer hop connections. By increasing the chance of acceptance, the setup time of more longer hops becomes shorter. This results in assignment of longer hop connections. In this situation, a number of wavelength-discontinuous routes may be generated.

If the traffic load for longer hops in the network is lower, preventing shorter hop connections from sharing some wavelengths will result in poor wavelength utilization and unnecessary degradation in the performance of shorter hop connections. A connection with the longest route among the designated shorter hop connections may acquire more penalties in comparison to a connection using the shortest route among the designated longer hop connections.

3.5.3 Limited Alternate Routing

In this method, shorter hop connections are provided with few alternate routes, whereas longer hop connections are provided with more alternate routes. By limiting the number of alternate routes for shorter hop connections, more longer hop connections are accommodated [39–41]. It provides moderate setup time and also increases fairness. In case of sparse network having high traffic load, the fairness may deteriorate. When traffic load is low, the provision of additional routes assigns wavelengths for both shorter and longer hop connections as the chance of finding a free wavelength on an additional route is enhanced. As traffic load increases, WA to longer hop connections is an important issue. Additional alternate routes may not provide significant improvement because additional routes have many hops sharing many links resulting in the use of more wavelengths for a connection request.

3.5.4 Static Priority Method

In this method, the available wavelengths are divided into two subsets – F set and P set. Accessing the wavelengths from F set is free, whereas accessing the wavelengths from P set is prioritized [42–44]. Every connection request (sd) has priority associated with it fixed a priori. This method first searches wavelengths from F set. If no free wavelength is available from F set, then the connection registers its priority on the links of the route and tries to get wavelengths from P set as per the priority level along the route. WRSV fails if higher-priority connections are registered.

The wavelength utilization increases as no request is prevented from using any wavelength. In fact, the longer hop connections have more chances to get wavelengths from P set.

The disadvantages of this method are various connection requests need to be determined and the higher the P set, higher is the penalty of shorter hop connections. The connection setup time is more because of the registration of priorities.

3.5.5 DYNAMIC PRIORITY METHOD

This method overcomes the shortcomings of static priority method by updating the priority of various streams. The priority of traffic stream at an instant of time depends on the past information – average, minimum, and maximum of estimated values of the streams in the network [44]. Every node can estimate the performance of the outgoing streams in terms of their BP. The average, minimum, and maximum of estimated values are transmitted periodically to other nodes in the network. Reservation is done in the same way as done in the static priority method. Since the priorities are estimated on the performance, shorter hop connections are not over penalized. If the performance degrades for some connections, the priorities of these connections should be increased. In turn, the performance increases.

In this case, the wavelengths are divided into two sets – free access set (F set) and priority access set (P set). A connection request is assigned with free wavelength from F set on the link of the route, whereas it can access a wavelength from P set as per priority estimated on the basis of information updated periodically. Every node keeps a table of pre-computed candidate routes. The priority information on a link includes a number of computing requests with their priority level. So the connection_id includes source, destination, and seq_no. The following control messages are used by the routing protocol.

- F_Res is used to reserve wavelengths in F set on the link of the routes. It consists of connection_id, route, and S_{free}. Since the message does not reserve any wavelength, S_{free} initially contains all the wavelengths. It travels along the route from the source node to the destination node.
- F_Res_Succ indicates the successful reservation of wavelengths in F set on the link of the routes. It consists of connection_id, route, and w_{res}. w_{res} represents the free wavelength that is assigned to the connection request. It travels back along the route from the destination node to the source node.
- F_Res_Fail indicates the unsuccessful reservation of wavelengths in F set on the link of the routes. It consists of connection_id and route. It travels back along the route from an intermediate node to the source node.
- Rel is sent to release the light path on some wavelength. It consists of only connection_id. It travels along the route from the source node to the destination node.
- Pri_Reg is used to register the priority of the connection request on the links of a route. It consists of connection_id, route, and pri_level. The pri_level indicates the priority level of the connection request. It travels along the route from the source node to the destination node.

- Pri_Cancel is used to cancel the priority of the connection request registered earlier by the request on the links of a route. It carries connection_id, route, and pri_level. The pri_level indicates the priority level of the connection request. It also travels along the route from the source node to the destination node.
- P_Res is used to reserve wavelengths in P set on the link of the routes. It consists of connection_id, route, S_{free}, and pri_level. It travels along the route from the source node to the destination node.
- P_Res_Succ indicates the successful reservation of wavelengths in P set on the link of the routes. It consists of connection_id, route, and w_{res}. It travels back along the route from the destination node to the source node.
- P_Res_Fail indicates the unsuccessful reservation of some wavelengths in P set on the link of the routes. It consists of connection_id and route. It travels back along the route from the intermediate node to the source node.
- Blk_Info carries information about the blocking performance of different traffic streams originating from a particular node. It broadcasts this information to other nodes in the network along the links of a pre-computed spanning tree over the shadow network. This message is sent periodically from every node to all the other nodes for computing priority level.

Protocol Mechanism

The procedure of the protocol for light path establishment and release is given below [44].

Step-1: When a request arrives at node s for connection to some other node d, the source node s prepares an F_Res message with the required information. A new sequence number seq_no is obtained. S_{free} has a maximum of MAX_WL wavelengths. The control message is transmitted through the first candidate route on the control channel.

Step-2: When the message is at intermediate node i, it selects the outgoing route from the route information. The set S_{free} is updated by considering the set of common available wavelengths with outgoing link from F set. If the new S_{free} has available wavelengths from F set, the node forwards the message to the next neighbor node with updated S_{free} along the outgoing link chosen from the route information. If the F_Res message reaches the destination node d, the reservation is successful, and node d transmits an F_Res_Succ. A wavelength in S_{free} is chosen and assigned to the field w_{res}. This message is sent back to the neighbor along the incoming link. This moves toward the source, updating table entries at the intermediate nodes and releasing wavelengths reserved except the selected wavelength. Otherwise it goes to the next step.

Step-3: If S_{free} has no available wavelengths at node i, the F_Res_Fail message is prepared to be transmitted and sent to the neighbor along the incoming link determined from the route information and ultimate sent back to the source node. When the source node receives the F_Res_Fail message, it goes to the next step.

Step-4: It creates a new S_{free} and again sends the F_Res along the route. If all the wavelengths in F set have been searched, then step-2 and step-3 are repeated for the remaining candidate routes one by one up to a maximum of MAX_TRIES leaving the GAP. If everything fails, it goes to the next step.

Step-5: The F_Res starts searching for a free wavelength from P set. The source node generates Pri_Reg message which includes its estimated priority level and sends it along the candidate routes. This message registers the priority on the links by incrementing the count associated with the appropriate priority level. Finally, it reaches the destination. The source node sends P_Res along the route. When an intermediate node i receives the same, it first determines the outgoing link by examining the number of competing requests registered on the outgoing link at the priority level higher than that mentioned in the received message. If no such registration exists at the higher priority levels and the intersections of S_{free} with the set of wavelengths from P set on the outgoing link is not empty, then the message is forwarded to the next intermediate node after performing necessary updates in the tables. Finally, it reaches the destination node d, and after it reaches there, the destination node transmits the P_Res_Succ message toward the source node along the same route passing through the same intermediate node. After receiving this message, the source node transmits Pri_Cancel along the same route before the transmission of the date through the same route. If the reservation fails, then it goes to the next step.

Step-6: The node i generates a P_Res_Fail message to be sent to the previous node along the incoming link, and the same message propagates toward the source through various intermediate nodes updating the state information in the tables. If the source node receives the same, it creates a new S_{free} and again sends the P_Res message along the route. If all the wavelengths in P set have been searched, then the same procedure for MAX_TRIES is repeated a number of times with the GAP between two successive tries. If everything fails, the connection request is rejected, and the source node generates the Pri_Cancel message which is sent along the routes.

When data transmission on an assigned light path is completed, the source node s generates a Rel message carrying the route and wavelength information used by the light path up to the destination. When the nodes receive the release messages, the messages are updated in the table. The proposed protocol and various control messages are shown in Figure 3.10. Figure 3.10a shows successful reservation of a wavelength in F set, whereas Figure 3.10b shows successful reservation of a wavelength in P set and Figure 3.10c represents reservation failure.

Priority Computation

Since the wavelength reservation is also made on the basis of the priority level of sd pair from P set, it is required to know how the priority level of sd pair is computed [41–44]. This method uses three priority levels – low, medium, and high. The priority levels are computed dynamically by estimating periodically the global network

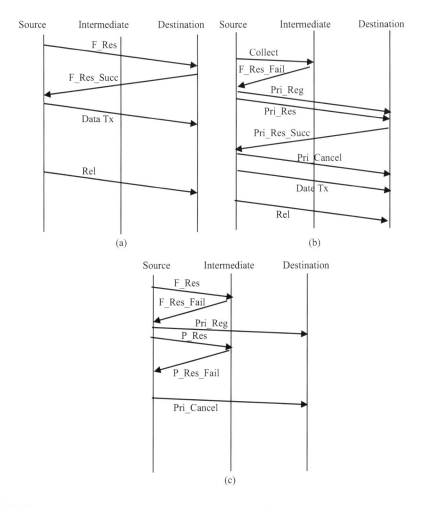

FIGURE 3.10 Dynamic priority reservation: (a) successful reservation of wavelength in F set, (b) successful reservation of wavelength in P set, and (c) failure of WRSV.

performance of the traffic streams. Every node keeps track of the BP performance of the traffic streams originating from it and destined to other nodes. This is measured with reference to a predefined start time. This information is updated at regular intervals of time. Thus, each node keeps track of the minimum (min), maximum (max), and average (avg) of the blocking probabilities of the traffic streams originating from it. At regular intervals of UPDATE (a protocol design parameter) time units, the values of min, max, and avg are updated. These values are kept in the Blk_Info message broadcasted to other nodes through links of a pre-computed spanning tree over the shadow network having the same number of nodes.

When a node receives a Blk_Info message, it computes a new estimate of the global minimum (gmin), global maximum (gmax), and global average (gavg) of the blocking performance of various traffic streams. These values are used to determine the priority level sd pairs. Two threshold values, Lthr (lower threshold) and Uthr (upper threshold), are computed as functions of gmin, gmax, and gavg. If the observed BP of an sd pair is above Uthr, its priority level is set to be high. If the observed BP of an sd pair is below Lthr, its priority level is set to be low. If the observed BP of an sd pair is in between Uthr and Lthr, its priority level is set to be medium. Thus, priority levels are changed adaptively. This provides several advantages: First it avoids over penalizing for shorter hop connections where for the degradation of the performance of shorter hop connections, the priority level increases.

3.6 MATHEMATICAL FORMULATION OF RWA

The physical topology of the network has been modeled as a unidirectional graph G (V, E), where V is the set of nodes and E is the set of links between nodes in network. We consider each link consists of two unidirectional fibers (two fibers carrying the same wavelengths in the reverse direction). It is assumed that each fiber link can carry the same number of wavelengths. We first introduce the following notation:

- i, j represent the end points of the physical link that occurs in the route of a connection.
- l is used as an index for the link number, where $l = 1, 2, 3, ..., L$; L = total number of links.
- w is used as an index for the wavelength number, where $w = 1, 2, 3 ... W$; W = total number of wavelengths carried by each fiber link.

The following inputs are supplied to the problem:

- An $N \times N$ distance matrix, where N = total number of nodes in the network and $d_{ij} = (i, j)$th element of distance matrix = distance between i and j nodes. Note that $d_{ij} = d_{ji}$ if and only if there exists a physical fiber link between i and j nodes and $d_{ij} = \infty$ if there is no fiber link.
- An $N \times N$ traffic matrix where the (i, j)th element, $\alpha_{i,j}$, is the traffic flow rate from i node to j node.

The time delay (T_d) experienced by the traffic is a combination of propagation time delay (T_p) and queuing time delay (T_q). The propagation time delay can be computed as follows [3]:

$$Tp = \sum_{i \neq j} \sum_{x,y} \sum_{w=1}^{W} \frac{\alpha_{i,j} \ P_{i,j,w}^{x,y} \ d_{x,y}}{L_v} \tag{3.9}$$

where $P_{i,j,w}^{x,y}$ = light path-wavelength-link indicator

$\quad$ = 1, if there is a light path from node i to node j and it uses wavelength w on a physical link from node x to node y

$\quad$ = 0, otherwise

$\quad L_v$ = velocity of light.

The queuing delay can be computed as follows:

$$Tq = \sum_{i \neq j} \frac{\alpha_{i,j}}{C_m - \alpha_{i,j}} \tag{3.10}$$

where C_m = maximum capacity per light path.

$$\alpha_{i,j} = \sum_{s,d} \alpha_{i,j}{}^{s,d} \quad \forall \quad i, j \tag{3.11}$$

Equation (3.11) shows that the traffic offered in the light path is the sum of the traffic on to the light paths due to all node pairs.

Minimize

$$T_d = T_q + T_p \tag{3.12}$$

3.6.1 Traffic Flow Constraints

The traffic flow constraints pertain to the traffic routed over the light paths in the network [3].

$$\alpha_{i,j} \leq C_m \quad \forall (i, j) \tag{3.13}$$

$$\alpha_{i,j}^{s,d} \leq P_{i,j}\, t^{s,d} \tag{3.14}$$

where $P_{i,j}$ = light path indicator
    = 1, if there is a light path from node i to node j
    = 0, otherwise.
$t^{s,d}$ = average traffic flow rate from source node s to destination node d

$$\sum_{j} \alpha_{i,j}^{s,d} - \sum_{j} \alpha_{j,i}^{s,d} = t^{s,d}, \text{if } s = i$$

$$= -t^{s,d}, \text{if } d = i \tag{3.15}$$

$$= 0, \text{if } s \neq i \text{ and } d \neq i.$$

Equation (3.13) defines network congestion; i.e., the component of traffic on a light path due to a node pair can be at the most the amount of traffic flow between the nodes in a pair. Equation (3.14) shows that the traffic can flow only through the existing light path. Equation (3.15) expresses the conservation of traffic flow at the end nodes of a light path.

3.6.2 WAVELENGTH CONSTRAINTS

The wavelength constraints pertain to the assignment of wavelengths to the light paths [1,3,45].

$$P_{i,j} = \sum_{w=0}^{W-1} P_{i,j,w} \quad \forall (i,j) \tag{3.16}$$

$$P_{i,j,w}^{x,y} \le P_{i,j,w} \quad \forall (i,j),(x,y),w \tag{3.17}$$

$$\sum_{i,j} P_{i,j,w}^{x,y} \le 1 \quad \forall (x,y),w \tag{3.18}$$

$$\left. \begin{array}{r} \displaystyle\sum_{w=0}^{W-1}\sum_{x} P_{i,j,w}^{x,y} l^{x,y} - \sum_{w=0}^{W-1}\sum_{x} P_{i,j,w}^{y,x} l^{y,x} = P_{i,j}, \text{if } y = j \\ = -P_{i,j}, \text{if } y = i \\ = 0, \text{if } y \ne i \text{ and } y \ne j. \end{array} \right\} \tag{3.19}$$

Equation (3.16) shows that the wavelength used by a light path is unique. Equation (3.17) expresses the wavelength-continuity constraint. Equation (3.18) shows that two light paths cannot use the same wavelength on a link. Equation (3.19) expresses the conservation of wavelengths at the end nodes of physical links on a light path.

3.7 PRIORITY-BASED RWA

To reduce the BP in the network, a priority-based RWA (PRWA) scheme [44,45] is considered in which connection requests for RWA are served according to their priority. The priority order of each connection request is determined on the basis of the following two criteria: type of path (direct link or indirect link physical path) and volume of traffic. Using these criteria, direct link connection requests are assigned with higher priority compared to connection requests having indirect link. Then after, the connection requests with direct or indirect link are served in the descending order of their traffic volume. Our goal is to reduce the overall blocking probability (BP) and hence to enhance the effective utilization of a given capacity optical network. To achieve our goal, we consider the type of path and traffic volume as the criteria for priority ordering of connection requests, which is required due to the wavelength-continuity constraint of the network. The wavelength-continuity constraint needs the use of the same wavelength on all hops in the end-to-end path of a connection. Use of a conventional RWA approach under the wavelength-continuity constraint may lead to a situation where wavelengths may be available but connection requests cannot be established due to the unavailability of the required wavelengths. Therefore, if the priority order of connection requests is estimated using these criteria, blocking of connection requests due to the wavelength-continuity constraint can be reduced to a great extent, which will in turn lead to better performance of the network in terms of lower BP. The overall concept of the scheme is explained in Figure 3.11. In the figure, random connection requests arrive at the system based on Poisson process. Then, the connection requests are enqueued into the priority queue to estimate their priority order. Finally, connection requests are served based on the RWA approach according to their priority order. If the connection request is not served within the holding time (t_H), it is treated as a blocked connection. The detailed procedure of the proposed scheme is given in Algorithm 1, and the functionality of the algorithm is explained by considering an example of the National Science Foundation Network (NSFNET) (Figure 3.12) and also assumes a few connection requests that are shown in Table 3.3.

According to Algorithm 1, two clustered sets of connection requests (R' and R'') [44] are estimated such that

$$R' = \{r^{WA.CA_2}, r^{WA.CA_1}, r^{WA.IL}\} \text{ and}$$

$$R'' = \{r^{WA.NY}, r^{WA.NE}, r^{WA.GA}, r^{WA.UT}, r^{WA.NJ}, r^{WA.PA}, r^{WA.MI}, r^{WA.TX}, r^{WA.CO}\} \quad (3.20)$$

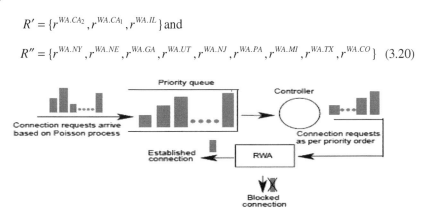

FIGURE 3.11   Priority-based WRA concept.

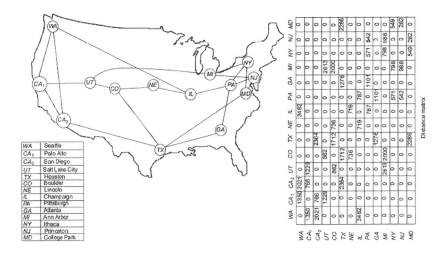

FIGURE 3.12   NSFNET T1 optical backbone with distance matrix.

TABLE 3.3
Connection Requests and Traffic Volumes

Connection Request	Traffic (kbps)	Connection Request	Traffic (kbps)	Connection Request	Traffic (kbps)
$r^{WA.CA_2}$	2500	$r^{WA.CA_1}$	3500	$r^{WA.UT}$	40500
$r^{WA.CO}$	20500	$r^{WA.TX}$	22500	$r^{WA.NE}$	48500
$r^{WA.IL}$	800	$r^{WA.PA}$	25800	$r^{WA.GA}$	45500
$r^{WA.MI}$	24300	$r^{WA.NY}$	70300	$r^{WA.NJ}$	30700

TABLE 3.4
Connection Requests with their Priority Order

Connection Request	Order	Connection Request	Order	Connection Request	Order
$r^{WA.CA_2}$	1st	$r^{WA.CA_1}$	2nd	$r^{WA.UT}$	3rd
$r^{WA.CO}$	4th	$r^{WA.TX}$	5th	$r^{WA.NE}$	6th
$r^{WA.IL}$	7th	$r^{WA.PA}$	8th	$r^{WA.GA}$	9th
$r^{WA.MI}$	10th	$r^{WA.NY}$	11th	$r^{WA.NJ}$	12th

Then, the priority order of each connection request is estimated and is given in Table 3.4. Finally, connection requests are served based on the RWA approach according to their priority order.

Algorithm 1: PRWA [44]

Input: Network configuration and a set of connection requests
Output: WA with total number of successful and unsuccessful connections in the network

Step-1: Enqueue all the connection requests in the priority queue to estimate their priority order.
Step-2: Cluster all the connection requests into two categories – direct physical link connection requests and indirect physical link connection requests.

$$R' = \left\{ r_{D,1}^{S_1,d_1}, r_{D,2}^{S_2,d_2}, \ldots, r_{D,x}^{S_x,d_x} \right\}$$

$$R'' = \left\{ r_{I,1}^{S_1,d_1}, r_{I,2}^{S_2,d_2}, \ldots, r_{I,Y}^{S_Y d_Y} \right\}$$

such that

$$\mathrm{Vol}(r_{D,1}^{S_1,d_1}) \geq \mathrm{Vol}(r_{D,2}^{S_2,d_2}) \geq .... \geq \mathrm{Vol}(r_{D,x}^{S_x,d_x})$$

$$\mathrm{Vol}(r_{I,1}^{S_1,d_1}) \geq \mathrm{Vol}(r_{I,2}^{S_2,d_2}) \geq ... \geq \mathrm{Vol}(r_{D,Y}^{S_Y,d_Y})$$

where R' and R'' are the two ordered sets of connection requests having direct physical link (D) and indirect physical link (I), respectively. X and Y are the total numbers of connection requests having direct and indirect physical links, respectively. The priority order of each connection request is assigned according to its position, i.e., either in R' or in R''. Connection requests in R' have higher priorities compared to those in R''. $\mathrm{Vol}(r_{D,1}^{S_1,d_1})$, $\mathrm{Vol}(r_{I,1}^{S_1,d_1})$ indicate the volumes of traffic for the connection requests of $r_{D,1}^{S_1,d_1}$, $r_{I,1}^{S_1,d_1}$, respectively.

Step-3: Compute K numbers of shortest paths (including primary path) using Dijkstra's algorithm for each of the connection requests on the basis of link state information.

Step-4: For each of the connection requests in R' and R'', selected based on their priority order, perform the following in the given sequence:

a. First, try to assign a wavelength according to wavelength constraints [2] to the primary path based on FF method.

b. If no WA is possible in step-4(a), consider the alternate paths in the ascending order of their light path distance for assigning a wavelength (with a constraint on wavelength similar to that in step-4(a)) till an alternate path is assigned a wavelength.

c. If no WA is possible in either step-4(a) or step-4(b) within tH, the connection request is treated as a blocked one. Otherwise, add the established connection to the total number of established connections in the network.

d. Drop the connection request from the network.

Mathematical Formulation

We model the physical topology of an optical network as a directed connected graph $G(V,E,W)$, where V is the set of nodes and E is the set of bidirectional optical fiber links or edges of the network. Here each link $e \in E$ has a finite number of wavelengths, W. In the network, a non-negative cost (distance between adjacent nodes) $C(e)$ is assigned for every e.

The cost between nodes a and b is considered to be 1 if there exists no link between a and b. The following assumptions are considered in the model [44,45]:

- Each fiber link can carry an equal number of wavelengths, and the network is without wavelength conversion capabilities.
- All the light paths using the same fiber link must be allocated distinct wavelengths.
- Each node can work as both an access node and a routing node.
- Each node is equipped with a fixed number of tunable transceivers.

- Each node is capable of multiplexing/demultiplexing as many connection requests (having the same SD pair) within the channel capacity.
- All the channels have the same bandwidth.
- The connection requests arrive in the system randomly based on a Poisson process.
- The holding times of all connection requests having the same sd pair are equal.

The following notations are used in this formulation:

- N and E are the total numbers of nodes and links, respectively, in the network. s and d are the source and destination of a connection request.
- A is the total number of different sd pairs for all connection requests ($A = N(N-1)$).
- W is the total number of wavelengths per fiber link.
- L is the number of links between an sd pair.
- Z is the total number of connection requests in the network.
- Y is the total number of groomed connection requests in the network ($Y \leq Z$).
- $\alpha^{s,d}$ is the total volume of traffic for a connection request between the source and the destination in an sd pair.
- $\alpha_{i,j}^{s,d}$ is the component of traffic due to an sd pair on a light path from node i to node j.
- $\alpha_{i,j}$ is the total amount of traffic on a light path from node i to node j.
- $C_B^{s,d}$ is the maximum bandwidth of a connection request between the source and the destination in an sd pair.
- C_B is the maximum bandwidth or capacity of a channel.
- t_H is the holding time of a connection request $C(s,d)$.
- K is the number of alternate paths.
- $P_{i,j,\lambda}^{x,y}$ is the light path-wavelength-link indicator.

   $P_{i,j,\lambda}^{x,y} = 1$, if there exists a light path from node i to node j and it uses wavelength λ, on a physical link between node x and node y
   $= 0$, otherwise

- $P_{i,j,\lambda}$ is the light path-wavelength indicator.
   $P_{i,j} = 1$, if there exists a light path from node i to node j
   $= 0$, otherwise
- $P_{i,j}$ is the light path indicator.
   $P_{i,j} = 1$, if there exists a light path from node i to node j
   $= 0$, otherwise
- $f_{\max}$ is the maximum traffic flow on any light path in the network.

Constraints

The wavelength constraints used in priority RWA are the same as those used in Sections 3.6.1 and 3.6.2 (equations 3.14–3.19)

Traffic Flow Constraints

The traffic flow constraints related to the traffic routed over the light paths in the virtual topology are given below [44]:

$$\alpha_{i,j} = \sum_{s,d} \alpha_{i,j}^{s,d} \quad \forall (i,j) \tag{3.21}$$

$$\alpha_{i,j}^{s,d} \le f_{\max} \tag{3.22}$$

Equation (3.21) states that the traffic on a light path is the total traffic due to all the node pairs. The congestion of the network is expressed by using equation (3.22).

Bandwidth Constraints

The bandwidth constraints related to the bandwidth of a connection request and the maximum capacity of a channel in the network are [44]

$$\alpha_{i,j} \le C_B(s,d) \tag{3.23}$$

$$\alpha_{i,j}^{s,d} \le C_B . P_{i,j} \tag{3.24}$$

Equation (3.23) expresses the fact that the traffic flow between the source and the destination in an sd pair cannot exceed the maximum bandwidth of the connection request. Equation (3.24) expresses the fact that the total amount of traffic on a light path from node i to node j cannot exceed the capacity of the light path from i to j.

In order to control wavelength routing and multiplexing or demultiplexing the signal in the optical network, a network node is being designed according to the PRWA algorithm [44] (which will be discussed later in this chapter). Figure 3.13 shows the logical architecture of the network node which uses a number of devices such as F WDMs/wavelength-division demultiplexers (WDDMs) [46], W thermo-optic switches (TOSWs) [47,48], W transceivers, W add–drop multiplexers (ADMs) [49,50], a SONET STS-192 multiplexer/SONET STS-192 demultiplexer [51], and a wavelength router based on the PRWA algorithm. In the figure, initially, a number of connection requests arrive at the system randomly based on a Poisson process. The connection requests (as per bandwidth) having the same sd pair are then groomed/grouped with the hierarchical time division multiplexer SONET STS-192. (For example, if the connection requests of bandwidth 622.08 Mbps are groomed, then a maximum number of 16 connection requests are accommodated with SONET STS-192.) The groomed connection requests are assigned the wavelengths using the PRWA algorithm. The signals of the assigned wavelengths are sent by using transmitters and added to TOSWs through ADMs. Then, the wavelengths are switched by TOSWs [48,49] and finally are multiplexed by WDMs to the output fiber link specified by the wavelength router based on the PRWA algorithm to deliver to the destination node. Further, the wavelengths from input fiber links are demultiplexed by WDDMs which are then switched by TOSWs

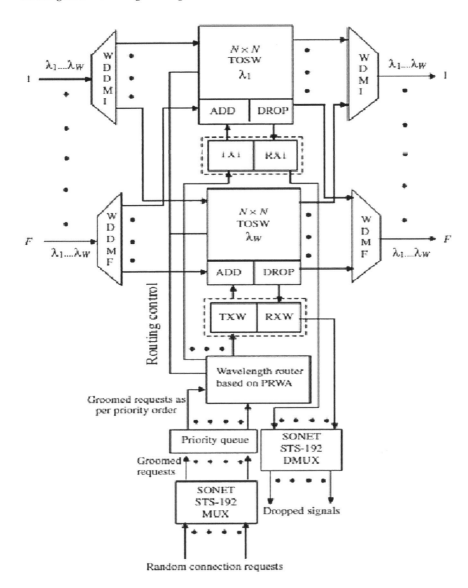

FIGURE 3.13 Network node architecture (WDDM: wavelength-division demultiplexer, WDM: wavelength-division multiplexer, TOSW: thermo-optic switch, MUX: multiplexer, DMUX: demultiplexer, TX: transmitter, RX: receiver, ADD–DROP: add–drop multiplexer, PRWA: priority-based routing and wavelength assignment).

and multiplexed to the corresponding output fiber link to deliver the signals to the destination node. The wavelength carrying the signal for the node itself is dropped through the ADM and demultiplexed by the SONET STS-192 demultiplexer to send the signals to the users.

3.8   COMPARATIVE STUDY OF DIFFERENT RWA
ALGORITHMS ON NSFNET T1 BACKBONE

Figure 3.14 shows BP versus the number of wavelengths (W) for different routing algorithms such as FR, fixed alternate path routing (FAR), and alternate path routing (AR) using FF method with 100,000 connection requests [5]. In this simulation, we do not consider low cost routing (LCR) because from the literature survey, it is seen that the performance of LCR in terms of BP is almost the same as that of FAR [5]. For comparison purpose, in this figure, we have included the result of PRWA scheme based on FF method with the same number of connection requests. It is seen from the figure that in all routing algorithms, BP decreases with the increase of number of wavelengths due to the establishment of more light paths, but the rate of decrease of BP for AR is more than that for other routing algorithms. This is because, in AR, all the possible routes ($K > 2$) are considered between the source and destination in an SD pair on the basis of link state information. Further, it is observed that BP for PWRA scheme ($K = 2$) is less than FR and FAR due to the incorporation of the prioritization concept with AR ($K = 2$). It is also seen from the figure that the BP of FAR is less than that of FR due to the consideration of alternate paths ($K = 2$) for establishing a connection request. We have also seen that the BP of AR with FF method is less than that of PRWA scheme based on FF method, but the average setup time of AR is much more higher compared to that of PRWA. This is mainly because, as the number of paths increases, the average setup time also increases (which is shown in the inner graph of Figure 3.7). AR algorithm considers all the possible routes/paths (in our simulation which is >2) for RWA, whereas in PRWA scheme, only two paths are considered for each connection request. Therefore, we incorporate prioritization concept with alternate path routing ($K = 2$) for the study of different WA schemes (Figure 3.15).

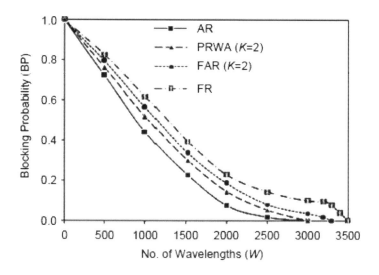

FIGURE 3.14   BP versus number of wavelengths for FR, AR, FAR, and PRWA.

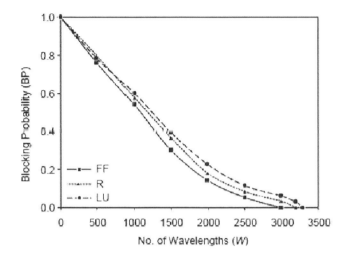

FIGURE 3.15   BP versus number of wavelengths for FF, LU, and random WA.

BP performance of different WA heuristics are compared with respective traffic loads using simulation on NSFNET T1 optical backbone as shown in Table 3.5 [5]. Each link in the network contains M fibers, and each fiber supports W wavelengths. Z is the total number of connection requests, L_2 is the total number of links, L_1 is the length of longest route of any node pairs, and N is total number of nodes in the network. The MP approach and LL approach are exclusively used in the network having multiple fibers in each link, whereas other approaches are used in both multi-fiber-link and single-fiber-link networks. Under high loads, MS, LL, and RCL perform well in terms of BP, whereas LU, MP, and MU show less BP at low traffic loads. Although the average setup time of the random approach is the lowest compared to that of other approaches, the BP is higher than that of others under both high and low traffic loads. Both BP and average setup time of PRWA are lower than those for FF approach under high traffic load condition, but under low load condition, FF approach performs better than PRWA. The time complexity of PRWA is higher than that of others, because of more number of loops in PRWA having K number of alternate paths for each sd pair.

Computational Complexity of Heuristics

The computational complexities of various RWA approaches are compared [1–5]. Random and FF have less computational complexity than others, and their running times are $\sim O(L_1 W Z)$. LU and MU are more complex than Random and FF, and time complexities of LU and MU are $\sim O(L_1 L_2 NZ)$. MP and LL are both used in multi-fiber networks. MP estimates $\prod_{l \in \pi(p)} D_{lj}$ for all W wavelengths and selects the wavelength by minimizing the product. The number of links on a path is limited by $O(N)$. Hence, the time complexity of computation in MP is $\sim O(L_{1M} W N Z)$, whereas that in LL is $\sim O(L_{1M} W N Z)$.

Max-Sum and RCL are costly. The worst-case running time of these approaches is of the order of $O(N^2)$. To find the capacity of each path, all the links along that

TABLE 3.5

Comparison between Different Approaches Such as MS, RCL, MP, LL, LU, LU, Random, FF, and PRWA

	Performance Analysis		No. of Fibers	Average Setup Time
Approach	Blocking Probability	Time Complexity		
MS	In multi-fiber networks, it performs better than the others for high traffic loads	$\sim O(L_1 \, W \, N^2 \, Z)$	Single/ multiple	Medium value
RCL	In single fiber networks, it performs well in comparison to others for high traffic loads	$\sim O(L_1 \, W \, N^2 \, Z)$	Single/ multiple	Medium value
MP	In multi-fiber networks, it performs well in comparison to others for low traffic loads	$\sim O(L_{1M} \, W \, N \, Z)$	Multiple	Medium value
LL	In multi-fiber networks, it performs well in comparison to others for high traffic loads	$\sim O(L_1 \, M, \, W \, N \, Z)$	Multiple	Slightly more than that of MP
LU	In both cases, it performs well for low traffic loads	$\sim O(L_1 \, L_2 N \, Z)$	Single/ multiple	Almost the same as that of LL
MU	In both cases, it performs well for low traffic loads	$\sim O(L_1 \, L_2 N \, Z)$	Single/ multiple	Almost the same as that of LL
Random	More BP than FF but almost same as that of others for low loads	$\sim O(L_1 \, W \, Z)$	Single/ multiple	Lowest among all
FF	Less BP than LU, MU, and random	$\sim O(L_1 \, W \, Z)$	Single/ multiple	More than that of random
PRWA	Less BP than LU, MU, FF, and random	$\sim O(L_1 \, L_2 \, K \, W \, Z)$	Single/ multiple	Less than that of FF under high load

MP – Min. Product, LL – Least Loaded, LU – Least Used, MU – Most Used, FF – first fit, PRWA – priority-based RWA.

path are examined for the minimum number of available wavelengths. The number of links on a path is limited by $O(K)$. Hence, in the worst case, time complexity is obtained as $O(L_1 \, W \, N^2 \, Z)$.

The time to perform PRWA for Z number of connection requests using K alternate paths is $O(L_1 L_2 \, W \, K \, Z)$ which is more than that of the FF and random approaches. The combined RWA problem is formulated as an ILP, which is NP-complete.

SUMMARY

This chapter discusses long-haul wide-coverage optical networks based on arbitrary (mesh) topology in which nodes employ wavelength-routing switches (or optical cross-connects (OXCs)), which establish WDM channels, called light paths, between node pairs. We have discussed different static and dynamic RWA approaches. We have also mentioned LP formulation along with different constraints used in the implementation of these approaches. For the analysis of these approaches, first we have discussed routing algorithms/schemes, and then we have mentioned different WAs along with wavelength searching and WRSV used in these approaches. We have also discussed different priority schemes used for WA for performance improvement.

We have made a comparative study of performances of different RWA algorithms in optical network along with time complexity.

EXERCISES

3.1. Given a graph $G = (V,E)$, define

$$k = \max_{1 \leq i \leq n}\left(1 + \deg_{\triangleleft v_1, v_2, \ldots, v_n \triangleright}(v_i)\right)$$

where $v_1, v_2, ..., v_n \in V$ and $<v_1, v_2, ...,v_n>$ is a vertex ordering. Define an SL vertex ordering $<v_1, v_2, ..., v_n>$ such that $\deg(v_{i+1}) \leq \deg(v_i)$, for $1 \leq i \leq n$. Show that over all the $n!$ possible vertex orderings, the SL vertex ordering minimizes the value of k.

3.2. Consider the network shown in Figure Exercise 3.1. Let the connection requests be as follows:

B-H, A-E, B-D, D-F, B-F, C-E, C-H, A-G, A-C.

Set up light paths to satisfy the above connection requests using at most three wavelengths per link. Assume no wavelength conversion.

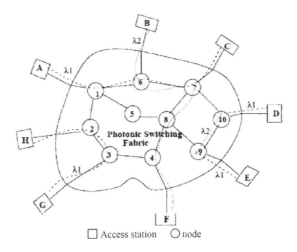

FIGURE EXERCISE 3.1   Sample network. ☐ Access station ○ node.

3.3. We know that the SLE problem is NP-complete. Show a simple transformation that transforms an SLE problem into a graph-coloring problem.

3.4. Consider the network in Figure Exercise 3.1 and the following light paths:
   a. C-7-8-9-E
   b. A-1-5-8-9-E
   c. H-2-1-5-8-7-C
   d. B-6-7-8-9-E
   e. A-1-6-7-10-D
   f. G-3-2-1-6-B
   g. H-2-3-4-F.
   Color the light paths using the minimum number of wavelengths.

3.5. Consider the NSFNET physical topology shown in Figure 3.12. Remove the nodes CA_2 and TX. Consider the connection requests shown in Figure Exercise 3.2. What is the minimum number of wavelengths needed to satisfy all the connection requests?

3.6. Compare the characteristics of various routing schemes.

3.7 Consider the Indian network physical topology shown in Ex Figure 3.3. What is the minimum number of wavelengths needed to satisfy all the connection requests? Assume that number of connection requests for each sd pair is the same and ~5000.

3.8. Consider the Indian network physical topology shown in Ex Figure 3.3. If node-2 and node-3 are removed or they fail, what is the minimum number of wavelengths needed to satisfy all the connection requests? Assume that the number of connection requests for each sd pair is the same and ~5000.

3.9. Compare the characteristics of different WAs heuristics.

3.10. Find the routing table for node-1, node-2, and node-4 for the following network by using Dijkstra's algorithm. The cost of each link is also shown in Ex Figure 3.4

3.11. Find the routing table for node-1, node-2, and node-3 for the network in Ex Figure 3.4 by using Dijkstra's algorithm if node-4 fails to work.

3.12. Find the routing table for node-1, node-2, and node-4 for the following network by using Bellman–Ford's algorithm. The cost of each link is also shown in Ex Figure 3.5.

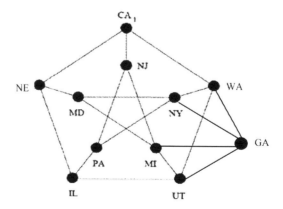

FIGURE EXERCISE 3.2 Sample network having 11 nodes.

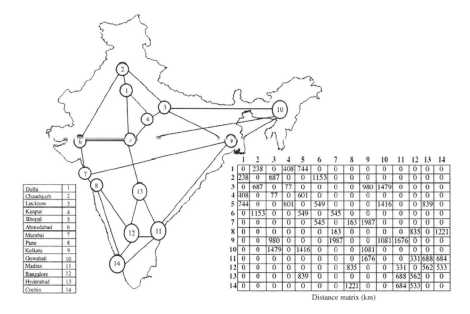

City index:

Delhi	1
Chandigarh	2
Lucknow	3
Kanpur	4
Bhopal	5
Ahmedabad	6
Mumbai	7
Pune	8
Kolkata	9
Guwahati	10
Madras	11
Bangalore	12
Hyderabad	13
Cochin	14

Distance matrix (km)

	1	2	3	4	5	6	7	8	9	10	11	12	13	14
1	0	238	0	408	744	0	0	0	0	0	0	0	0	0
2	238	0	687	0	0	1153	0	0	0	0	0	0	0	0
3	0	687	0	77	0	0	0	0	980	1479	0	0	0	0
4	408	0	77	0	601	0	0	0	0	0	0	0	0	0
5	744	0	0	601	0	549	0	0	0	1416	0	0	839	0
6	0	1153	0	0	549	0	545	0	0	0	0	0	0	0
7	0	0	0	0	0	545	0	163	1987	0	0	0	0	0
8	0	0	0	0	0	0	163	0	0	0	0	835	0	1221
9	0	0	980	0	0	0	1987	0	0	1081	1676	0	0	0
10	0	0	1479	0	1416	0	0	0	1081	0	0	0	0	0
11	0	0	0	0	0	0	0	0	1676	0	0	331	688	684
12	0	0	0	0	0	0	0	835	0	0	331	0	562	533
13	0	0	0	0	839	0	0	0	0	0	688	562	0	0
14	0	0	0	0	0	0	1221	0	0	0	684	533	0	0

FIGURE EXERCISE 3.3 Indian network connecting major cities of India.

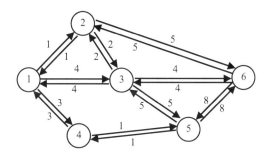

FIGURE EXERCISE 3.4 Sample network having 6 nodes and 9 bidirectional links.

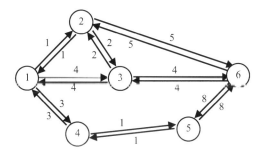

FIGURE EXERCISE 3.5 Sample network 6 nodes and 8 bidirectional links.

3.13. Find the routing table for node-1, node-2, and node-4 for the following network by using genetic algorithm. The cost of each link is also shown in Ex Figure 3.5.

3.14. Find the routing table for node-1 and node-2 for the network in Ex Figure 3.5 by using simulated-annealing algorithm.

3.15. Update the routing table for node-1 and node-2 for the network in Ex Figure 3.5 by using Bellman–Ford's algorithm if node-4 fails.

REFERENCES

1. B. Mukherjee, *Optical WDM Networks*, Springer-Verlag, 2006.
2. I. Chlamtac, A. Ganz, and G. Karmi, "Lightnets: Topologies for high speed optical networks," *IEEE/OSA Journal of Lightwave Technology*, vol. 11, pp. 951–961, 1993.
3. H. Zang, J. P. Jue, and B. Mukherjee, "A review of routing and wavelength Assignment approaches for wavelength-routed optical WDM networks," *SPIE Optical Networks Magazine*, vol. 1, pp. 47–60, 2000.
4. H. Zang, J. P. Jue, and B. Mukherjee, "Capacity allocation and contention resolution in a photonic slot routing all-optical WDM mesh network," *IEEE/OSA Journal of Lightwave Technology*, vol. 18, no. 12, pp. 1728–1741, 2000.
5. B. C. Chatterjee, N. Sarma, and P. P. Sahu, "Review and performance analysis on routing and wavelength assignment approaches for optical networks", IETE Technical Review, vol. 30, pp. 12–23, 2013.
6. R. Ramaswami and K. Sivarajan, "Optimal routing and wavelength assignment in all-optical networks," *IEEE/ACM Transactions on Networking*, vol. 3, pp. 489–500, 1995.
7. R. Ramaswami and K. Sivarajan, "Design of logical topologies for wavelength-routed all-optical networks," *Proceedings, IEEE INFOCOM'95*, Boston, MA, pp. 1316–1325, April 1995.
8. R. Ramaswami and K. N. Sivarajan, "Design of logical topologies for wavelength-routed optical networks," *IEEE Journal on Selected Areas in Communications*, vol. 14, no. 5, pp. 840–851, 1996.
9. P. Raghavan and C. D. Thonlpson, "Randomized rounding: A technique for provably good algorithms and algorithmic proofs," *Combinatorica*, vol. 7, no. 4, pp. 365–374, 1987.
10. C. Ou, H. Zang, N. Singhal, B. Mukherjee, et al., "Sub-path protection for scalability and fast recovery in optical WDM mesh networks," *IEEE Journal on Selected Areas in Communications*, vol. 22, no. 11, pp. 1859–1875, 2004.
11. D. W. Matula, "k-components, clusters and slicings in graphs," *SIAM Journal of Applied Mathematics*, vol. 22, pp. 459–480, 1972.
12. M. Berkelaar, "lpsolve: Readme file," Documentation for the lP Solve program, 1994.
13. CPLEX, http://www.ilog.com
14. D. W. Matula, G. Marble, and J. D. Isancson, "Graph coloring algorithms," *Graph Theory and Computing* (R. C. Read, ed.), New York and London: Academic Press, 1972.
15. K. M. Chan and T. S. Yum, "Analysis of least congested path routing in WDM light-wave networks," *Proceedings, IEEE INFOCOM'94*, Toronto, Canada, pp. 962–969, June 1994.
16. R. Bhandari, *Survivable Networks: Algorithms for Diverse Routing*, Kluwer Academic Publishers, 1999.
17. W. Stalling, *Data and Computer Communication*, PHI, 1999.
18. J. J. Garcia-Luna-Aceves, "Distributed routing with labeled distances," *Proceedings, IEEE INFO COM'92*, Florence, Italy, pp. 633–643, May 1992.

19. S. Rarnamurthy, "Optical design of WDM network architectures," Ph.D. Dissertation, University of California, Davis, 1998.

20. R. Libeskind-Hadas, "Efficient collective communication in WDM networks with a power budget," *Proceedings, Ninth, IEEE International Conference on Computer Communications and Networks (ICCCN)*, Las Vegas, Nevada, pp. 612–616, October 2000.

21. B. Rarnamurthy and B. Mukherjee, "Wavelength conversion in optical networks: Progress and challenges," *IEEE Journal on Selected Areas in Communications*, vol. 16, pp. 1040–1050, 1998.

22. Y. Huang, J. P. Heritage and B. Mukherjee, "Connection provisioning with transmission impairment consideration in optical WDM networks with high-speed channels," *IEEE/OSA Journal of Lightwave Technology*, vol. 23, no. 3, pp. 982–993, 2005.

23. S. Ramamurthy and B. Mukherjee, "Survivable WDM mesh networks, part I -protection," *Proceedings, IEEE INFOCOM '99*, New York, pp. 744–751, March 1999.

24. L. Li and A. K. Somani, "Dynamic wavelength routing using congestion and neighborhood information," *IEEE/ACM Transactions on Networking*, vol. 7, no. 5, pp. 779–786, 1999.

25. D. Banerjee and B. Mukherjee, "Practical approaches for routing and wavelength assignment in large all-optical wavelength-routed networks," *IEEE Journal on Selected Areas in Communications*, vol. 14, pp. 903–908, 1996.

26. D. Banerjee and B. Mukherjee, "Wavelength-routed optical networks: Linear formulation, resource budgeting tradeoffs, and a reconfiguration study," *IEEE/ACM Transactions on Networking*, vol. 8, no. 5, pp. 598–607, 2000.

27. S. Subramaniam, R.A. Barry, "Wavelength assignment in fixed routing WDM networks," *IEEE International Conference on Communications*, 1997, pp. 406–410.

28. C.R. Reeves, *Modern Heuristic Techniques for Combinatorial Problems*, McGraw-Hill, 1995.

29. V.J. Rayward-Smith, I.H. Osman, C.R. Reeves, and G.D. Smith, *Modern Heuristic Search Methods*, John Wiley & Sons, 1996.

30. A. Hertz and D. de Werra, "Using Tabu search techniques for graph coloring", *Computing*, vol. 39, pp. 345–351, 1987.

31. R. Ramaswami, K. N. Sivarajan, Routing and wavelength assignment in all-optical networks, Tech Rep. RC 19592, IBM Research Report, 1994.

32. Y. Sun, J. Gu, and D. H. K. Tsang, "Multicast routing in all optical wavelength routed networks," *Optical Networks Magazine*, vol. 2, pp. 101–109, 2001.

33. G. Jeong and E. Ayanoglu, "Comparison of wavelength interchanging and wavelength-selective cross-connects in multiwavelength all-optical networks," *Proceedings, IEEE INFOCOM'96*, San Francisco, CA, pp. 156–163, March 1996.

34. E. Karasan and E. Ayanoglu, "Effects of wavelength routing and selection algorithms on wavelength conversion gain in WDM optical networks," *IEEE/ACM Transactions on Networking*, vol. 6, no. 2, pp. 186–196, 1998.

35. R. A. Barry and S. Subramaniam, "The MAX-SUM wavelength assignment algorithm for WDM ring networks," *Proceedings, OFC'97*, Dallas, TX, pp. 121–122, February 1997.

36. X. Zhang and C. Qiao, "Wavelength assignment for dynamic traffic in multi-fiber WDM networks," *Proceedings, 7th International Conference on Computer Communications and Networks*, Lafayette, LA, pp. 479–485, October 1998.

37. A. Birman and A. Kershenbaum, "Routing and wavelength assignment methods in single-hop all- optical networks with blocking," *Proceedings, IEEE INFOCOM'95*, Boston, MA, pp. 431–438, April 1995.

38. X. Yuan, R. Melhem, R. Gupta, Y. hlei, and C. Qiao, "Distributed control protocols for wavelength reservation and their performance evaluation," *Photonic Network Communications*, vol. 1, no. 3, pp. 207–218, 1999.

39. N. Charbonneau and V. M. Vokkarane, "A survey of advance reservation routing and wavelength assignment in wavelength-routed WDM networks," *IEEE Communications Surveys & Tutorials*, vol. 14, pp. 1037–1064, 2012.

40. N. Fazlollahi and D. Starobinski, "Distributed advance network reservation with delay guarantees," *IEEE International Symposium on Parallel Distributed Processing (IPDPS)*, pp. 1–12, 2010.

41. C. Xie, H. Alazemi, and N. Ghani, "Routing and scheduling in distributed advance reservation networks," Proc. IEEE GLOBECOM, 2010.

42. B. C. Chatterjee, N. Sarma, and P. P. Sahu, "Priority based routing and wavelength assignment with traffic grooming for optical networks," IEEE/OSA Journal of Optical Communication and Networking, vol. 4, no. 6, pp. 480–489, 2012.

43. B. C. Chatterjee, N. Sarma, and P. P. Sahu, "Priority based dispersion-reduced wavelength assignment for optical networks", IEEE/OSA Journal of Lightwave Technology, vol. 31, no. 2, pp. 257–263, 2013.

44. D. M. Shan, K. C. Chua, G. Mohan, and M. H. Phunq, "Priority-based offline wavelength assignment in OBS networks," *IEEE Transactions on Communications*, vol. 56, no. 10, pp. 1694–1704, 2008.

45. B. C. Chatterjee, N. Sarma, and P. P. Sahu, "A heuristic priority based wavelength assignment scheme for optical networks," Optik –International Journal for Light and Electron Optics, vol. 123, no. 17, pp. 1505–1510, 2012.

46. P. P. Sahu, "Compact optical multiplexer using silicon nano-waveguide," *IEEE Journal of Selected Topics in Quantum Electronics*, vol. 15, no. 5, pp. 1537–1541, 2009.

47. P. P. Sahu, "Thermooptic two mode interference photonic switch," *Fiber and Integrated Optics*, vol. 29, pp. 284–293, 2010.

48. P. P. Sahu and A. K. Das, "Polarization-insensitive thermo-optic Mach Zehnder based on silicon oxinitride waveguide with fast response time" *Fiber and Integrated Optics*, vol. 29, no. 1, pp. 10–20, 2010.

49. P. P. Sahu, "Tunable optical add/drop multiplexers using cascaded Mach Zehnder Coupler," *Fiber and Integrated Optics*, vol. 27, no. 1, pp. 24–34, 2008.

50. P. P. Sahu "Polarization insensitive thermally tunable Add/Drop multiplexer using cascaded Mach Zehnder coupler," *Applied Physics: Lasers and Optics*, vol. B92, pp. 247–252, 2008.

51. G. Keiser, *Optical Fiber Communication*, McGraw Hill, 2002.

4 Virtual Topology

Design of next generation optical Wide-Area Networks (WANs) based on Wavelength-Division Multiplexing (WDM) are done for nationwide and global coverage. Considering wavelength multiplexers and optical switches (cross-connects) in routing nodes, there is a need to design a virtual topology for a given physical fiber networks [1–50]. The virtual topology having a set of "light paths" is set up for transmitting packets of information in the optical domain using optical circuit switching, but packet forwarding from one light path to another light path is also carried out by optical switching [1]. Each light path in the virtual topology is established by using the Routing and Wavelength Assignment (RWA) techniques mentioned in Chapter 2. In this chapter, we will discuss the virtual topology having a constraint and additional relaxations. If no wavelength converter is used in the optical cross-connect (OXC) (discussed in the previous chapter), the wavelength of the light path obeys wavelength continuity constraint. This network architecture depends on a combination of well-known "single- hop" and "multi-hop" approaches, and here attempts are made to consider the characteristics of both. A "light path" in this architecture provides "single-hop" communication between any two nodes. Within a limited number of wavelengths, "light paths" between all source-destination pairs are obtained. As a result, multi-hopping between "light paths" is needed. In addition, when the existing traffic pattern changes, a different number of "light paths" making a different "multi-hop" virtual topology is required. A networking issue challenge is to make the necessary reconfiguration with minimal disruption of the network operations in network architecture where the use of wavelength multiplexers gives the advantage of much higher aggregate system capacity due to spatial reuse of wavelengths and supports a large number of users.

The design problem of virtual topology requires the optimization of one or two possible objective functions [4–6] which are for a given traffic matrix (intensities of packets flow between various pairs of nodes), minimization of the network-wide average packet delay (corresponding to a solution for present traffic demands) for a given traffic matrix (intensities of packets flow between various pairs of nodes) and maximization of the scale factor by which the traffic matrix can be scaled up (to provide the maximum capacity upgrade for future traffic demands). Here we need an iterative approach which combines "simulated annealing" (searching for a virtual topology) and flow deviation.

4.1 VIRTUAL TOPOLOGY ARCHITECTURE

We consider the NSFNET backbone (Figure 2.11) for virtual topology design [1,10]. Virtual topology information is transported over this backbone as packets of variable sizes. The store-and-forward packet switching is needed to be performed at the network nodes. Here a packet for a source-destination node pair passes through one more intermediate nodes, and at each such intermediate node, the packet has to be completely received (stored in memory), its header has to be processed by the inter-mediate node to determine the node's outgoing links for this forwarded packet, and the packet may wait at this node longer if that corresponding outgoing link is busy due to the transmission of other packets. Although a fiber connects the nodes, the fiber's tremendous transmission bandwidth is not used since data transmission on each fiberlink is limited on to a single wavelength.

There are requirements of concerns in the virtual design of network architec-ture [1–4] - such as upgradation of any future technology, how the WDM solution can be used to upgrade an existing ATM solution, and how the WDM solution can accommodate a variety of electronic interfaces or services and protocol transpar-ency property.

4.1.1 GENERAL PROBLEM STATEMENT

The problem of designing a virtual topology for a given physical topology (fiber net-work is formally stated below. The following are the input to the problem [1–3,6] are:

- A physical topology $G(V,E)$ as a weighted undirected graph, where V is the set of network nodes, and E is the set of links connecting nodes, as opposed to virtual links (or light paths) in a virtual topology, and E is the set of links connecting the nodes. In undirected states, each link is bidirectional. So, each link has two fibers. Links are assigned weights, corresponding to physical distances between nodes and traffic loads. A network node i is equipped with a $D_p(i) \times D_p(i)$ wavelength-routing switch (WRS), where $D_p(i)$ is physical degree of node i, and the number of physical fiberlink ema-nating out of node i.
- Number of wavelength channels of each fiber $= M$.
- $N \times N$ traffic matrix, where N is the number of network nodes, and the (i,j) th element is the average rate of packet traffic flow from node i to node j. The traffic flowing from node i to node j is different from the flow from node j to node i.
- The number of wavelength-tunable lasers (transmitters) and the number of wavelength-tunable filters (receivers) at each node depend on the number of wavelength channels per link.

Our goal is to determine the following area:

- To determine a virtual topology $G(V,E)$ in which the out-degree of a node is the number of transmitters at that node and the in-degree of a node is the number of receivers at that node. The nodes of the virtual topology cor-respond to the nodes in the physical topology. Each link between a pair of nodes in the virtual topology corresponds to a "light path" directly between the corresponding nodes in the physical topology.
- To determine the wavelength assignment for light paths such that if two light paths have a common physical link but necessarily different wavelengths.

- To determine the sizes and configurations of the WRSs at the intermediate nodes are known, the virtual topology is determined and the wavelength assignments have been performed.
- To setup communication is to set up by using a path having a sequence of light paths from the source node to the destination node on the virtual topology.

4.2 NSFNET OPTICAL BACKBONE: VIRTUAL TOPOLOGY

In NSFNET, two more fictitious nodes *AB* and *XY* are considered to make upgradation of the WDM-based existing fiber optic network as shown in Figure 4.1. A hypercube is embedded as a virtual topology over this physical topology of NSFNET seen in Figure 4.2. All light paths are undirected, comprising bidirectional paths [1,6].

The NSFNET backbone has originally 14 nodes. Optical node is based on wavelength-routing switch (WRS) from these nodes to one another and some links connecting to the outside world [1]. Two fictitious nodes *AB* and *XY* are added to find the effect of NSFNET's connections to Canada's communication network, CA-NET, and networks of other countries. Node *XY* is joined to Ithaca (NY) and Princeton (NJ) nodes of NSFNET, while node AB is joined to the Seattle (WA) and Salt Lake City (UT) nodes, where the last link is taken as a fictitious link to make the physical topology richer and fault tolerant. The electronic component is an electronic packet router, which serves as a store and forward electronic over lay on top of the optical virtual topology. Figure 4.3 represents a schematic diagram of the architecture of the Utah node (UT) of NSFNET topology [1]. An array of optical space-division switches, one per wavelength, between the demultiplexer (demux) and multiplexer (mux) stages, are considered in the Utah node. These switches can be reconfigured under electronic control, e.g., to adapt the network's virtual topology on demand.

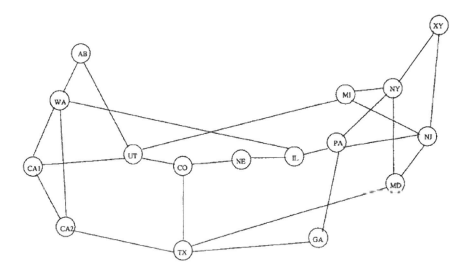

FIGURE 4.1   Modified NSFNET with nodes AB and XY [1].

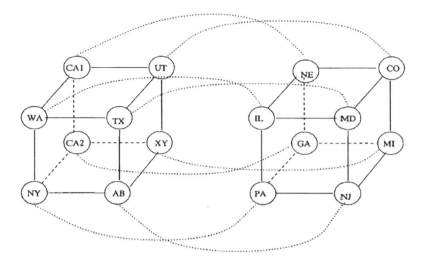

FIGURE 4.2   16 Node hypercube embedded on NSFNET topology [1].

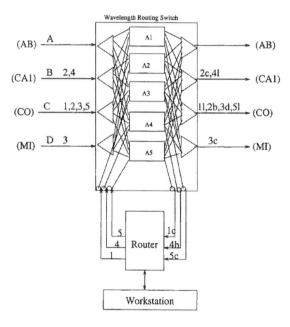

FIGURE 4.3   Node architecture of Utah [1].

The virtual topology is considered as a 16-node hypercube (as shown in Figure 4.2) [1]. One of several possible wavelengths is used on one of several possible physical paths [1], such as CA1-UT-CO-NE, or CA1-WA-IL-NE, or others as shown in Figure 4.3. The WRSs at the UT and CO nodes are configured to establish this CA1-NE light path. The switch at UT has wavelength 2 on its fiber to CA-1 connected to CO.

Figure 4.4 shows a solution requiring a maximum of five or seven wavelengths per fiber, by using shortest-path routing of light paths on the physical topology in Figure 4.1 [1]. Only one fiber connects the local node to the local WRS where each of the light paths originating from and terminating at that node has a different wavelength to avoid wavelength clash on the local fiber. This solution needs more wavelengths to be set in a virtual topology. If multiple fibers join the local node to the local WRS, multiple light paths transmit the same wavelength from a node. In Utah node, the switch has to operate with four incoming fibers plus four outgoing fibers, one each to nodes AB, CAI, CO, and MI, as per the physical topology shown in Figure 4.4.

The output fiber to CO indicates the UT-CO fiber using four wavelengths 1, 2, 3, and 5, with wavelengths 2 and 3 being "clear channels" (i.e., optical-circuit-switched channels) through the UT switch and directed to the physical neighbors CA1 and MI, respectively, while wavelengths 1 and 5 are used for two local lasers. However, this virtual topology rooted on NSFNET is an "incomplete" hypercube with nodes *AB* and *XY* considered non-existent. Hence some nodes have less than four neighbors. Three laser-filter pairs at UT node need to be operated – one on wavelength 1 (for connection to physical neighbor CO to get the light path UT–TX), another on wavelength 4 (for connection to physical neighbor CAI, to obtain the light path UT-CAI), and at wavelength 5 (for connection to physical neighbor CO, to obtain the light path UT-CO) in Figure 4.5 [1].

The WRS associated with the Utah switch is different from that in Figure 4.4. Since multiple fibers connect the electronic router to the WRS, the size of the WRS has a 7×7 switch instead of the 4×4 switch used in the solution found in Figure 4.5.

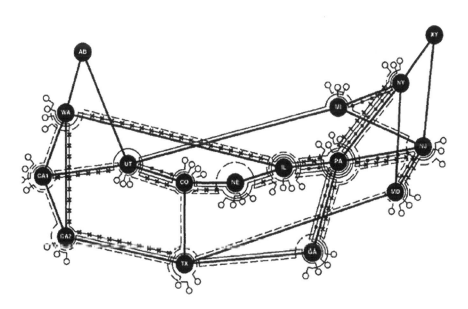

FIGURE 4.4 Physical topology with embedded wavelengths corresponding to an optimal solution (more than one transceiver at any node can be tuned to the same wavelength) [1].

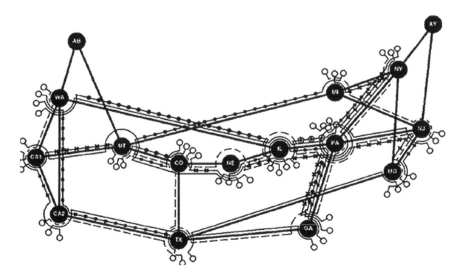

FIGURE 4.5   Modified virtual topology embedded on NSFNET T1 backbone [1].

However, since the solution corresponding to Figure 4.4 requires fewer wavelengths, the number of space-division switches inside the largest WRS decreases from seven to five.

4.2.1 Formulation of Virtual Topology

The problem is an optimization problem formulated by using principles from multicommodity flow for physical routing of light paths and traffic flow on the virtual topology and using the following notation [1,6]:

- s and d represent subscript or superscript, represent the source and destination, respectively.
- i and j indicate originating and terminating nodes, of a light path.
- m and n denote end-points of a physical link that might occur in a light path.

Parameters
- N = number of nodes in the network.
- M = maximum number of wavelengths per fiber.
- Physical topology P_{mn}, where $P_{mn} = P_{nm} = 1$ if and only if there exists a direct physical fiberlink between nodes m and n, where $m, n = 1, 2, 3, \ldots, N$; $P_{mn} = P_{nm} = 0$ otherwise (i.e., fiberlinks are assumed to be bidirectional).
- Distance matrix, viz., fiber distance d_{mn}, from node m to node n. For simplicity in expressing packet delays, d_{mn} is expressed as a propagation delay (in time units).
    $d_{mn} = d_{nm}$ since fiberlinks are bidirectional, and $d_{mn} = 0$ if $P_{mn} = 0$.

- T_i = number of transmitters at node i ($T_i \geq 1$).
- Number of receivers at node $i = R_i$ ($R_i \geq 1$).
- Traffic matrix λ_{sd} shows mean/average rate of traffic flow from s node to d node with $\lambda_{ss} = 0$ for $s,d = 1, 2, ..., N$. Additional assumptions: packet inter-arrival duration at node s and packet lengths are exponentially distributed. M/M/1 queuing results can be applied to each network link (or "hop") by using the independence assumption on interarrival and packet lengths due to traffic multiplexing at intermediate hops. Mean packet length (in bits per packet), λ_{sd}, can be expressed in packets per second.
- Capacity of each channel = C in packets per second.

Variables

– Virtualtopology [1]:

Variable $V_{ij} = 1$ if there exists a light path from node i to node $j = 0$ otherwise. The light paths are not considered to be bidirectional, i.e., $V_{ij} = 1 \nRightarrow V_{ji} = 1$. For multiple light paths between node pairs, $V_{ij} > 1$.

– Traffic routing [1]: The variable λ_{ij}^{sd} is the traffic flowing from node s to node d and considering V_{ij} as an intermediate virtual link between i and j nodes. The traffic from node s to node d may be "bifurcated" with different components (or fractions) taking different sets of light paths.

Physical-topology route:

$P_{mn}^{ij} = 1$ if the fiberlink P_{mn}^{ij} is present in the light path for virtual link $V_{ij} = 0$ otherwise.

Wavelength color [1]:

$C_k^{ij} = 1$ if a light path from node i to node j is assigned the color k, where $k = 1,2,..., M = 0$ otherwise.

Constraints

On virtual-topology connection matrix V_{ij} [1]

$$\sum_j V_{ij} \leq T_i \quad \forall i$$

$$\sum_i V_{ij} \leq R_j \quad \forall j$$

(4.1)

The equalities in equation (4.1) are satisfied if all transmitters are at node i and all receivers are at node j.

On physical route variables:

$$\left.\begin{array}{c} P_{mn}^{ij} \leq P_{mn} \\[2mm] P_{mn}^{ij} \leq V_{ij} \end{array}\right\}$$

(4.2)

$$\left.\begin{aligned}
\sum_{m} P_{mk}^{ij} &= \sum_{n} P_{kn}^{ij} \quad \text{if } k \neq i, j \\
\sum_{m} P_{mk}^{ij} &= V_{ij} \\
\sum_{n} P_{in}^{ij} &= V_{ij}
\end{aligned}\right\} \qquad (4.3)$$

On virtual topology traffic variables λ_{ij}^{sd}:

$$\left.\begin{aligned}
\lambda_{ij}^{sd} &\geq 0 \\
\sum_{j} \lambda_{sj}^{sd} &= \lambda_{sd} \\
\sum_{i} \lambda_{id}^{sd} &= \lambda_{sd} \\
\sum_{i} \lambda_{ik}^{sd} &= \sum_{j} \lambda_{kj}^{ij} \quad \text{if } k \neq i, j \\
\sum_{s,d} \lambda_{ij}^{sd} &\leq V_{ij} * C
\end{aligned}\right\} \qquad (4.4)$$

On virtual topology coloring variables [1]:

$$\sum_{k} C_{k}^{ij} = V_{ij} \qquad (4\ 5)$$

$$\sum_{ij} P_{mn}^{ij} C_{k}^{ij} \leq 1 \quad \forall m,n,k \qquad (4.6)$$

Objective
   Minimizing delay [1]:

$$\sum_{ij} \left[\sum_{sd} \lambda_{ij}^{sd} \left(\sum_{mn} P_{mn}^{ij} d_{mn} + \frac{1}{C - \sum_{sd} \lambda_{ij}^{sd}} \right) \right] \qquad (4.7)$$

Maximizing offered load (equivalent to minimizing maximum flow in a link) [1]:

$$\min\left(\max\sum_{sd}\lambda_{ij}^{sd}d_{mn}\right)=\max\frac{C}{\max\sum_{sd}(\lambda_{ij}^{sd})} \tag{4.8}$$

Equations (4.7) and (4.8) obey conservation of flows and resources (transceivers, wavelengths, etc.) as well as conflict-free routing. Equation (4.1) shows that the number of light paths originating and terminating at a node's out-degree and in-degree, respectively. Equation (4.2) indicates that P_{mn}^{ij} can exist only if there is a physical fiber along with a corresponding light path. Equation (4.3) represents the multi-commodity for the routing of a light path from its origin to its termination. Equation (4.4) represents the routing of packet traffic on the virtual topology (considering no packet dropping in the router), making sure that the combined traffic is transmitted through a channel within the channel capacity. Equation (4.5) assures that a light path is of only one color, whereas equation (4.6) assures that the colors considered in different light paths are mutually exclusive over a physical link.

Equations (4.7) and (4.8) are objective functions. In equation (4.7), the innermost brackets, the first component corresponds to the propagation delays on the links mn which is basically the light path ij, while these components indicate delay due to queuing and packet transmission on light path ij (using an M/M/l queuing model for each light path). For shortest-path routing of the light paths over the physical topology, P_{mn}^{ij} values can be estimated. For very small queuing delays, the optimization problem in equation (4.7) becomes:

$$\text{Minimize}\sum_{sd}\sum_{mn}\sum_{ij}\lambda_{ij}^{sd}P_{mn}^{ij}d_{mn}$$

where the variables V_{ij} and C_k^{ij} need to have integer solutions [1]. The objective function in equation (4.8) represents the maximum amount of traffic transmitted through any light path corresponding to obtain a virtual topology maximizing the offered load to the network.

4.2.2 ALGORITHM

4.2.2.1 Subproblems

This optimization of virtual topology is NP-hard, since its subproblems are also NP-hard. The problem of optimal virtual-topology design consists of the following four independent subproblems [1] on how to:

1. derive a virtual topology where nodal transmitters are directly connected to nodal receivers.
2. find the light paths over the physical topology.
3. allot wavelengths optimally to the various light paths (this problem has been shown to be NP-hard in [2]).
4. route packet traffic on the virtual topology (in any packet-switched network).

Subproblem 1 is about deriving the optimal use of the limited number of available transmitters and receivers. Subproblem 2 is about proper usage of the limited number of available wavelengths, and this subproblem was discussed as an RWA problem in Chapter 2. Subproblem 4 is about decreasing the effect of store-and-forward (queuing and transmission) delays at intermediate electronic hops.

The problem of optimal design of virtual topologies is also mentioned in [15]. It accommodates many of the physical connectivity constraints. For resolution, exhaustive search/heuristic approaches are used in [15]. Some algorithms for obtaining a hypercube virtual topology are mentioned in [2,10]. In these works, the minimization of average hop accommodates more network traffic. The works in [12,13] consider the physical topology as a subset of the virtual topology with algorithms for maximizing the throughput. The previous work in [16] considers an adaptation mechanism to follow the changes in traffic without a prior knowledge of the future traffic pattern by utilizing the measured Internet backbone traffic characteristics. In this case, virtual topology is redesigned as per the expected traffic pattern. The key issue is to adapt the optical connectivity by estimating the actual traffic load on light paths and by either adding or deleting one or more light paths at a time due to fluctuations in the traffic.

For subproblems 1 and 4, the number of available wavelengths are considered a constraint. An iterative approach uses "simulated annealing" to obtain a virtual topology (subproblem 1) with the "flow-deviation" algorithm for optimal (possibly "bifurcated") routing of packet traffic on the virtual topology (subproblem 4). As the virtual topology has an undirected graph, light paths are bidirectional.

A virtual-topology-based random configuration is derived by using simulated annealing via node-exchange techniques that are similar to branch-exchange [17] techniques. Then, the traffic matrix is scaled up to determine the maximum throughput required for the virtual topology, using flow deviation for packet routing over the virtual topology. For a given traffic matrix, using the flow-deviation algorithm, the network-wide packet delay is minimized by properly distributing the flows on the virtual links (to reduce the effect of large queuing delays). Our objective is to devise the virtual topology that provides the maximum traffic scale-up and also maximum throughput expected from the current fiber network if it were to support WDM and if future traffic characteristics were to model present-day traffic.

4.2.2.2 Simulated Annealing

Simulated annealing provides the solutions to complex optimization problems [18] for virtual topology design. In the simulated-annealing process, the algorithm begins with an initial random configuration for the virtual topology. Node-exchange operations are made at neighboring configurations. In a node-exchange operation, adjacent nodes in the virtual topology are used for swapping wherein node i is connected to nodes j, a, and b, while node j is linked to nodes p, q, and i in virtual topology; after the node-exchange operation between nodes i and j, node i is linked to nodes p, q, and j, while node j is connected to nodes a, b, and i. Solutions are accepted with a certain probability which is estimated by a system control parameter, but it decreases as the algorithm progresses in time so as to obtain the cooling rate associated with annealing. The probability of acceptance is inversely proportional to the difference between the current solutions. As time progresses, the probability of accepting worse solutions decreases, and the algorithm resolves the problem after several iterations [18].

4.2.2.3 Flow-Deviation Algorithm

The flow-deviation algorithm [19] is an optimal algorithm for minimizing the network-wide average packet delay. The different fractions of traffic get a path along in order to reduce the packet delay. The balancing of flows is affected due to excessive loading of a particular channel which causes congestion (which leads to large delays) in that channel and thus has a negative influence on the network-wide average packet delay. The algorithm having shortest-pathflows first calculates in a channel. The cost rates use a shortest-pathflows (which can be solved using one of several well-known algorithms such as Dijkstra's algorithm or Bellman–Ford's algorithm). An iterative algorithm is estimated by how much the original flow has been deviated.

Both simulated-annealing algorithm as well as the flow-deviation algorithm are used to get the results for the virtual-topology design problem, viz., to study subproblems 1 and 4 mentioned in Section 4.2.2.1. There are two matrixes for simulation in the network – traffic matrix and distance matrix. The traffic matrix used for the mapping is derived from an actual measurement of the traffic on the NSFNET backbone for a 15-minute period (11:45 p.m. to midnight on January 12, 1992, EST [10]). The distance matrix is basically derived from the actual geographical distances connecting major cities in USA. The number of wavelength channels per fiber is considered to be sufficient enough so that all possible virtual topologies are derived.

Since the aggregate capacity for the carried traffic is set by the number of links in the network, decreasing the average hop distance can provide higher values of load that the network can support. The queuing delay is normally estimated considering a standard M/M/l queuing system with a mean packet length determined from the measured traffic (133.54 bytes per packet) as shown in Table 4.1. The "cooling" parameter for the simulated annealing is updated after every 100 acceptances using a geometric parameter of value 0.9. A state is to be "frozen" when there is no improvement over 100 consecutive trials.

TABLE 4.1
Maximum Scale-Up of NSFNET T1 Backbone Considering 133.54 Bytes per Packet

Parameter	Physical Topology (No WDM)	Multiple Point-to-Point Links (No WRS)	Arbitrary Virtual Topology (Full WDM)
Maximum scale-up	49	57	106
Avg. packet delay	11.739 ms	12.3342 ms	17.8146 ms
Avg. propagation delay	10.8897 ms	10.9188 ms	14.4961 ms
Queuing delay	0.8499 ms	1.6353 ms	3.31849 ms
Avg. hop distance	2.12186	2.2498	1.71713
Max. link loading	98%	99%	99%
Min. link loading	32%	23%	71%

Physical Topology as Virtual Topology (No WDM)

We considered a network having no WDM but having electronic hardware, packet-switched network nodes with fiberlinks, and point-to-point connections having one bidirectional light path channel per fiberlink. The maximum scale-up achieved was found to be 49 using flow deviation. (Only integer values of the scale up were considered.) The link with the maximum traffic (WA-IL) was loaded at 9896, while the link with the minimum traffic (NY-MD) was at 32%. These values are treated as a basis for comparison in terms of throughput by adding extra WDM optical hardware such as tunable transceivers and wavelength routing switches at nodes.

Multiple Point-to-Point Links with No WRS

We considered network that had no WRS, and had WDM on some links but did not use switching capability at any node. The network had 21 bidirectional links in the NSFNET physical topology. Using extra transceivers at the nodes, extra links on the paths NE-CO, NE-IL, WA-CA2, CAI-UT, MI-NJ, and NY-MD were taken to accommodate nodes of other networks [1,10]. Different combinations were taken, and the choice of channels providing the maximum scale-up were chosen. For 14 nodes, each with a nodal degree of four, 28 channels were used. The GA node was joined only to TX and PA, both of which were physically connected to four nodes already. After the inclusion of six new channels, the maximum scale-up was found to be 57. Two NY-MD channels with minimum load of only 23% were obtained, while the UT-MI channel took a maximum load of 99%.

Arbitrary Virtual Topology

We considered a network with full WDM with all nodes having WRSs; the number of wavelengths supported in each fiber were restricted for possible virtual topologies, and also, all light paths were found over the shortest path on the physical topology. Beginning with a random initial topology, we considered simulated annealing to get the best virtual topology [10]. The best virtual topology shown in Table 4.1, provides a maximum scale-up of 106. Clearly, the increasing scale-up provides the benefits of the WDM-based virtual-topology approach. Now, the minimum loading is obtained on link UT–TX as 71%, while all the other links have loading above 98%.

Comparisons

The delay characteristics include overall average packet delay (FD) and average propagation delay encountered by each packet (PD), average queuing delay experienced by each packet (QD), and the mean hop distance (HD) and traffic matrix depends on scale-up (throughput) for the three types of network structures mentioned above. The scale-up determines the throughput in the network. Here the propagation delay is the dominant component of the packet delay. Also, at light loads, the average propagation delay faced by packets in NSFNET is a little over 9 ms (for the given traffic matrix), and this serves as a lower bound on the average packet delay. The coast-to-coast, one-way propagation time in the U.S. is nearly 23 ms. On an average, each packet travels about 40% of the coast-to-coast distance (as shown in Table 4.2).

The full WDM is considered (with a WRS at each node). As the nodal degree is enhanced to 5 and 6, the maximum scale-up increases nearly proportionally with

TABLE 4.2

Virtual Topology for Nodal Degree $P = 4$ [1]

Source	Neighbors
WA	CA1, CA2, MI, UT
CA1	WA, CO, IL, TX
CA2	WA, PA, NE, GA
UT	WA, TX, IL, MD
CO	CA1, MD, NE, GA
TX	CA1, UT, GA, NJ
NE	CA2, CO, IL, MI
IL	CA1, UT, NE, PA
PA	CA2, IL, NY, NJ
GA	CA2, CO, TX, NY
MI	WA, NE, NY, NJ
NY	PA, GA, MI, MD
NJ	TX, PA, MI, MD
MD	CO, NY, NJ, UT

increasing nodal degree. Minimizing hop distance is an important optimization problem and is considered for the study later in this chapter. Although no constraints on wavelengths per fiber were considered in this study, we also examined the wavelength requirements to set up a virtual topology using shortest-path routing of light paths on the physical topology. Considering no limit on the supply of wavelengths, but with the wavelength constraints outlined before, the maximum number of wavelengths needed for obtaining the best virtual topology (which provided the maximum scale-up) with degree $P = 4$, 5, and 6 was found to be 6, 8, and 8, respectively in NSFNET [1]. The virtual topology having shortest-path routing of light paths, and wavelength constraints, is obtained if there is no link on the physical topology that uses all of the required wavelengths [16].

4.3 ADVANCED VIRTUAL TOPOLOGY OPTIMIZATION

Here the design of a light path-based virtual topology is considered as a linear optimization problem, and the problem formulation is derived for an exact optimal virtual network design [1,16]. This formulation consists of non-linear equations which cannot resolve the problem. The objective function is to minimize the average packet hop distance which is inversely proportional to the network throughput under balanced network flows. By making a shift in the objective function to minimal hop distance and by relaxing the wavelength-continuity constraints (i.e., assuming wavelength converters at all nodes), the entire optical network design can be linearized, simplified, and hence resolved optimally. The network throughput is bounded by

$T \leq \dfrac{CL}{H}$, where C = capacity, the number of light paths is L, and the average packet

hop distance is H. Therefore, minimizing H and maximizing the network throughput are asymptotic problems, while the equality can be satisfied. A linear program (LP) formulation is used to minimize H, the average packet hop distance, for a virtual-topology-based, wavelength-routed network. The LP gives a complete specification to the virtual-topology design, routing of the constituent light paths, and intensity of packet flows though the light paths [16]. The LP formulation requires a balanced network to be designed where the utilization of both transceivers and wavelengths are optimized, thus reducing cost of the network t. We make trade-offs in the budgeting of resources (transceivers and switch sizes) in the optical network.

Considering some of our underlying assumptions reduces running time of the solution. There are two simple heuristics with fast running times whose performance compares favorably with the performance of the optimized solution. Heuristics become important when the size of the problem becomes larger than what an LP solver can handle [1,16].

4.3.1 Problem Specification of LP

The notation and constraints used in Section 4.2. The same are considered here again. There are additional constraints/notations which include a new linear objective function, a constraint to bound the light path length, a constraint to bound the maximum loading per channel, a constraint to incorporate the physical topology as part of the virtual topology [1,16], and all of the simplifying assumptions in Section 4.2.1.

4.3.1.1 Linear Formulation

The problem is designated as an optimization problem, derived from multi-commodity flow for physical routing of light paths and traffic flow on the virtual topology [1], and the following notation is used:

- s and d, subscript or superscript, indicate source and destination of a packet, respectively.
- i and j indicate initial and final nodes, respectively, in a light path.
- m and n indicate endpoints of a physical link that might occur in a light path.
  Consider the following:
  Number of nodes in the network = N.
  Maximum number of wavelengths per fiber = W.

In physical topology, P_{mn} denotes the number of fibers interconnecting node m and node n.

$P_{mn} = 0$ for nodes which are not physically adjacent to each other.

$P_{mn} = P_{nm}$ indicating that there are an equal number of fibers joining two nodes in different directions. There may be more than one fiberlink connecting adjacent nodes in the network.

$$\sum_{m,n} P_{mn} = M$$

= the total number of fiberlinks in the network.

For fiber distance d_{mn} from node m to node n, when $P_{mn} = 0$, $d_{mn} = d_{nm}$ and $d_{mn} = \infty$.

Shortest-path delay matrix is denoted as D where D_{sd} denotes the delay (sum of propagation delays only) of the shortest path between nodes s and d. Light path length bound α, $1 \leq \alpha < \infty$, limits the delay over a light path between two nodes i and j, with respect to the shortest-path delay D_{ij} between them; i.e., the maximum acceptable propagation delay over the light path between the two nodes i and j is D_{ij}.

Number of transmitters at node $i = T_i$ ($T_i > 1$). Number of receivers at node $i = R_i$ ($R_i > 1$). In general, we assume that $T_i = R_i$, Vi, although this is not a strict requirement [1]. Traffic matrix λ_{sd} denotes the average rate of traffic flow in packets per second from node s to node d, with $\lambda_{ss} = 0$ for $s, d = 1, 2, \ldots, N$. Capacity of each channel $= C$ (normally expressed in bits per second but converted to packets per second by knowing the mean packet length).

Maximum loading per channel $= \beta$, $0 < \beta < 1$. The β restricts the queuing delay on a light path from getting unbounded by avoiding excessive link congestion [1].

4.3.1.2 Variables

Virtual topology: The variable V_{ij} represents the number of light paths from node i to node j in the virtual topology. Since light paths are not necessarily assumed to be bidirectional, $V_{ij} = 0$ does not indicate that $V_{ji} = 0$. There may be multiple light paths between the same source-destination pair, i.e., $V_{ij} = 1$, for the case when traffic between nodes i and j is greater than a single light path's capacity (C).

Traffic routing: The variable λ_{ij}^{sd} shows the traffic flowing from node s to node d, and V_{ij} represents an intermediate virtual link.

Physical-topology route: The variable P_{mn}^{ij} represents the number of light paths between nodes i and j being routed though fiberlink mn. For example, since the light path from CA1 to TX passes through CA2, the variable

$$P_{CA1CA2}^{CA1TX1} = 1$$

4.3.1.3 Objective: Optimality Criterion

$$\frac{1}{\sum\limits_{sd} \Lambda_{sd}} \sum\limits_{ij} \sum\limits_{s,d} \lambda_{ij}^{sd}$$

The objective function minimizes the average packet hop distance in the network. $\sum\limits_{ij} \sum\limits_{s,d} \lambda_{ij}^{sd}$ is a linear sum of variables, while $\sum\limits_{sd} \Lambda_{sd}$ is a constant for a given traffic matrix. The two objective functions are minimization of the average packet delay over the network, and maximization of the scale factor by which the traffic matrix can be scaled up.

4.3.1.4 Constraints

On virtual-topology [1,2] connection matrix V_{ij}:

$$\sum_j V_{ij} \leq T_i \quad \forall i$$

$$\sum_i V_{ij} \leq R_j \quad \forall j$$

(4.9)

The above equation shows that the number of light paths emerging from a node is constrained by the number of transmitters at that node, while the number of light paths terminating at a node are constrained by the number of receivers at that node. V_{ij} variables can only hold integer values [1]. $V_i > 1$ indicates that there is more than one light path between the particular source-destination pair.

On physical route [1,2], P_{mn}^{ij} variables:

$$\left.\begin{array}{l} \sum_m P_{mk}^{ij} = \sum_n P_{kn}^{ij} \ if \ k \neq i, j \\[2ex] \sum_n P_{in}^{ij} = V_{ij} \\[2ex] \sum_m P_{im}^{ij} = V_{ij} \\[2ex] \sum_{ij} P_{im}^{ij} = WP_{mn} \end{array}\right\}$$

(4.10)

On virtual topology [1,2] traffic variables:

$$\left.\begin{array}{l} \sum_j \lambda_{sj}^{sd} = \lambda_{sd} \\[2ex] \sum_i \lambda_{id}^{sd} = \lambda_{sd} \\[2ex] \sum_i \lambda_{ik}^{sd} = \sum_j \lambda_{kj}^{ij} \quad if \ k \neq i, j \\[2ex] \lambda_{ij}^{sd} \leq \lambda_{sd} \cdot V_{ij} \\[2ex] \sum_{s,d} \lambda_{ij}^{sd} \leq \beta \cdot V_{ij} \cdot C \end{array}\right\}$$

(4.11)

The above equations show the multi-commodity flow controlling the routing of light paths from source to destination [1,16]. The number of light paths passing through a fiberlink does not exceed *W*. The equations do not show the wavelength-continuity constraint (under which the light path is allotted with the same wavelength on all the fiberlinks through which it passes). In the absence of wavelength-continuity constraints in the equations, the solution obtained from our current formulation requires the network having wavelength converters.

Optional constraint [1,2]:

As physical topology is a subset of the virtual topology,

$$P_{mn} = 1 \Rightarrow V_{mn} = 1, P_{mn}^{mn} = 1 \tag{4.12}$$

Bound light path length:

$$\sum_{mn} P_{mn}^{ij} d_{mn} \leq \alpha.D_{ij} \text{ for } K \text{ alternative paths} \tag{4.13}$$

Equations (4.12) and (4.13) are incorporated to get bounded packet delays in this approach. These equations reduce the solution size of the problem. Equation (4.12) indicates every link in the physical topology as a light path in the virtual topology, in addition to which there are light paths which span multiple fiberlinks, Figure 4.6 demonstrates a virtual topology for a two-wavelength solution [1]. The light paths of the physical topology are used to find a path for transmission of network-control messages efficiently.

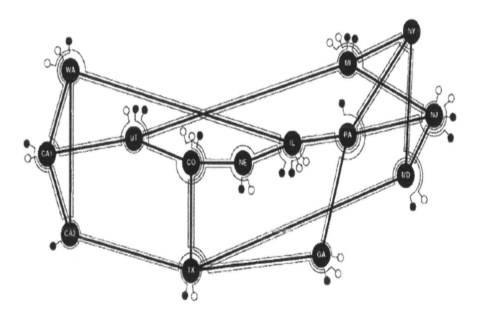

FIGURE 4.6 Optimal NSFNET virtual topology with two wavelengths per link [1].

Equation (4.13) restricts the list of variables to be only among those present in K alternate shortest paths from i to j, where $K > 1$. Equations (4.12) and (4.13) prevent long and convoluted light paths, i.e., light paths with an unnecessarily long route instead of a shorter route, from occurring. The value of K determined by the network design is normally 2.

Equations (4.12) and (4.13) represent the virtual-topology problem which is difficult to solve. The following simplifying assumptions are needed to make the problem solvable.

- The total number of light paths considered through a fiber is less than or equal to W. Adding wavelength-continuity constraints to equations (4.12) and (4.13) significantly increases the complexity of the problem, e.g., if the variable $C_k^{ij} = 1$ indicates that a light path from node i to node j is allotted with the wavelength k (where $k = 1, 2, \ldots, W$), the simplifying equations are as follows:

$$\sum_k C_k^{ij} = V_{ij} \tag{4.14}$$

$$\sum_k P_{mn}^{ij} C_k^{ij} = 1 \tag{4.15}$$

Equation (4.15) involves the product of two variables where the wavelength assignment of light paths is ignored in the current problem formulation, considering that the wavelength-assignment problem will be resolved separately, as per the light path routes obtained through this formulation or the availability of wavelength converters at the routing nodes [1].

Queuing delay simplifies the objective function, and the propagation delay dominates the overall network delay in nationwide optical networks. The exact optimization function for delay is as follows [1]:

$$\sum_{ij} \left[\sum_{sd} \lambda_{ij}^{sd} \left(\sum_{mn} P_{mn}^{ij} d_{mn} + \frac{1}{C - \sum_{sd} \lambda_{ij}^{sd}} \right) \right] \tag{4.16}$$

This is a non-linear equation involving the product of two variables λ_{ij}^{sd} and P_{mn}^{ij} because the term $\dfrac{1}{C - \sum_{sd} \lambda_{ij}^{sd}}$ is non-linear in λ_{ij}^{sd}.

The time complexity in the original problem formulation is $O(N^4)$. For reduction of time complexity, we reduce the number of constraints in the problem and consider a limited number of alternate shortest paths, denoted by K, between source-destination pairs, such that the selected routes are within a constant factor $(\alpha > 1)$ of the shortest-path distance between the given source-destination pair. The traffic flow is made by only the light paths connecting the nodes present in these alternate paths. Since these assumptions are considered during the generation of the problem formulation, the total number of equations and variables are decreased. The value of K is a function of the size of the problem that can be solved by the chosen LP solver [20]. If we consider two alternate shortest paths between any source-destination pair, then the two alternate paths from node (CA) to node (IL) in NSFNET are CAI-UT-CO-NE-IL and CAI-WA-IL. Then, the list of variables for light path routing are as follows [1]:

$$P_{CA1,UT}^{CA1,IL}, P_{UT,CO}^{CA1,IL}, P_{CO,NE}^{CA1,IL}, P_{NE,IL}^{CA1,IL}, P_{CA1,WA}^{CA1,IL}, P_{WA,NE}^{CA1,IL}, P_{NE,IL}^{CA1,IL},$$

Likewise, the enumerated variables for packet routing are as follows [1]:

$$\lambda_{CA1,UT}^{CA1,IL}, \lambda_{UT,CO}^{CA1,IL}, \lambda_{CO,NE}^{CA1,IL}, \lambda_{NE,IL}^{CA1,IL}, \lambda_{CA1,WA}^{CA1,IL}, \lambda_{WA,NE}^{CA1,IL}, \lambda_{NE,IL}^{CA1,IL},$$

This formulation provides bifurcated routing of packet traffic. To specify non-bifurcated routing of traffic, new variables γ_{ij}^{sd} are permitted to take only binary values, and the equations are suitably modified as follows [1]:

$$\left. \begin{aligned} & \lambda_{s'd'} = Max(\lambda_{sd}) \\ & \gamma_{ij}^{sd} \in (0,1) \\ & \sum_j \gamma_{sj}^{sd} = 1 \\ & \sum_i \gamma_{id}^{sd} = 1 \\ & \sum_i \gamma_{ik}^{sd} = \sum_j \gamma_{kj}^{jj} \quad \text{if } k \neq i,j \\ & \gamma_{ij}^{sd} \leq .V_{ij} \\ & \sum_{s,d} \gamma_{ij}^{sd} \lambda_{sd} \leq \beta.V_{ij}.C \end{aligned} \right\} \qquad (4.17)$$

The objective function [1] becomes

$$\text{Minimize} \ \frac{1}{\displaystyle\sum_{s,d} \lambda_{sd}} \left(\sum_{ij} \sum_{sd} \gamma_{ij}^{sd} \lambda_{sd} \right) \qquad (4.18)$$

Bifurcated routing is used. In non-bifurcated routing, packet traffic significantly increases the running time of the optimization solution. The increase in running time is due to the computation of the product terms in equation (4.17).

4.3.2 HEURISTIC APPROACHES

There are two heuristic approaches mainly used to resolve big problems in the virtual-topology design in order to minimize the average packet hop distance – maximizing single-hop traffic and maximizing multi-hop traffic. Formulations of these problems are made to increase the physical size of the network, and their solutions by traditional LP methods [20] are obtained using computational constraints.

For the optimization formulation [20], we do not include wavelength-continuity constraints in light path routing, since the heuristics accommodate this feature without any sacrifice in their running time.

1. Maximizing single-hop traffic

   This simple heuristic sets up light paths between source-destination pairs with the highest λ_{sd} values with constraints on the number of transceivers at the two end nodes and the availability of a wavelength in some path connecting two end nodes. The procedure for this heuristic is as follows [1],

   Procedure MaxSingleHop(void)s

   While (not done)

   $\lambda_{s'd'} = \lambda_{s'd'} - C$

   if

   $\qquad \lambda_{s'd'} = \lambda_{s'd'} - C$

   ((free transmitter available at s')

   AND

   if (free receiver available at d') AND

   (free wavelength available in any alternate path from s' to d'))

   $\qquad\qquad$ begin

   $\qquad\qquad$ Establish light path between s' and d'

   $\qquad\qquad\qquad$ end

   $\qquad\qquad$ endif

   $\qquad\qquad$ endwhile

2. Maximizing multi-hop traffic

   In a packet-switched network, the traffic carried by a link includes forwarded traffic as well as traffic originating from that node. So, any light path-establishment heuristic having the forwarded traffic performs better than a heuristic maximizing the single-hop traffic. The heuristic begins with the physical topology as the initial virtual topology and adds more light paths one by one. H_{sd} is the number of electronic hops needed to send a packet from source s to destination d. The heuristic establishes light paths in the decreasing order of $\lambda_{sd}(H_{sd} - 1)$, making with constraints on the number of transceivers at the two end nodes and the availability of a wavelength in some path connecting the two end nodes. After each light path is set up, H_{sd} values are estimated, as traffic flows are changed due to the new light path, in order to minimize the average packet hop distance. This algorithm allows only a single light path to be established between any source-destination pair (but this can be generalized). The procedure for this heuristic is provided below [1].

Procedure MaxMultiHop(void)
  Initial Virtual Topology = Physical Topology
while (not done) Compute $H_{sd} \; \forall s, d$
Find $\lambda_{sd}(H_{sd} = 1) = \text{Max} \; (\lambda_{sd}(H_{sd} - 1))$
if ((free transmitter available at s')
      AND
  (free receiver available at d')
      AND
  (free wavelength available in any alternate path from s' to d'))
begin
  Establish light path between s' and d'
      end
    endif
end while

4.4   NETWORK DESIGN: RESOURCE BUDGETING AND COST MODEL

This section discusses some of the virtual-topology design principles of a network derived from LP formulation. A cost model is required for the network design, in terms of the costs for the transmission as well as the switching equipment, to derive a minimum-cost solution [1].

4.4.1   BUDGETING

In a network having a very large number of transceivers per node with very few wavelengths per fiber and few fibers between node pairs, a large number of transceivers are used because some light paths are set up due to wavelength constraints. In a network with few transceivers, a large number of wavelengths are also unutilized due to the network transceiver constraint. These unutilized transceivers and wavelengths influence the cost of the network directly, as the number of wavelengths and number of transceivers in the network determine the cost of the switching devices and terminating devices. These network resources are optimized in order to maximize the utilization of both the transceivers and the wavelengths in the network. In this direction, resource budgeting becomes important to optimize network design with cost constraints. A very simple analysis is needed to resolve the resource-budgeting problem [1].

For a physical topology, and a routing algorithm of light paths, the average length of a light path is estimated as a function of the number of fiberlinks traversed by a light path averaged overall source-destination pairs in the network. The average length of a light path is indicated by H_p. If there are M fiberlinks in the network, accommodating W wavelengths, then the maximum number of light paths is MW/H_p, considering uniform utilization of wavelengths on all fiberlinks.

Therefore, the approximate number of transceivers per node in a balanced network [1] is obtained as

$$T_i = R_i = \frac{M.W}{N.H_p} \tag{4.19}$$

Cost Model

Budgeting in the network influences the cost of operation of the network. The cost model of the WRS is required taking the prototype which consider OXCs and transmission equipment at a node integrate together to form the corresponding WRS [1]. The following are the notations:

- C_t is the cost of a transceiver.
- C_m is the cost of a multiplexer or a demultiplexer.
- C_x is the cost of a 2×2 optical cross-point switching element.

Then, the aggregate network-wide equipment cost for transceivers is given by

$$C_t \left(\sum_i T_i + \sum_i R_i \right) \tag{4.20}$$

The aggregate cost of multiplexers/demultiplexers in the network is obtained as

$$C_m \left(2M + \sum_{i=0}^{N} (T_i/W) + \sum_{i=0}^{N} (R_i/W) \right) \tag{4.21}$$

$\sum_{i=0}^{N} (T_i/W) + \sum_{i=0}^{N} (R_i/W)$ represents the cost of providing (de)multiplexers for the local, access ports needed to launch or terminate light paths. The cost of a switch with q input and q output ports of 2×2 optical switching elements in a multi-stage interconnection network (MINN) is written as $C(q) = C_x.q \log(q/2)$. There is a MINN switch per wavelength in a WRS; hence, the cost for node m with degree $q_m = \sum_{n=1}^{N} P_{mn} + T_m/W$. The total network cross-switching cost is obtained as

$$W \sum_{m=1}^{N} C(q_m)$$

The total cost of the network is derived as [1]

$$C_{\text{total}} = C_t \left(\sum_i T_i + \sum_i R_i \right) + C_m \left(2M + \sum_{i=0}^{N} (T_i/W) + \sum_{i=0}^{N} (R_i/W) \right) + W \sum_{m=1}^{N} C(q_m)$$

$$\tag{4.22}$$

The total cost in the above equation does not include the wavelength-converter cost. The cost of a wavelength converter is also included in the current model; we consider to ignore the converter cost as less number of converters are used.

4.5 RECONFIGURATION OF VIRTUAL TOPOLOGY

An optical network requires reconfiguration of its virtual topology to be adapted as per changing traffic patterns. Some reconfigurations have been reported by previous authors [21–23]. These studies consider that the new virtual topology is based on the cost and sequence of branch-exchange operations for the conversion from the current virtual topology to the new virtual topology. The new virtual topology is designed considering the optimization of a given objective function, as well as reduction of the changes required in the current virtual topology [24]. The LP formulates new virtual topologies [1] from existing virtual topologies. For a given small change in the traffic matrix, the new virtual topology should be similar to the current virtual topology, in terms of the constituent light paths and the routes for these light paths, to minimize the changes in the number of WRS configurations required for shifting from the existing virtual topology to the new virtual topology [24].

Two traffic matrices λ_{sd}^1 and λ_{sd}^2 are considered at two but not-too-distant time instants. There is a correlation between these two traffic matrices. For a certain traffic matrix, there may be many different virtual topologies, each of which has the same optimal value with regard to the objective function in equation (4.18).

Virtual topology is designed using reconfiguration algorithm corresponding to λ_{sd}^2 which is close to the virtual topology corresponding to λ_{sd}^1.

4.5.1 RECONFIGURATION ALGORITHM

The steps of this algorithm are as follows:

- Do linear formulations $F(1)$ and $F(2)$, corresponding to traffic matrices λ_{sd}^1 and λ_{sd}^2, respectively, based on the formulation in Section 4.2.
- Estimate the solutions $S(1)$ and $S(2)$, corresponding to $F(1)$ and $F(2)$, respectively. The variables in $S(1)$ are indicated as $Q(1)$, $V_{ij}(1)$, $P_{mn}^{ij}(1)$, and $\lambda_{ij}^{sd}(1)$, and those in $S(2)$ are indicated as $Q(2)$, $V_{ij}(2)$, $P_{mn}^{ij}(2)$, and $\lambda_{ij}^{sd}(2)$, respectively. The values of the objective function considered for $S(1)$ and $S(2)$ are OPT_1 and OPT_2, respectively [1,51].
- Adjust $F(2)$ to $F'(2)$ by adding the new constraint,

$$\frac{1}{\sum_{s,d} \lambda_{sd}} \left(\sum_{ij} \sum_{sd} \gamma_{ij}^{sd} \lambda_{sd} \right) = \text{OPT}_2 \tag{4.23}$$

This ensures that all the virtual topologies made by $F'(2)$ are optimal as per the optimization of objective function.

- The new objective function for $F'(2)$ is [1,51,52]

$$\sum_{ij}\sum_{mn}\left|P_{mn}^{ij}(2)-P_{mn}^{ij}(1)\right| \hspace{3cm} (4.24)$$

Minimize

$$\sum_{ij}\left|V_{ij}(2)-V_{ij}(1)\right|$$

The mod operation, $|x|$, is a non-linear function. If P_{mn}^{ij} and V_{ij} have only binary values, then the above equations are linear. If $V_{ij}(1)=1$, then $\left|V_{ij}(2)-V_{ij}(1)\right|=(1-V_{ij}(2))$. If $V_{ij}(1)=0$, then $\left|V_{ij}(2)-V_{ij}(1)\right|=V_{ij}(2)$. $F'(2)$ may be solved directly using an LP solver. $\left|P_{mn}^{ij}(2)-P_{mn}^{ij}(1)\right|$ also follows the same assumptions [1].

4.5.2 NSFNET VIRTUAL TOPOLOGY DESIGN

In this section, we discuss virtual-topology network-design problem of the NSFNET backbone consisting of 14 nodes connected in a mesh network [1]. Each link is made of a pair of unidirectional fibers which transmit in opposite directions and which join physically adjacent nodes, i.e., $P_{mn}=P_{nm}=1$. Each node is based on a WRS along with multiple transceivers for origination and termination of light paths. The number of transmitters is considered to be equal to the number of receivers, and it is the same for all nodes. The traffic matrix is randomly made, considering a certain fraction F of the traffic uniformly over the range $[0,C/a]$ and the remaining traffic uniformly over range $[0,CT/a]$, where C is the light path channel capacity, a is an arbitrary integer which is 1 or greater, and T indicates the average ratio of traffic intensities between node pairs with high traffic values and node pairs with low traffic values. This model produces traffic patterns with varying characteristics. The average hop distance in the network is a function of the number of light paths which in turn depends on the number of transceivers and wavelengths. The average hop distance decreases with an increase of the number of transceivers and wavelengths. The increase of transceivers without adding extra wavelengths gives marginally better results. Transceiver utilization decreases with the decrease of number of wavelengths/increase of the number of transceivers, whereas the wavelength utilization decreases. It is necessary to obtain the correct balance between transceivers and wavelengths in the system for efficient utilization of both these expensive resources. As a cost constraint, resource-budgeting trade-off is needed to avoid under utilization of transceivers and wavelengths in the system.

There are 14 nodes in the network, and five transceivers are taken per node. So a maximum of $14 \times 5 = 70$ light paths are set up in the network. In the two-wavelength solution, only 59 light paths are set up, out of which 42 light paths are used in the physical topology and embedded as a virtual topology. In the five-wavelength solution, all of the 70 light paths are established, so that the transceiver utilization is 100% as opposed to a transceiver utilization of less than 85% for the two-wavelength case [1,52].

4.6   VIRTUAL-TOPOLOGY ADAPTATION
WITH DYNAMIC TRAFFIC

Considering the measured Internet backbone traffic characteristics, an adaptation mechanism [1] is needed to get accustomed with the changes in traffic without prior knowledge of the future traffic pattern. An adaptation mechanism redesigns the virtual topology according to an expected (or known) traffic pattern and then modifies the connectivity to achieve the target topology. The key step of our approach is to get accustomed to the underlying optical connectivity by measuring the actual traffic load on light paths continuously in a time interval and reacting promptly to the load imbalances caused by fluctuations in the traffic, by either adding or deleting one or more light paths at a time.

4.6.1   PROBLEM DEFINITION

We consider a virtual topology having the set of all such light paths in a network in which virtual topology is used by an Internet Service Provider (ISP) or a large institutional user of bandwidth to connect its end equipment (e.g., IP routers) by using wavelength channels from the network operator. The traffic rates between node pairs fluctuate distinguishably over time [53–55] which is an important obstacle in fixed virtual-topology design. A typical traffic measurement can be seen in Figure 4.7 [53] in which the measurements for both directions of a link on the Abilene network are displayed as two profiles over a 33-hour period, beginning at 9:00 a.m. on one day and ending a little after 6:00 p.m. the next day. A virtual topology is optimized for a specific traffic demand which does not respond with equal efficiency to a different traffic demand.

In this section, the problem of online redesign of the virtual topology is resolved in WDM mesh networks where the traffic load changes dynamically over time. Reconfiguration of optical networks has been studied by many authors both for broadcast optical networks [56] and for wavelength-routed networks [57]. The solution of the problem needs two-phase operation where the first step is virtual-topology design under the new traffic conditions and the second step is the transition from the old virtual topology to the newly designed one. It requires dynamic virtual-topology

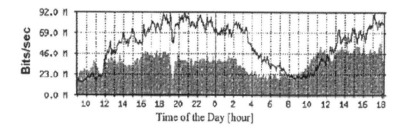

FIGURE 4.7   Traffic measurements on a link in the Abilene network during a 33-h period from 9:00 a.m. on day 1 to 6:00 p.m. on day 2. The two profiles correspond to the two directions of traffic on a link [1].

design. The light paths involved in transition are used by the ongoing traffic. To minimize the disruption to the ongoing traffic [58], all network elements are reconfigured concurrently or step-by-step changes are applied until the new virtual topology is designed [58]. But it is difficult to remove the traffic disruption because the transition step cannot be removed.

According to the methodology, the transition between topologies is considered by first setting up all new links without eliminating any link. The links of the old topology are removed only when the traffic passes through the links of the new topology. As traffic changes over time are monitored systematically, the virtual topology is reconfigured accordingly. The reconfiguration process is started as a continuous measurement made, instead of waiting in system. A new light path is added when congestion occurs. A light path is deleted if it is underutilized. In previous studies on virtual-topology design, many light paths were set up as far as possible from each other, and reconfiguration did not change the number of light paths [22].

Network Model

The arbitrary physical topology of a network having nodes is designed by connecting the nodes with bidirectional optical links. Each optical link has W wavelengths, and node i has T_i transmitters and R_i receivers. Each node consists of n OXC which should have high wavelength-conversion capability needed for efficient virtual-topology reconfiguration so that light paths are set up between any node pair if the resources (an optical transmitter at source, an optical receiver at destination, and at least one wavelength on each fiberlink) are available along the path. Virtual-topology design for wavelength conversion is not simple. Each OXC has an edge device called IP router, which controls a source or a destination of a traffic flow.

In this model, a centralized approach is considered for the virtual-topology reconfiguration with the following considerations:

1. Each router processes all packet traffic flowing through it and taking the amount of traffic on its outgoing light paths.
2. A central manager will collect the virtual-link-usage information from routers at the end of every observation period.
3. The link-usage information needed to make a reconfiguration decision is about which links are overloaded, which links are under loaded, and what the intensities of end-to-end packet traffic flowing through the overloaded links are.
4. The decision for a topology change will then be made by the central manager, and a signaling system will be modified if a light path addition or deletion is required **as** a result of the decision algorithm.
5. An implicit assumption here is that the observation period is much longer (typically hundreds of seconds or longer) than the time it takes for control signals to propagate from various nodes to the central manager.

We use shortest-path routing for routing light paths on the physical topology and the first-fit scheme for wavelength assignment. A shortest-path (minimum-hop) routing scheme is used for the efficient usage of network.

Problem formulation [1]

- The network graph is represented as $G(V, E_p)$ where V is the set of nodes and E_p is the set of links connecting the nodes. Graph nodes correspond to network nodes with OXCs, and links correspond to the fibers between nodes [22].
- Number of wavelength channels carried by each fiber.
- Number of transmitters and receivers at each node.
- Current virtual topology $V(V, E_v)$ is represented as a graph where the nodes correspond to the nodes in the physical topology. Each link in E_v corresponds to a direct light path between the nodes.
- Current traffic load transmitted by each light path.

Problem dealing [1]

- The current virtual topology of the network should be efficient for the current traffic.
- A change should be made in the virtual topology.
- Light paths should be added and/or deleted if necessary.

Steps in solving a problem [1]

- Traffic should be monitored continuously to provide adequate information to the reconfiguration system.
- A decision mechanism triggers a virtual-topology change if the current topology is not convenient.
- Finally, the exact modification to the topology should be determined.

Virtual-Topology Adaptation

The formulation of virtual topology is obtained by mixed integer linear program (MILP). This formulation on several backbone networks considers the amount of traffic between node changes in a smooth and continuous manner [53] as seen in Figure 4.7 in which long-term variations have time-of-the-day characteristics where traffic intensities change in terms of hours [59].

Formulation of Adaptation using an MILP

High wavelength-conversion capability is required at each node in this formulation. We use the following notations:

- s and d indicate *source* and *destination* of a traffic flow.
- i and j indicate originating and terminating nodes of a light path, respectively.
- m and n indicate the end points of a physical link. H and W_L denote the high and low water marks which are used to detect the link-usage efficiencies in a network

At any step of the adaptation, one of these three decisions is taken:

- addition of a light path
- deletion of a light path
- no change to the virtual topology.

The highest and the lowest light path loads select the proper action for the best light path having maximum linkload of local optimization ("local" with respect to time). This problem uses the adaptation method [59].

Given
- Number of nodes in the network = N.
- Physical topology of the network $P = P_{mn}$, where P_{mn} represents the number of fibers between nodes m and n, and $P_{mn} = P_{nm}$ for $m = 1, 2, 3, \ldots, N$ and $n = 1, 2, 3, \ldots, N$.
- Current traffic matrix $\lambda = \lambda_{sd}$ indicates the average traffic rate (in bits/s) measured during the last observation period between every node pair, with $\lambda_{ss} = 0$ for $s = 1, 2, 3, \ldots, N$.
- Current virtual topology $V = V_{ij,q}$ where $V_{ij,q}$ is a binary value indicating the qth light path between nodes i and j, and $V_{ii,q} = 0$. If there is no light path from node i to node j, $V_{ij,1} = V_{ij,k-1} = 1$, and $V_{ij,k} = 0$ if there are k light paths from node i to node j. Since light paths are not necessarily bidirectional, $V_{ij,q} = 0$, but $V_{ji,q}$ need not necessarily be 0.
- Number of wavelengths on each fiber = W
- Capacity of each wavelength channel = C bps.
- Number of transmitters and receivers at node i are considered to be T_i and R_i, respectively.
- High watermark value = W_H where $W_H \in (0, 1)$, e.g., $W_H = 0.8$ implies that a light path is considered to be overloaded when its load exceeds $0.8C$.
- Low watermark value = W_L where $W_L \in (0, 1)$.
- Highest and lowest light path loads measured during the observation period are L_{max}^P bps and L_{min}^P bps, respectively.

Variables
- Physical routing binary variable $P_{mn}^{ij,q} = 1$ if the qth light path from node i to node j is routed through the physical link (m, n).
- New virtual topology: $V' = V'_{ij,q}$ where $V'_{ij,q}$ is written as same as $V_{ii,q}$.
- Traffic routing: The binary variable $\rho_{sd}^{ij,q} = 1$ when the traffic flowing from node s to node d traverses light path $V'_{ij,q}$ and 0 otherwise. $\rho_{sd}^{ij,q} = 0$ by definition. The traffic from s to d is not bifurcated; i.e., all traffic between s and d will flow through the same path.
- Load of maximally loaded light path in the network = L_{max}

Objective
- Minimize L_{max}
- The objective function gives the load of the maximally loaded light path in the network. The network load is balanced in the new virtual topology by addition or deletion of the best possible light path.

Constraints
- On physical topology [1]:

$$\forall i,j,q \quad \sum_{n} P_{in}^{ij,q} = V_{ij,q}' \tag{4.25}$$

$$\forall i,j,q \quad \sum_{n} P_{nj}^{ij,q} = V_{ij,q}' \tag{4.26}$$

$$\forall i,j,l,q \quad \sum_{n} P_{nl}^{ij,q} - \sum_{n} P_{ln}^{ij,q} = 0, \quad i \neq l \neq j \tag{4.27}$$

$$\forall m,n \sum_{i} \sum_{j} \sum_{q} P_{mn}^{ij,q} \leq W * P_{mn} \tag{4.28}$$

$$\forall m,n,i,j,q \quad P_{mn}^{ij,q} \leq V_{ij,q}' \tag{4.29}$$

Equation (4.25) indicates that only one outgoing physical link of the source node is assigned to a light path, whereas equation (4.26) states that only one incoming physical link at the destination node is allotted to a light path. Equation (4.27) confirms the number of incoming and outgoing links set aside for a light path at any intermediate node. The total number of wavelengths used between two nodes is limited by equation (4.28). We consider wavelength conversion capability on network nodes, and we use the wavelength channels on different fibers as non-distinguishable entities. Equation (4.29) indicates that a physical link is allotted only if the light path exists.

- On virtual-topology connections [1]:

$$\sum_{i} \sum_{j} \sum_{q} V_{ij,q}' = \sum_{i} \sum_{j} \sum_{q} V_{ij,q} + k_H - (1 - k_H) * k_L \tag{4.30}$$

where

$$k_H = \left[L_{max}^P / C - W_H \right], k_L = \left[W_L - L_{max}^P / C \right]$$

$$\forall i,j,q \quad [1 + 2 * (k_H - 1) * k_L] * \left(V_{ij,q}' - V_{ij,q} \right) \geq 0 \tag{4.31}$$

The values of k_H and k_L are binary and are estimated by using the maximum and the minimum light path loads measured in the last observation period, watermark values, and channel capacity. $k_H = 1$ indicates that one or more light paths take a heavy load and results in the addition of a new light path to the virtual topology, whereas $k_H = 1$ indicates that one or more light paths has a load below the low watermark. $k_H = 0$ shows that none of the light paths in the virtual topology is heavily loaded and results in the deletion of one light path. Here a higher priority is assigned to a light path addition than light path deletion to accommodate more traffic. Equation (4.30) indicates the total number of light paths in the new virtual topology where a light path should be added or deleted. Equation (4.31) makes sure that the new virtual topology has the same light paths of the old virtual topology except one light path added/deleted.

- On virtual-topology traffic routing variables [1]:

$$\forall s,d,l,q \quad \sum_q \rho_{il,q}^{sd} - \rho_{li,q}^{sd} = \begin{cases} 1 & l = d \\ 0 & l \neq s \\ -1 & l = s \end{cases} \tag{4.32}$$

$$\forall s,d,i,j,q \quad \rho_{ij,q}^{sd} \leq V_{ij,q}' \tag{4.33}$$

$$L_{\max} \leq C * W_H \tag{4.34}$$

$$\forall i,j,q \quad \sum_s \sum_d \lambda_{sd} * \rho_{ij,q}^{sd} \leq L_{\max} \tag{4.35}$$

Equation (4.32) indicates a multi-commodity-flow controlling the routing of packet traffic on virtual links. Equation (4.33) indicates traffic flowing only through an existing light path, and equation (4.34) indicates the capacity constraint for any light path. Equation (4.35) shows the load constraint on any light path to be lower than or equal to the maximum load $L_{\max}$.

On transmitters:

$$\forall i \quad \sum_j \sum_q V_{ij,q}' \leq T_i \tag{4.36}$$

On receiver:

$$\forall \sum_i \sum_q V_{ij,q}' \leq R_j \tag{4.37}$$

Equations (4.36) and (4.37) provide the constraints to the total number of light paths originating from and terminating at a node and to the total number of transmitters and receivers at that node. The above formulation gives the best selection for a virtual-topology adjustment of one light path [1].

4.6.2 ADAPTATION WITH MINIMAL LIGHT PATH CHANGE

The number of light paths (i.e., resource usage cost) and the number of changes need to be minimized by reconfiguring topology with operation cost [1,54]. The new adaptation approach is inexpensive. The comparison is based on solving MILP formulations of both methods by using the standard solver CPLEX [60]. The reconfiguration of virtual topology is done in the following manner:

- Start with initial virtual topology.

Every time interval Δ seconds do:

- Determine the optimal virtual topology for the new traffic pattern.
- Determine the virtual topology requiring minimum number of changes from the previous topology.

The solution of this MILP is found by substituting the objective function to minimize the total number of light paths in the network [54]

$$\text{Minimize} \sum_i \sum_j \sum_q V_{ij,q} \leq L_{\max} \qquad (4.38)$$

The maximum load $L_{\max}$ to W_H should be considered. The optimum virtual topology is determined from this formulation providing the minimum number of light paths. The MILP can then be modified as follows:

1. The objective function can be modified to assure that the new virtual topology should be as close as possible to the previous one.
2. A new constraint is added to the formulation where the new virtual topology will have q light paths.

The second step considers the virtual topology to be a feasible topology having exactly q light paths carrying the given traffic demand. The optimal adaptation method is compared with the full reconfiguration method in which the total number of light paths in the network and the number of light path additions and deletions are taken. The performance in terms of light path depends on the variation of traffic load for 24 hours. Under both heavy and low traffic loads, the adaption method requires more light paths in comparison to the optimal method.

The design of virtual topology is essential for next-generation optical WANs, using WDM targeting nationwide coverage. WDM-based network architecture can provide a high aggregate system capacity due to spatial reuse of wavelengths. Our objective was to find the overall design, analysis, upgradeability, and optimization of a nationwide WDM network consistent with device capabilities [1].

SUMMARY

This chapter begins with the basic formulation of virtual topology. We have made simulations on NSFNET T1 optical backbone. We have discussed different algorithms for virtual topology design. Then advanced virtual-topology optimizations are mentioned in which an LP formulation to derive an exact minimal-hop-distance solution to the virtual-topology design problem in a wavelength-routed optical network is presented in the absence of wavelength-continuity constraints. The problem formulation derives a complete virtual-topology solution, including the choice of the constituent light paths, routes for these light paths, and intensity of packet flows through these light paths. Resource-budgeting trade-offs are required in the allocation of transceivers per node and wavelengths per fiber. An MILP formulation is presented here for the selection of the light path to be added or deleted to minimize the maximum link load in the network for the adaptation algorithm. The performance of the adaptation scheme is comparable to the optimal reconfiguration in terms of number of light paths and much better in terms of the cumulative number of changes.

EXERCISES

4.1. For the physical NSFNET topology, find a logical ring configuration which uses only one wavelength.

4.2. Why is a virtual topology embedded on a physical topology?

4.3. Consider the NSFNET physical topology, and remove nodes WA, UT and GA.
  a. Design the new physical topology.
  b. Set up light paths on the new topology that result in the virtual topology. In your virtual topology, what is the maximum number of wavelengths used on any link in the network?
  c. Show the details of the CO switch.
  d. Assuming a uniform traffic matrix, i.e., equal amount traffic between any two nodes and packets routed via the shortest path, find the average packet hop distance when packets are routed over the physical topology.

4.4. Explain why the average packet hop distance is used as the objective function.

4.5. Draw the virtual topology of the network shown in Figure Exercise 4.1. Calculate the number of light paths per link in this network.

4.6. Consider the network shown in Figure Exercise 4.1 with two transmitters and two receivers per node, two wavelengths, capacity of each wavelength equal to 10 units, and the following traffic matrix:

$$\begin{pmatrix} 0 & 2 & 3 & 4 & 5 & 6 \\ 2 & 0 & 5 & 9 & 4 & 2 \\ 5 & 3 & 0 & 2 & 1 & 8 \\ 7 & 3 & 2 & 0 & 2 & 9 \\ 3 & 1 & 3 & 3 & 0 & 1 \\ 5 & 3 & 2 & 1 & 2 & 0 \end{pmatrix}$$

Determine a set of light paths using the max single-hop heuristic.

4.7. Assume that the network equipment cost budget is $1,000,000. Using Table 9.1, find the network configurations (i.e., number of transceivers and wavelengths) that can be supported. Which network configuration maximizes the total network throughput?

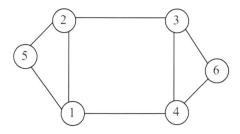

FIGURE EXERCISE 4.1   Physical network topology.

4.8. Assume the Petersen graph as the physical topology. Assume that the network can support 10 wavelengths. Now assume that we embed a complete graph on 10 nodes as the virtual topology. (Use shortest-path routing to embed the virtual topology.) Use the cost model described.

REFERENCES

1. B. Mukherjee, *Optical WDM Networks*, Springer-Verlag, 2006.
2. I. Chlamtac, A. Ganz, and G. Karmi, "Lightnets: Topologies for high speed optical networks," *IEEE/OSA Journal of Lightwave Technology*, vol. 11, pp. 951–961, 1993.
3. H. Zang, J. P. Jue, and B. Mukherjee, "A review of routing and wavelength assignment approaches for wavelength-routed optical WDM networks," *SPIE Optical Networks, Magazine*, vol. 1, pp. 47–60, 2000.
4. B. Ranmmurthy and A. Ramakrishnan, "Virtual topology reconfiguration of wavelength-routed optical WDM networks," *Proceedings, IEEE Globecom'00*, San Francisco, CA, pp. 1269–1275, November/December 2000.
5. N. Sreenath, C. S. R. Murthy, B. H. Gurucharan, and G. Mohan, "A two-stage approach for virtual topology reconfiguration of WDM optical networks," *SPIE Optical Networks Magazine*, vol. 2, no. 5, pp. 58–71, May 2001.
6. J.-F. P. Labourdette, G. W. Hart, and A. S. Acampora, "Branch-exchange sequences for reconfiguration of lightwave networks," *IEEE Transactions on Communications*, vol. 42, pp. 2822–2832, 1994.
7. G. N. Rouskas and M. H. Ammar, "Dynamic reconfiguration in multihop WDM networks," *Journal of High Speed Networks*, vol. 4, no. 3, pp. 221–238, 1995.
8. J.-F. P. Labourdette and A. S. Acampora, "Logically rearrangeable multihop lightwave networks," *IEEE Transactions on Communications*, vol. 39, pp. 1223–1230, 1991.
9. J. A. Bannister, L. Fratta, and M. Gerla, "Topological design of the wavelength-division optical network," *Proceedings, IEEE INFOCOM-90*, San Francisco, CA, pp. 1005–1013, June 1990.
10. B. Mukherjee, D. Banerjee, S. Ramamurthy, and A. Mukherjee, "Some principles for designing a wide-area optical network," *IEEE/ACM Transactions on Networking*, vol. 4, pp. 684–696, 1996.
11. R. Ramaswami and K. Sivarajan, "Design of logical topologies for wavelength-routed all-optical networks," *Proceedings, IEEE INFOCOM-95*, Boston, MA, pp. 1316–1325, April 1995.
12. R. Ramaswami and K. N. Sivarajan, "Design of logical topologies for wavelength-routed optical networks," *IEEE Journal on Selected Areas in Communications*, vol. 14, no. 5, pp. 840–851, 1996.
13. R. Ramaswami, and K. Sivarajan, *Optical Networks: A Practical Perspective*, 2nd ed., Morgan Kaufmann Publishers, 2001.
14. Z. Zhang and A. Acampora, "A heuristic wavelength assignment algorithm for multihop WDM networks with wavelength routing and wavelength reuse," *IEEE/ACM Transactions on Networking*, vol. 3, pp. 281–288, 1995.
15. J. A. Bannister, L. Fratta, and M. Gerla, "Topological design of the wavelength-division optical network," *Proceedings, IEEE INFOCOM-90*, San Francisco, CA, pp. 1005–1013, June 1990.
16. A. Gencata and B. Mukherjee, "Virtual-topology adaptation for WDM mesh networks under dynamic traffic," *IEEE/ACM Transactions on Networking*, vol. 11, no. 2, pp. 236–247, 2003.
17. J.-F. P. Labourdette and A. S. Acampora, "Logically rearrangeable multihop lightwave networks," *IEEE Transactions on Communications*, vol. 39, pp. 1223–1230, 1991.

18. A. Aarts and J. Korst, *Simulated Annealing and Boltzmann Machines*, New York: John Wiley & Sons, 1989.
19. L. Fratta, M. Gerla, and L. Kleinrock, "The flow deviation method: An approach to store- and-forward communication network design," *Networks*, vol. 3, pp. 97–133, 1973.
20. M. Berkelaar, "lpsolve: Readme file," Documentation for the lpSolve program, 1994.
21. R. Dutta and G. N. Rouskas, "Optical WDM networks: Principles and practice," *Design of Logical Topologies for Wavelength Routed Networks,* Norwell, MA: Kluwer, pp. 79–102, 2000.
22. J.-F. P. Labourdette, G. W. Hart, and A. S. Acampora, "Branch-exchange sequences for reconfiguration of lightwave networks," *IEEE Transactions on Communications*, vol. 42, pp. 2822–2832, 1994.
23. G. N. Rouskas and M. H. Ammar, "Dynamic reconfiguration in multihop WDM networks," *Journal of High Speed Networks*, vol. 4, no. 3, pp. 221–238, 1995.
24. D. Banerjee and B. Mukherjee, "Wavelength-routed optical networks: Linear formulation, resource budgeting tradeoffs, and a reconfiguration study," *IEEE/ACM Transactions on Networking*, vol. 8, no. 5, pp. 598–607, 2000.
25. H. Zang, J. P. Jue, and B. Mukherjee, "A review of routing and wavelength assignment approaches for wavelength-routed optical WDM networks," *SPIE Optical Networks Magazine*, vol. 1, no. 1, pp. 47–60, 2000.
26. S. Subramaniam, and R. A. Barry, "Wavelength assignment in fixed routing WDM networks," *IEEE International Conference on Communications*, pp. 406–410, 1997.
27. C. R. Reeves, *Modern Heuristic Techniques for Combinatorial Problems*, McGraw-Hill, 1995.
28. V. J Rayward-Smith, I. H. Osman, C. R. Reeves and G. D. Smith, *Modern Heuristic Search Methods*, John Wiley & Sons, 1996.
29. A. Hertz and D. de Werra, Using Tabu search techniques for graph coloring. *Computing* 39, pp. 345–351, 1987.
30. S. Subramaniam and R. A. Barry, "Wavelength assignment in fixed routing WDM networks," *Proceedings, IEEE ICC-97*, Montreal, Canada, pp. 406–410, June 1997.
31. Y. Sun, J. Gu, and D. H. K. Tsang, "Multicast routing in all optical wavelength routed networks," *Optical Networks Magazine*, vol. 2, pp. 101–109, 2001.
32. G. Jeong and E. Ayanoglu, "Comparison of wavelength interchanging and wavelength-selective cross-connects in multiwavelength all-optical networks," *Proceedings, IEEE INFOCOM-96*, San Francisco, CA, pp. 156–163, March 1996.
33. E. Karasan and E. Ayanoglu, "Effects of wavelength routing and selection algorithms on wavelength conversion gain in WDM optical networks," *IEEE/ACM Transactions on Networking*, vol. 6, no. 2, pp. 186–196, 1998.
34. R. A. Barry and S. Subramaniam, "The MAX-SUM wavelength assignment algorithm for WDM ring networks," *Proceedings, OFC-97, Dallas*, TX, pp. 121–122, February 1997.
35. X. Zhang and C. Qiao, "Wavelength assignment for dynamic traffic in multi-fiber WDM networks," *Proceedings, 7th International Conference on Computer Communications and Networks*, Lafayette, LA, pp. 479–485, October 1998.
36. A. Birman and A. Kershenbaum, "Routing and wavelength assignment methods in single-hop all-optical networks with blocking," *Proceedings, IEEE INFOCOM-95*, Boston, MA, pp. 431–438, April 1995.
37. X. Yuan, R. Melhem, R. Gupta, Y. hlei, and C. Qiao, "Distributed control protocols for wavelength reservation and their performance evaluation," *Photonic Network Communications*, vol. 1, no. 3, pp. 207–218, 1999.
38. N. Charbonneau and V. M. Vokkarane, "A survey of advance reservation routing and wavelength assignment in wavelength-routed WDM networks" *IEEE Communications Surveys & Tutorials*, vol. 14, pp. 1037–1064, 2012.

39. N. Fazlollahi and D. Starobinski, "Distributed advance network reservation with delay guarantees," *IEEE International Symposium on Parallel Distributed Processing (IPDPS)*, pp. 1–12, 2010.

40. C. Xie, H. Alazemi, and N. Ghani, "Routing and scheduling in distributed advance reservation networks," *Proc. IEEE GLOBECOM*, 2010.

41. B. C. Chatterjee, N. Sarma, and P. P. Sahu, "Priority based routing and wavelength assignment with traffic grooming for optical networks," *IEEE/OSA Journal of Optical Communication and Networking*, vol. 4, no. 6, pp. 480–489, 2012.

42. B. C. Chatterjee, N. Sarma, and P. P. Sahu, "Priority based dispersion reduced wavelength assignment for optical networks," *IEEE/OSA Journal of Lightwave Technology*, vol. 31, no. 2, pp. 257–263, 2013.

43. D. M. Shan, K. C. Chua, G. Mohan, and M. H. Phunq, "Priority-based offline wavelength assignment in OBS networks," *IEEE Transactions on Communications* vol. 56, no. 10, pp. 1694–1704, 2008.

44. B. C. Chatterjee, N. Sarma, and P. P. Sahu, "A heuristic priority based wavelength assignment scheme for optical networks," *Optik – International Journal for Light and Electron Optics*, vol. 123, no. 17, pp. 1505–1510, 2012.

45. P. P. Sahu, "Compact optical multiplexer using silicon nano-waveguide" *IEEE Journal of Selected Topics in Quantum Electronics*, vol. 15, no. 5, pp. 1537–1541, 2009.

46. P. P. Sahu, "Thermooptic two mode interference photonic switch" *Fiber and Integrated Optics*, vol. 29, pp. 284–293, 2010.

47. P. P. Sahu and A. K. Das, "Polarization-insensitive thermo-optic Mach Zehnder device based on silicon oxinitride waveguide with fast response time" *Fiber and Integrated Optics*, vol. 29, no. 1, pp. 10–20, 2010.

48. P. P. Sahu, "Tunable optical add/drop multiplexers using cascaded Mach Zehnder coupler" *Fiber and Integrated Optics*, vol. 27, no. 1, pp. 24–34, 2008.

49. P. P. Sahu "Polarization insensitive thermally tunable Add/Drop multiplexer using cascaded Mach Zehnder coupler," *Applied Physics: Lasers and Optics*, vol. B92, pp. 247–252, 2008.

50. G. Keiser, *Optical Fiber Communication*, McGraw Hill, 2002.

51. D. Bienstock and O. Gunluk, "A degree sequence problem related to network design," *Networks*, vol. 24, pp. 195–205, 1994.

52. B. Ranmmurthy and A. Ramakrishnan, "Virtual topology reconfiguration of wavelength-routed optical WDM networks," *Proceedings, IEEE Globecom-00*, San Francisco, CA, pp. 1269–1275, November/December 2000.

53. Abilene Network Traffic Statistics, http://www.abilene.iu.edu.

54. R. Dutta and G. N. Rouskas, "A survey of virtual topology design algorithms for wavelength routed optical networks," *Optical Networks Magazine,* vol. 1, no. 1, pp. 73–89, Jan. 2000.

55. Cooperative Association for Internet Data Analysis, http://www.caida.org.

56. I. Baldine and G. N. Rouskas, "Traffic adaptive WDM networks: A study of reconfiguration issues," *IEEE/OSA Journal of Lightwave Technology*, vol. 19, no. 4, pp. 433–455, 2001.

57. K. Bala et al., "Toward hitless reconfiguration in WDM optical networks for ATM transport," *Proceedings, IEEE GLOBECOM-96*, London, UK, pp. 316–320, November 1996.

58. A. Narula-Tam and E. Modiano, "Dynamic load balancing for WDM based packet networks," *Proceedings, IEEE INFOCOM-00*, Tel Aviv, Israel, pp. 1010–1019, March 2000.

59. J. A. Bannister, L. Fratta, and M. Gerla, "Topological design of the wavelength-division optical network," *Proceedings, IEEE INFOCOM-90*, San Francisco, CA, pp. 1005–1013, June 1990.

60. CPLEX, http://www.ilog.comf

5 Wavelength Conversion in WDM Networks

Wavelength conversion (WC) can be included in WDM networks to increase link utilization/efficiency [1,2]. Wavelength converters [3–6] were already mentioned in Chapter 2 of volume 1. This chapter deals with the network performance issues when WC is incorporated. Various analytical models used to assess the performance benefits in a wavelength-convertible network will be reviewed. To establish a lightpath, a wavelength is allocated on all of the links in the path considering the wavelength-continuity constraint [1]. This constraint differentiates the wavelength-routed network from a "circuit-switched" network which has no such constraints since the latter blocks a connection only when there is no capacity along any of the links in the path assigned to the connection.

Figure 5.1a represents wavelength routing and wavelength continuity with $W = 2$ wavelengths per link. There are two lightpaths established – (1) between node 1 and node 2 on wavelength λ_1 and (2) between node 2 and node 3 on wavelength λ_2. Now, we consider that a new lightpath between node 1 and node 3 needs to be set up. But it is impossible to establish such a lightpath; all the links along the path from node 1 to node 3 do not have the same wavelength. This is because the available wavelengths on any two links are *different*. Thus, a wavelength-routed network with

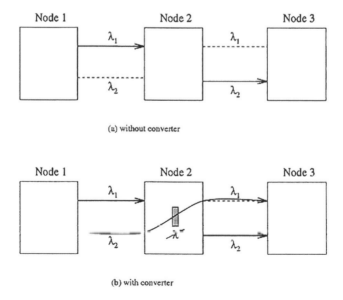

(a) without converter

(b) with converter

FIGURE 5.1 Wavelength-continuity constraint in WRS [1].

177

wavelength-continuity constraint has higher blocking probability (BP) in comparison to a circuit-switched network [2].

The wavelength-continuity constraint limitation is removed if we can use wavelength to convert the data arriving on one wavelength along a fiber link into another wavelength at an intermediate node and transmit it along the next fiber link. This technique is known as wavelength conversion (WC) as shown in Figure 5.1b, by converting from wavelength λ_2 to λ_1. Thus, WC enhances link utilization and reduces BP in the network by resolving the wavelength conflicts of the lightpaths [1].

5.1 BASICS OF WC

5.1.1 WAVELENGTH CONVERTERS

A wavelength converter is used to convert a signal's input wavelength to a different output wavelength among the N wavelengths in the system [1] (as shown in Figure 5.2). In this figure, λ_i indicates the input signal wavelength; λ_2, the converted wavelength; λ_p, the pump wavelength; f_i, the input frequency; f_o, the converted frequency; f_p, the pump frequency; and CW, the continuous wave generated as the signal.

Following are the characteristics [4] of an ideal converter used in the network:

- transparent toward bit rates and signal formats
- high-speed establishment of output wavelength
- conversion of both shorter and longer wavelengths
- moderate input power levels
- insensitive to input signal polarization
- large signal-to-noise ratio
- simple operation.

The details of WC devices are provided in Chapter 2 of volume 1.

5.1.2 SWITCHES

One of the most important things is to place the wavelength converter in the network. It is positioned in the switches (i.e., cross-connects) of network nodes. An architecture of a fully wavelength-convertible switching node has a dedicated wavelength-convertible switch shown in Figure 5.3 [7]). In this architecture, each wavelength along each output link in a switch requires a full wavelength converter. Each wavelength after demultiplexing goes to the desired output port via the non-blocking optical switch. The output signal has its wavelength changed by a wavelength converter. Finally, various wavelengths combine to form a multiplexed signal which is sent to the outbound link. The dedicated wavelength-convertible switch is not very costly since all the wavelength converters are not needed always. An effective method for

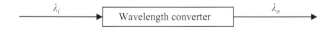

FIGURE 5.2 Operation principle of a wavelength converter.

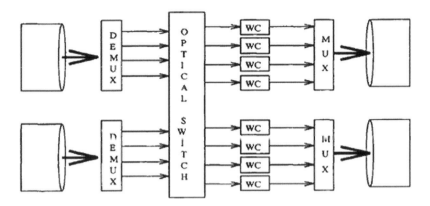

FIGURE 5.3 Fully dedicated wavelength-convertible switch node [1].

reduction of the costs is the sharing of the converters. There are two architectures used for nodes having sharing wavelength converters shared per node wavelength convertible architecture as shown in Figure 5.4a and shared per wavelength node wavelength convertible architecture as shown in Figure 5.4b. In Figure 5.4a, all the converters at the switching node are combined in the wavelength converter block having a collection of a few wavelength converter, each of which is assumed to have identical characteristics and can convert any input wavelength to any output wavelength. This block can be connected with any of the incoming lightpaths by appropriately configuring the larger optical switch. In this architecture, only the wavelengths which require conversion are directed to the converter block. The converted wavelengths are then sent to the appropriate outbound link by the second optical switch. In Figure 5.4, each outgoing link is provided with a dedicated converter using only those lightpaths available on that particular outbound link.

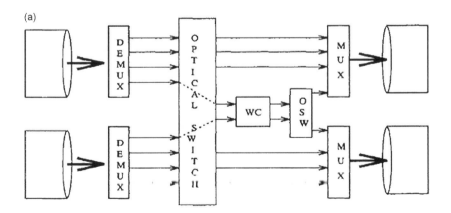

FIGURE 5.4 (a) Shared-per-node wavelength-convertible node architecture [1]. (b) Shared-per-link wavelength-convertible node architecture [1].

(*Continued*)

(b)

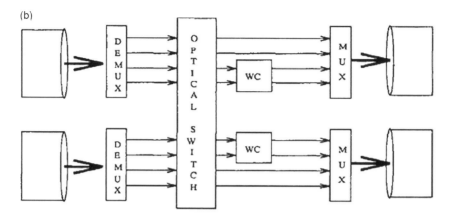

FIGURE 5.4 (CONTINUED)   (a) Shared-per-node wavelength-convertible node architecture [1]. (b) Shared-per-link wavelength-convertible node architecture [1].

5.2 OPTICAL NETWORK DESIGN, CONTROL, AND MANAGEMENT WITH WAVELENGTH CONVERSION

5.2.1 OPTICAL NETWORK DESIGN WITH WAVELENGTH CONVERTER

An efficient optical network is designed with the inclusion of WC effectively. Network designers consider various conversion techniques among the several switch architectures mentioned in Section 5.2. Following are the disadvantages of using WC technology.

1. **Sharing of converters:** To make cost-effective and efficient utilization of WC in switch architectures, converters are shared among various signals [1] at a switch of a node. The performance of such a network is not improved even if the number of converters at a switch increases beyond a certain threshold.

2. **Limitation in availability of wavelength converters at the nodes:** As wavelength converters are expensive; it is difficult to use WC at all the nodes in the network. Sparse conversion (i.e., having only a few converting switches in the network) can be used [2].

3. **Limited-range WC:** It is difficult to use WC for a wide range of wavelengths. It restricts the number of wavelength channels used in a link on the other hand. Mainly, four-wave-mixing-based all-optical wavelength converters are used, but they have a limited-range conversion capability. Analysis shows that networks employing such devices, however, compare favorably with those utilizing converters with full-range capability, under certain conditions [8].

The wavelength converters have disadvantages – especially the wavelength converter based on SOA in XGM mode degrades the signal quality during WC. Signal quality is degraded after many such conversions in cascade. There are efficient wavelength-convertible switch architectures, and their optimal placement and several other design techniques provide better link utilization and hence improve throughput in the network. In this direction, networks having multiple fibers on each link have been considered for potential gains [9] in wavelength-convertible networks. One important problem is the design of a fault-tolerant wavelength-convertible network.

5.2.2 CONTROL OF OPTICAL NETWORKS WITH WAVELENGTH CONVERTERS

Network management is required for the efficient use of network resources. The control algorithm for this provides the path to the connection requests while maximizing throughput. There are two mechanisms of routing control – *static* and *dynamic* – and they are described below.

1. **Dynamic routing:** In a wavelength-routed optical network, connection requests of lightpaths between source–destination (SD) pairs arrive randomly at the source node and while holding time is over, it is blocked. To establish the connection dynamically between SD pairs, it is required to determine a path through the network connecting the source to the destination and then allot a free wavelength to this path having a common wavelength to maintain wavelength continuity. If there is no common wavelength, a wavelength converter is used for assigning a wavelength to the node without maintaining wavelength continuity. Routing algorithms are required for allotting a wavelength in wavelength-convertible networks [7]. The routing algorithm makes an estimate of the cost function of routing as the sum of individual costs due to the use of channels and wavelength converters. An auxiliary graph is drawn [10], and the shortest-path algorithm is applied to the graph to determine the route. In [11], an algorithm has been provided for such a technique. There is a fixed path between every SD pair in the network. But the advantages of WC can also be better realized by using alternate-path routing algorithms.
2. **Static routing:** In contrast to the dynamic routing problem described above, the static routing and wavelength assignment (RWA) problem requires that all the lightpaths used in the network are pre-fixed as traffic arrivals are assumed to remain constant. Here, the main aim is to maximize the overall throughput in the network. An upper bound on the carried traffic per available wavelength has been obtained for a network with and without wavelength conversion by relaxing the corresponding integer linear programming (ILP) [12]. Several heuristic-based approaches have been reported for solving the static RWA problem in a network without WC [1].

5.2.3 NETWORK MANAGEMENT

The main management issue regarding the use of WC is to promote interoperability across subnetworks managed by independent operators. WC supports the distribution of network control and management functionalities by allowing flexible wavelength assignments within each subnetwork. As shown in Figure 5.5, network operators 1, 2, and 3 manage their own WC in the lightpath [1].

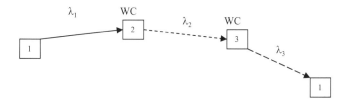

FIGURE 5.5 WC management in a lightpath.

5.3 BENEFIT ANALYSIS OF WAVELENGTH CONVERSION

As mentioned earlier, WC violates the wavelength-continuity constraint in wavelength-routed WDM networks. The wavelength assignment algorithm becomes insignificant in these networks because all the wavelengths are given equal importance. The converters may permit many long-distance connection requests which were blocked earlier due to the wavelength-continuity constraint, and in addition to this, they increase fairness. The attempts involve either probabilistic approaches or deterministic algorithms for benefit analysis in specific network topologies [13].

5.3.1 A PROBABILISTIC APPROACH TO WC BENEFITS' ANALYSIS

A probabilistic model for benefit analysis estimates the benefits such as reduction in BP by using WC. We consider W wavelengths per fiber link and p as the probability that a wavelength is used on any fiber link. Since pW is the expected number of busy wavelengths on any fiber link, there is an H-link path for a connection from node A to node B to be set up. The probability P_L that a connection request from A to B is blocked is the probability that, along this H-link path, there exists a fiber link with all of its W wavelengths in use. If there are no wavelength converters, the probability P_b that the connection request from A to B is blocked is the probability that, along this H-link path, each wavelength is used on at least one of the H links, so that

$$P_b = 1 - \left(1 - \rho^W\right)^H \tag{5.1}$$

We consider q to be the target utilization for a given BP in a network with WC; for small values of P_b/H, q is written as

$$q = 1 - \left(1 - P_b^{1/H}\right)^{1/W} \approx \left(P_b/H\right)^{1/W} \tag{5.2}$$

So the achievable utilization is inversely proportional to the length of the lightpath connection (H) for large H. The probability P_b' that the connection request from A to B is blocked is the probability that, along this H-link path, each wavelength is used on at least one of the H links.

We consider p to be the target utilization for a given BP in a network without WC. So p can be written as

$$p = 1 - (1 - P_b^{/1/H})^{1/W} \approx -\frac{1}{H}\ln\left(1 - P_b^{/1/W}\right) \tag{5.3}$$

An approximation is made for large values of H for small correlation of successive link utilizations. We consider $G = q/p$ to be a measure of the benefit of WC.

$$G \approx -H^{1-1/W}\frac{P_b^{1/W}}{\ln\left(1 - P_b^{1/W}\right)} \tag{5.4}$$

where $P_b = P_b^{/}$ and W is moderate. If $H = 1$ or $W = 1$, then $G = 1$; i.e., there is no difference between networks with and without wavelength converters in these cases.

The gain increases as the BP decreases, and as W increases, G also increases until it peaks around $W = 10$ [14]. After peaking, G decreases but very slowly. Generally, it is found that, for a moderate to a large number of wavelengths, the benefits of WC increase with the length of the connection and decreases (slightly) with an increase in the number of wavelengths [14].

5.3.2 A REVIEW OF BENEFIT-ANALYSIS STUDIES

Wavelength-convertible optical networks are analyzed by considering various techniques: converter placement, allocation, conversion router architectures, and sparse WC. Here, we consider the optimal placement of a wavelength converter in limited number of nodes of the network to reduce the number of wavelength converters for blocking performance improvement.

5.3.2.1 Bounds on RWA Algorithms with and without Wavelength Converters

Upper bounds on the carried traffic in a wavelength-routed WDM network have already been reported in a previous study [12]. The problems of both static and dynamic RWA are resolved using integer linear programs (ILPs) with the objective of maximizing the number of lightpaths successfully routed. The formulation is same as that of a multi-commodity flow problem with integer flows through the links. The upper bound is obtained by relaxing the wavelength-continuity constraints for networks with full WC at all nodes. The bound is determined asymptotically by a fixed RWA algorithm. A heuristic shortest-path RWA algorithm for dynamic routing is used with a set of shortest paths, and it assigns the first available free wavelength to the lightpath requests. The wavelength reuse factor is the maximum traffic per wavelength for which the BP can be reduced either by using sufficiently large number of wavelengths or by using wavelength converters in large networks.

5.3.2.2 Probabilistic Model Not Based on Link-Load Assumption

An approximate model [13] is designed for a deterministic fixed-path wavelength-routed network with an arbitrary topology with and without WC. In this model, the following assumptions are considered:

1. The traffic loads on the different links are not dependent on each other.
2. Wavelength occupancy probabilities on the links are also not dependent on each other.
3. In a wavelength-assignment strategy, a lightpath is allotted a wavelength randomly from the available wavelengths in the path.

The performance of the network is based on the BP of the lightpaths. The benefits of WC are obtained in networks such as non-blocking centralized-switch topology and ring topology where WC significantly improves performance.

5.3.2.3 Probabilistic Model Based on Link-Load Assumption

The following assumptions are considered in this model [14]:

1. The link loads are not dependent on each other.
2. A wavelength used on successive links is not dependent on other wavelengths.
3. The interference length L (the expected number of links shared by two sessions) which share at least one link, is introduced.
4. Analytical expressions for link utilization and BP are obtained by considering an average path which spans H (average hop distance) links in networks with and without WC.

The gain (G) in using WC is written as the ratio of the link utilization with WC to that without WC for the same BP. The gain is directly proportional to the effective path length.

5.3.2.4 Probabilistic Model for a Class of Networks

The model [15] provides an approximate process for estimating the BP in a wavelength-routed network with the following assumptions:

a. Traffic arrival follows Poisson distribution.
b. It is considered to be a Markov chain model with state-dependent arrival rates.

There are two different routing schemes considered: fixed routing and alternate-path routing. Analyses and simulations are carried out by using fixed routing for networks of arbitrary topology with paths of length having three hops at the most and by using least-load routing (LLR) for fully connected networks with paths of one or two hops.

5.3.2.5 Multi-Fiber Networks

Multi-fiber links in the network are used to reduce the gain due to the use of WC where the number of fibers is more important than the number of wavelengths for a network [9]. A heuristic is used to resolve the capacity assignment problem in a wavelength-routed network having no WC where the multiplicity of the fibers is considered to be minimized. A mesh network provides higher utilization gain with WC for the same traffic demand than a ring or a fully connected network. The benefits of WC in a network with multiple fiber links also need to be studied by extending the analysis presented in [1] to multi-fiber networks.

5.3.2.6 Sparse Wavelength Conversion

The sparse WC also reduces the cost of the network. In this case, only a few nodes in the network are capable of full WC, and the remaining nodes do not support any conversion at all [2,16]. The analytical model in [2] improves on the model proposed in [13] by relaxing the wavelength-independence constraint while retaining the link-load independence assumption, thus incorporating, to a certain extent, the correlation between the wavelengths used in successive links of a multi-link path.

5.3.2.7 Limited-Range WC

The limited-range WC also achieves gains in the performance of a network [16,17]. The model uses the functionality of certain all-optical wavelength converters (e.g., those based on four-wave mixing) whose conversion efficiency drops with increasing range. The analytical model considers both link-load independence and wavelength independence. The wavelength-assignment algorithm considering fixed routing minimizes the number of converters by choosing the input wavelength with the lowest index at the converters. The simulation results show significant improvement in the blocking performance of the network obtained for limited-range wavelength converters [2].

5.3.3 Benefits of Sparse Conversion

There are three degrees of sparseness to reduce WC [2,13–17]:

- **Sparse nodal conversion:** in a few nodes of the network with "full" conversion capabilities
- **Sparse switch-output conversion:** limiting the number of wavelength converters at each node
- **Sparse- or limited-range conversion:** limiting the conversion capabilities both in range and in the number of nodes in the network [17].

To realize the advantages of sparse conversion, some degree of sparse WC is used without the degradation of the performance of the network. The following strategies may be considered:

- For "full" WC at a few nodes, it is required to find the "best" nodes at which wavelength converters are placed.
- With different and effective wavelength-converting-switch designs, one needs to utilize fewer wavelength converters effectively, to determine which switch design should be implemented in the network.
- In order to avoid under- or overutilization of the wavelength converters at each node, it is required to have many wavelength converters at selected nodes.
- It is required to find whether wavelength converters offer significant benefits to optical networks, in terms of the reduction in BPs justifying the increase in costs due to the deployment of wavelength converters.
- It is required to investigate how different traffic loads affect the need for or desirability of wavelength converters.

Proper simulation is required to investigate these requirements of sparse conversion.

5.4 RWA WITH ALL THE NODES FULLY WAVELENGTH CONVERTIBLE

Before discussing RWA for the network having all the nodes fully wavelength convertible, the wavelength-convertible node architecture, mathematical formulation, and related constraints need to be mentioned [18,19].

5.4.1 FULLY WAVELENGTH-CONVERTIBLE NODE ARCHITECTURE

There are three kinds of WC switch architectures used for fully convertible nodes [18,19]:

  i. dedicated
 ii. shared-per-node
iii. shared-per-link.

In dedicated WC switch architecture, each wavelength is convertible by dedicated WC as seen in Figure 5.4a. This figure shows WC node for three ingoing (denoted as D links) and three outgoing (denoted as D links) fibers each having three wavelengths (denoted as W). It contains D number of 1XW wavelength multiplexers (MUXs) [20], same number of demultiplexers (DEMUXs) [20], and DW X DW switch [21,22].

In case of shared-per-node architecture, wavelength converters which converts each wavelength channels to others are used in a node and shared by all links in node as shown in Figure 5.4b. Although this architecture provides a reduction in the cost of a node, the BP may be increased due to the prior use of converters for conversion of wavelengths [20].

In case of shared-per-link architecture, the number of wavelength converters is equal to the number of links in which each converter converts all wavelength channels each to others per link and shared each converter per link in node as shown in

Figure 5.4c. This architecture increases the cost of a node slightly in comparison to shared-per-node architecture (Figure 5.4b), but the BP may be slightly reduced in comparison to shared-per-node architecture. Among these architectures, the dedicated architecture (Figure 5.4a) provides the best blocking performance.

5.4.2 MATHEMATICAL FORMULATION AND CONSTRAINTS

The physical topology of the network is represented as a unidirectional graph G (V,E,W) where V is the number of nodes, E is the number of the links in network, W = number of wavelengths per link. Each link consists of two unidirectional fibers. Each link has geographical distance between nodes. We consider the following notation [18,19]:

i, j denotes the end points (i.e., nodes) of physical link that might occur in the route of a connection.

e is used as an index for link number, where e = 1, 2, 3, ... E.

w is used as an index for the wavelength number of link e, where w = 1, 2, 3, ... W.

l is an index of a SD pair/connection where l = 1, 2, 3, ... L; L is the total number of possible SD pairs.

We require the following parameters for the solution of the RWA problem:

W_e is the number of wavelengths used in link e for protection.

S_e is the number of wavelengths used for traffic flow on link e.

$P_{ij}^{l,w,b}$ is an integer that takes a value 1 if the link ij of the route b is assigned a wavelength w for shared protection of the lth SD pair/connection; otherwise it is zero.

$R_{ij}^{l,w,b}$ is an integer that takes a value 1 if the link ij of the route b is assigned a wavelength w for restricted shared protection of lth SD pair/connection; otherwise it is zero.

$S_{ij}^{l,w,b}$ is an integer that takes a value 1 if the wavelength w is free on the link ij of the route b for the lth SD pair; otherwise it is zero.

$L_{ij}^{l,w,b}$ is an integer that takes a value 1 if the primary lightpath uses wavelength w on the link ij of the route b for the lth SD pair; otherwise it is zero.

The placement of wavelength converter in the network is represented by status vector

$$\{x_i\} = \{x_1, x_2, \dots x_N\}$$

where x_1, x_2,...x_N correspond to the statuses of node-1, node-2, ... node-N, respectively. The status of node i is defined as [19]

$$x_i = 1 \quad \text{if wavelength converter is place at node } i$$

$$= 0 \quad \text{otherwise}$$

Objective:

$$\text{Maximize} \quad \sum_{i=1}^{N} U_i . x_i \qquad (5.5)$$

where U_i is the utilization factor which is defined as the number of times the converter at node i is used for the connection of the SD pairs.

Constraint [19]:

$$\sum_{i=1}^{N} x_i = C \qquad (5.6)$$

where C = set of converters placed in the network:

$$C \leq N \qquad (5.7)$$

The other constraints for RWA (discussed in Chapter 4) are used for the assignment of wavelength in the shortest path (path of minimum delay).

5.4.3 ALGORITHM

We consider only the algorithm-based alternate routing in which free wavelength is not available for wavelength routing to improve the performance [19,23].

1. If any connection request comes for a SD pair, set protection backup lightpath for the following objective: Minimize $\sum_{e=1}^{E} W_e$ for all SD pairs.
2. Compute m number routes (R_i) on the basis of link state information in descending order of time delay (where $i = 1, 2, 3\ldots m$ correspond to 1st, 2nd, mth shortest time delay paths).
3. Choose the first shortest time delay path ($i = 1$) and select and assign the wavelength using parallel reservation scheme. If the wavelength is not available, then we use CONV_PLACE algorithm for establishing the connection of SD pair. If it is not assigned, go to the next step.
4. Try to assign a wavelength for the next alternative path in descending order of time delay. If it is not assigned to the wavelength, then we place converters in the same positions as those in step-3 for the connection of SD pair.
5. If it is not assigned, repeat step-4 k times; else go to next step.
6. The connection is blocked, and request is rejected.

CONV_PLACE algorithm:

We place the converter properly in such a way that the BP is reduced. The procedure of CONV_PLACE algorithm is given below [19,23]:

1. Fix the number of converters (C).
2. Choose a particular converter placement $\left(\left\{x_i^k\right\} = \left\{x_1^k, x_2^k, \dots x_N^k\right\}\right)$ and find $\sum_{i=1}^{N} U_i^k . x_i^k$ where U_i^k is the utilization factor of ith node in kth placement.
3. Repeat step-2 to maximize $\sum_{i=1}^{N} U_i^k . x_i^k$.
4. End.

5.4.4 SIMULATION

The simulation study requires the following assumptions [16,24]:

- There are enough transceivers at each node so that there is no constraint.
- The simulator is made flexible enough to test all possible aspects of wavelength converters, such as traffic model, (sparse) switch design, arbitrary network topology with arbitrary set of nodes with conversion capabilities, and arbitrary RWA algorithms. Our present study will assume.
- We consider Poisson arrivals, exponential holding times, and uniform (symmetric, balanced) traffic, and alternate-path routing (shortest path with respect to hops) for each connection, with one chosen randomly when multiple shortest paths exist.

5.5 RWA OF SPARSE WAVELENGTH CONVERTER PLACEMENT PROBLEM

Sparse wavelength converter placement is one of the most challenging problems under mesh topologies. In a sparse wavelength converter, one of the common objectives is to minimize the overall BP. There are number of approaches used for solving this problem–fixed alternate-path routing-first fit (FAR-FF) algorithm, least loading routing-first fit (LLR-FF) algorithm, and least loading routing-weighted maximum segment length (LLR-WMSL) algorithm. Before discussing these algorithms, one should know about the analytical model for the estimation of BP.

5.5.1 ANALYTICAL MODEL FOR THE ESTIMATION OF BLOCKING PROBABILITY

In Figure 5.6 for a path, we consider the source is i_0 and destination is i_1. So there is no scope to place a converter. So the success probability can be found out from the logic that if any wavelength can be found out, then we will get success or if all most

FIGURE 5.6   A path for converter placement.

all wavelengths are blocked, then blocking will be a success. The success probability is written as [19]

$$S\left(R_{sd}^{k}\right) = \left(1 - \rho_{0,1}^{W_{0,1}^{k}}\right).$$

However for two hop paths (suppose source i_0 and destination i_2), there is a scope to place a converter at node i_1, so the equation will have two terms, one with converter and another without converter. When there is no converter, the total path is only one segment of the equation and will be one hop path, but BP is different for different links. So the equation can be given as [19]

$$\left(1 - \left\{1 - \left(1 - \rho_{0,1}\right)\left(1 - \rho_{1,2}\right)\right\}^{W_{0,2}^{k}}\right)^{(1-x_1)}$$

Here this term is effective if $x_1 = 0$ (no converter at node 1). But if there is a converter, then we will get success if we are able to get any wavelength in the first link and any other wavelength in the second link. Or in another sense, the path will be blocked if all wavelengths are blocked in the first link and all wavelengths are blocked in the second link. The overall equation can be given as follows [19]:

$$S\left(R_{sd}^{k}\right) = \left(1 - \rho_{0,1}^{W_{0,1}^{k}}\right)^{x_1}\left(1 - \rho_{1,2}^{W_{1,2}^{k}}\right)\left(1 - \left\{1 - \left(1 - \rho_{0,1}\right)\left(1 - \rho_{1,2}\right)\right\}^{W_{0,2}^{k}}\right)^{(1-x_1)} \quad (5.8)$$

For multi-hop paths, the placement of the converter is given by a combined formula. If there are n nodes, then the total way of placing n converters is 2^n. We can fragment the path into so many individual parts or segments according to the converters placed. If there are converters at each node, then the equation takes the form

$$\left(1 - \rho_{0,1}^{W_{0,1}^{k}}\right)^{x_1}\left(1 - \rho_{1,2}^{W_{1,2}^{k}}\right)^{x_1 x_2}$$

$$\left(1 - \rho_{H-2,H-1}^{W_{H-2,H-1}^{k}}\right)^{X_{H-2}X_{H-1}}\left(1 - \rho_{H-1,H}^{W_{H-1,H}^{k}}\right)^{X_{H-1}}. \quad (5.9)$$

Here if x_i's of all nodes from i_0 to i_{H-1} are 1, then blocking is effective. So, there is scope to choose any wavelength in each link. If there is at least one converter, then an independent segment exists between two converters, between the first node and the first converter and between the last converter and the last node. If there is no converter, then there is only one segment from the first node to the last node. Formulas for different combinations of converters can be found with this approach so that wavelength continuity is maintained between the links (if multiple links are present) belonging to the same segment and any different or same wavelength between two segments.

The BP that a route R_{sd}^k for a SD pair is blocked is formulated as [19]

$$P\left(R_{sd}^k\right)=\left[1-S\left(R_{sd}^k\right)\right] \tag{5.10}$$

The BP of a connection considering alternate paths of the SD pair is written as [19]

$$P_{nt}=\prod_{k=1}^{K}\left[\left(1-S\left(R_{nt}^k\right)\right)\right] \tag{5.11}$$

and K is the maximum number of alternate paths considered for the establishment of lightpath for the SD pair.

Considering all SD pairs, we obtain the overall BP as [19]

$$P=\frac{\displaystyle\sum_{\forall s,d}\lambda_{sd}P_{sd}}{\displaystyle\sum_{\forall s,d}\lambda_{sd}} \tag{5.12}$$

Equation (5.12) gives the general formula for the overall BP of the network using alternate routing and partial placement of converters in the network.

5.5.2 FAR-FF ALGORITHM

This approach uses a heuristic algorithm based on maximum BP first (MBPF) which is employed for sparse wavelength converter placement [18]. In this case, each path is divided into segments in which one segment is the path length between two immediate neighboring convertible nodes along the selected path as shown in Figure 5.7.

Following are the parameters of this algorithm:

- M = number of wavelength converters placed in the network
- M_a = number of paths provided for ith node pair assigning wavelengths to that paths with the sequences $R_a^{(1)}, R_a^{(2)}, \dots R_a^{(M_i)}$

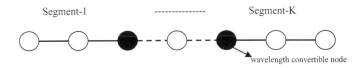

FIGURE 5.7 Segments of a path for converter placement where K = number of converters on the path.

The following assumptions have been considered for the estimation of BP:

- Connection requests arrive at a SD node pair a as per Poisson distribution. The call holding times are considered to be exponentially distributed with unit time.
- When a request arrives, paths are attempted sequentially from $R_a^{(1)}, R_a^{(2)}, \ldots R_a^{(M_i)}$ until a wavelength is allotted for a path from available wavelengths.
- If there are w_a^t wavelength converters on the path R_a^t excluding two end nodes, we can divide the path with $w_a^t + 1$ segments. Each segment has wavelength continuity. The kth segment is denoted by $R_a^{t,k}$, and the number of hop counts is represented as hR_a^t.
- The offered traffic is denoted as traffic arrived, and the carried traffic is already being set up successfully.

When a connection request for ith node pair arrives, a path is selected from M_a number of paths with sequences $R_a^{(1)}, R_a^{(2)}, \ldots R_a^{(M_i)}$ by using FAR-FF algorithm until a valid wavelength is assigned to the node pair. If no wavelength is available, the connection request is blocked. Once a connection request is established, the FF wavelength assignment is considered on each segment along the selected path where the free wavelength with the smallest label will be allotted to all the links in that segment. The selection of the best possible sparse converter placement is a difficult problem for a WAN. A heuristic algorithm of converter placement based on FAR-FF RWA [1] has been reported where the converters are placed one by one. In this case, we consider the placement of the converter in the node where the overall BP of the path passing through the node is more. This algorithm is called as MBPF, and its procedure is given below [25]:

Step-1: We find the paths $R_a^{(1)}, R_a^{(2)}, \ldots R_a^{(M_i)}$ for each node pair using the FAR algorithm. We put M converters into the network one by one.

Step-2: For each candidate node v, we first consider that a wavelength converter has been placed at that node, and then determine the corresponding BP using the analytical model discussed above. After calculating the BP of all candidate nodes, we place a wavelength converter at the node that can result in the minimum overall BP.

Step-3: If there are still wavelength converters remaining, then go to step-2.

The time complexity of this algorithm is estimated as $O(M \cdot N)$.

5.5.3   LLR-FF Algorithm

The LLR-FF algorithm provides less BP than the FAR-FF algorithm [1]. It needs proper sparse wavelength converter placement especially in an arbitrary mesh network. The following parameters are considered:

- The mesh network consists of N nodes and J fiber links, and each link has W wavelengths that are labeled form 1 to W.
- M is the number of wavelength converters. Our aim is to find proper converter placements to minimize of overall BP.

The following assumptions are considered:

- The arrival of connection requests at nodes follows a Poisson distribution with rate λ_{sd} (s–source and d–destination).
- The connection request holding time is exponentially distributed with one-unit time.
- There are M_a number of routes provided for node pair a, and the routes are denoted by $R_a^{(1)}, R_a^{(2)}, \dots R_a^{(M_i)}$.
- The notation of segments is as same as that in previous section. The number of free wavelengths of segment $R_a^{t,k}$ is indicated by $f(R_a^{t,k})$. For each segment $R_a^{t,k}$, the maximum segment length is the largest value of $h(R_a^{t,k})$ among the $w_a^t + 1$ segments and is denoted as $s(R_a^{t,k})$, and the number of free wavelengths is the smallest value of $f(R_a^{t,k})$ among those of all segments in kth path $R_a^{t,k}$.

Once a call request arrives at a node, a path should be selected for the assignment of wavelength on that path. Wavelength converters are placed in the network segment-wise with at least one converter placed in a segment. The following procedure is considered for wavelength assignment on a selected path.

Step-1: After arrival of a call request of node pair a, the states of free wavelengths are examined on M_a paths in node pair a. The path with maximum number of free wavelengths is chosen to set up a connection for the call. If two or more paths have the maximum number of free wavelengths, the path with the smallest label is selected. If no free wavelength is available, go to next step.

Step-2: Examine the availability of free wavelengths segment-wise on the paths. If free wavelengths are available, assign the free wavelength with the smallest label on each segment. Use a wavelength converter in the segment to convert the assigned wavelength to the free wavelength with the smallest label of the next segment, and continue the same till the request reaches the destination of node pair a. If the wavelength is not assigned by using wavelength converter, then call request is treated as blocked.

5.5.4 WMSL Algorithm

This is a heuristic algorithm for wavelength converter placement in which the converters are placed one by one sequentially. This approach allots a weight value to each candidate node denoting the weight of the node for converter placement. The length of the path is an important parameter that affects the BP of the path if there is no converter. The converters divide the path into several segments in which one converter is placed in each path. This leads to reduction in wavelength-continuity constraints. The traffic offered to the node pair a is distributed to all the provided routes evenly. The offered load for a path R_a^t is denoted as $\alpha(R_a^t)$ considering the above on the paths $R_a^{(1)}, R_a^{(2)}, \ldots R_a^{(M_i)}$.

The WMSL approach [1,12] tries to minimize the sum of the maximum segment lengths over the whole network considering the offered traffic to each path. The procedure of the approach is given below:

Step-1: Find the paths $R_a^{(1)}, R_a^{(2)}, \ldots R_a^{(M_i)}$ for each node pair a.
Step-2: Approximate the offered load

$$\alpha(R_a^t) = \frac{1}{M_a} A^a$$

where A^a = offered traffic in advance for a node pair a.
Step-3: Calculate the weight value $W(u)$ for each candidate node u. $s(R_a^{t,k})$ is the original maximum segment length of the path R_a^t after a converter is placed on a node u. The weight function $W(u)$ is then defined as

$$W(u) = \sum_{\substack{\text{all the routes that} \\ \text{transit through node } u}} \alpha\left(R_a^t\right)\left\{s\left(R_a^t\right) - s\left(R_a^t(u)\right)\right\}$$

After calculation over all the candidate nodes, we place a wavelength converter with maximum weight value.
Step-4: If there are still wavelength converters left, go to step-3.

The computation time complexity of the WMSL algorithm is as follows: There are M converters, and for each converter placement, each weight value can be calculated in $O(N2)$ for N nodes of the network. For M number of steps, the time complexity is written as $O(M \cdot N3)$ where M_a is less than N [12].

5.6   SIMULATION OF BENEFITS OF USING WAVELENGTH CONVERTERS

The different benefits of using wavelength converters are BP reduction, reduction of wavelength channels, and increase of wavelength channel utilization. These performances in the network ultimately reduce the cost of the network. Most of the simulations in this direction are mainly analyzed in terms of BP. The benefits of wavelength converters in the networks depend on the topology of the network. In this section, the benefits of using wavelength converters in the ring network, mesh–torus network [26,27], and NSFNET T1 backbone are mentioned [1].

Single Optical Ring

Figure 5.8 represents a unidirectional ring network having dynamic traffic for assessing the benefits of wavelength converters. Here, the nodes have full WC

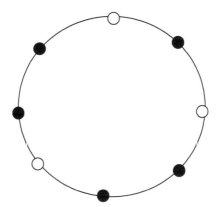

FIGURE 5.8 An eight-node optical ring with a sparse placement of converters. ○–
Converter in a node.

capabilities (i.e., any wavelength entering a node can exit on any free wavelength on
any output fiber that wavelength converters have limited usefulness in a single optical ring. There is negligible reduction of BP after the placement of certain number of
wavelength converters in the nodes in the ring. In the case of Figure 5.8, the optimal
number of converters is 3 [1].

Mesh–Torus Network

In the case of a mesh network, the benefits of using wavelength converters are
more than those in the case of a ring network. In a mesh–torus network, the density
of nodes is more [27] (Figure 5.9).

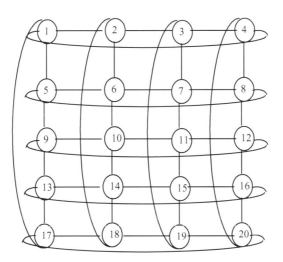

FIGURE 5.9 Torus network [1] with ten converters placed in nodes 2, 5, 7, 9, 10, 11, 13, 15,
17, and 19.

In the case of a 20-node torus, the reduction of BP stops after the placement of converters in ten nodes, which are the nodes 2, 5, 7, 9, 10, 11, 13, 15, 17, and 19.

NSFNET T1

NSFNET T1 backbone is nationwide arbitrary mesh network in which the sparse placement of wavelength converters provides improved performance in comparison to a ring network as the "mixing" of traffic is more [1,2]. There are many possibilities of sparse wavelength converter placements depending on traffic arrivals/traffic distribution in the network. The wavelength-assignment strategy used here is the FF algorithm, which uses the ordering of the wavelengths, and whenever wavelengths have to be chosen, it chooses the first available wavelength as per ordering. The exhaustive search of all combinations of wavelength-converter placements is based on the minimum BPs in the NSFNET. The optimal number of wavelength converters is eight; beyond this, the reduction of BP is very slow for any traffic load in a 14-node NSFNET. The placement of these eight wavelength converters in the NSFNET depends on traffic pattern between different nodes. A heuristic wavelength-converter-placement search is used to find the nodes in which wavelength converters may be placed. Therefore, a good heuristic for placing C sets of wavelength converters is to place them at the C nodes with the highest average output link congestion. Traffic load also influences the benefits of wavelength converters at light loads, there is not much need for WC as only few connections have a path to the irrespective destinations. So, the benefit of wavelength converters increases with the increase of traffic. The difference between the BPs with and without conversion becomes almost the same after the traffic load reaches a certain value. At heavy loads, using wavelength converters can establish new additional connections which no conversion is unable to set up. The rate of increase of overall blocking is less than that for a network without the placement of converters.

Few wavelength converters are required a teach node to obtain the minimum BP, and also most of the benefits are obtained with WC. We consider node-2 in NSFNET showing how its wavelength converters are utilized efficiently. Node-2 achieves WC 95% of the time utilizing three or fewer wavelength converters. Implementing a three-converter version of the switch (with three output fibers) would be very reasonable at node-2 [16,24].

SUMMARY

This chapter discusses the various aspects and benefits of WC in a network from its *incorporation* in a wavelength-routed network design to its *effect* on efficient routing and management algorithms to a *measurement* of its potential benefits under various network conditions. Some of the important results reported by our simulation-based case study of sparse WC are summarized below.

A network needs the mixing of traffic for wavelength converters to obtain benefits of WC. The mesh topology has higher connectivity benefits compared to the ring topology. Sparse nodal conversion or sparse output conversion provides the same benefits as a network having "full" conversion capabilities. Simple heuristics are used to efficiently place wavelength converters. The traffic load also affects the benefits of WC.

EXERCISES

5.1. There are three optical networks $N1$, $N2$, and $N3$ – $N1$ with one fiber between adjacent nodes in the physical topology and four wavelengths per fiber, allowing WC, $N2$ with four fibers between adjacent nodes in the physical topology and one wavelength per fiber, and $N3$ with full WC. We consider connection requests set up dynamically. Let p_1, p_2 and p_3 be the average BPs of networks $N1$, $N2$, and $N3$, respectively. How are p_1, p_2, and p_3 when compared with one another?

5.2. Given an optical network with the facility of recoloring existing lightpaths, show that such a network may block a connection request which could have been satisfied if WC was allowed.

5.3. Explain why employing multiple fibers between nodes is better (i.e., why it results in lower BPs) than increasing the number of wavelengths?

5.4. For a physical link having two wavelengths, show that the Indian network of dynamic connection setup requests in Figure Exercise 5.1 can be satisfied only with WC and cannot be satisfied otherwise.

5.5. Two networks $N1$ and $N2$ have the same physical topology and number of wavelengths per fiber. $N1$ has no WC, whereas there is WC in network $N2$. We consider the *least-congested-path* routing scheme to satisfy dynamic connection requests. The blocking probabilities for a sequence of connection requests S are p_1 and p_2 for the networks $N1$ and $N2$, respectively. Prove

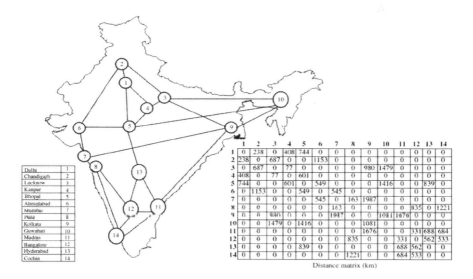

		1	2	3	4	5	6	7	8	9	10	11	12	13	14
Delhi	1	0	238	0	408	744	0	0	0	0	0	0	0	0	0
Chandigarh	2	238	0	687	0	0	1153	0	0	0	0	0	0	0	0
Lucknow	3	0	687	0	77	0	0	0	0	980	1479	0	0	0	0
Kanpur	4	408	0	77	0	601	0	0	0	0	0	0	0	0	0
Bhopal	5	744	0	0	601	0	549	0	0	0	1416	0	0	839	0
Ahmedabad	6	0	1153	0	0	549	0	545	0	0	0	0	0	0	0
Mumbai	7	0	0	0	0	0	545	0	163	1987	0	0	0	0	0
Pune	8	0	0	0	0	0	0	163	0	0	0	835	0	0	0
Kolkata	9	0	0	980	0	0	0	1987	0	0	1081	1676	0	0	0
Guwahati	10	0	0	1479	0	1416	0	0	0	1081	0	0	0	0	0
Madras	11	0	0	0	0	0	0	0	835	1676	0	0	331	688	684
Bangalore	12	0	0	0	0	0	0	0	0	0	0	331	0	562	533
Hyderabad	13	0	0	0	0	839	0	0	0	0	0	688	562	0	0
Cochin	14	0	0	0	0	0	0	0	1221	0	0	684	533	0	0

Distance matrix (km)

FIGURE EXERCISE 5.1   The Indian network.

that $p_1 > p_2$ for all S. Under this condition show $p_2 > p_1$ for a network topology and sequence of connections S.

5.6. Plot the percentage gain from using full conversion versus the network load for the Indian network. Explain the local maximum in the plot.

5.7. Consider a path consisting of five links. Each link supports up to five wavelengths, and average link utilization is 0.5. Calculate the BP with and without WC. What is the gain for a BP of 0.8?

5.8. Find out the benefits of sparse WC in the above-mentioned Indian network in terms of BP considering a sequence of traffic S (following Poisson's distribution for traffic arrival) in its nodes.

5.9. Find out the benefits of full WC in the above-mentioned Indian network in terms of BP considering a sequence of traffic S (following Poisson's distribution of arrival) in its nodes.

5.10. Find out the benefits of sparse WC with limited ranges in the above-mentioned Indian network in terms of BP considering a sequence of traffic S (following Poisson's distribution for arrival) in its nodes.

REFERENCES

1. B. Mukherjee, *Optical WDM Networks*, New York, Springer-Verlag, 2006.

2. S. Subramaniam, M. Azizoglu, and A. K. Somani, "All-optical networks with sparse wavelength conversion," *IEEE/ACM Transactions on Networking*, vol. 4, pp. 544–557, 1996.

3. H. Song, J. Lee, and J. Song, "Signal up-conversion by using a cross-phase-modulation in all-optical SOA-MZI wavelength converter," *IEEE Photonics Technology Letters*, vol. 16, no. 2, pp. 593–595, 2004.

4. T. Durhuus, et al., "All-optical wavelength conversion by semiconductor optical amplifiers," *IEEE/OSA Journal of Lightwave Technology*, vol. 14, pp. 942–954, 1996.

5. V. Eramo and M. Listanti, "Packet loss in a bufferless optical WDM switch employing shared tunable wavelength converters," *IEEE/OSA Journal of Lightwave Technology*, vol. 18, no. 12, pp. 1818–1833, 2000.

6. B. Mikkelsen, et al., "Wavelength conversion devices," *Proceedings, Optical Fiber Communication (OFC '96)*, San Jose, CA, vol. 2, pp. 121–122, 1996.

7. K.-C. Lee and V. O. K. Li, "A wavelength convertible optical network," *IEEE/OSA Journal of Lightwave Technology*, vol. 11, no. 516, pp. 962–970, 1993.

8. J. Yates, J. Lacey, D. Everitt, and M. Summerfield, "Limited range wavelength translation in all-optical networks," *Proceedings, IEEE INFOCOM '96*, San Francisco, CA, pp. 954–961, March 1996.

9. K. Sivarajan and R. Ramaswami, "Lightwave networks based on de Bruijn graphs," *IEEE/ACM Transactions on Networking*, vol. 2, no. 1, pp. 70-79. 1994.

10. K. Bala, T. E. Stern, and K. Bala, "Algorithms for routing in a linear lightwave network," *Proceedings, IEEE INFOCOM '91*, Bal Harbour, FL, vol. 1, pp. 1–9, April 1991.

11. I. Chlamtac, A. Faragó, and T. Zhang, "Lightpath (wavelength) routing in large WDM networks," *IEEE Journal on Selected Areas in Communications*, vol. 14, pp. 909–913, June 1996.

12. R. Ramaswami and K. Sivarajan, *Optical Networks: A Practical Perspective*, 2nd ed., San Francisco, CA, Morgan Kaufmann Publishers, 2001.

13. M. Kovacevic and A. S. Acampora, "Benefits of wavelength translation in all-optical clear-channel networks," *IEEE Journal on Selected Areas in Communications*, vol. 14, pp. 868–880, 1996.

14. R. A. Barry and P. A. Humblet, "Models of blocking probability in all-optical networks with and without wavelength changers," *IEEE Journal on Selected Areas in Communications*, vol. 14, pp. 858–867, 1996.

15. A. Sengupta, S. Bandyopadhyay, A. Jaekel, On the performance of dynamic routing strategies for all-optical networks, Proc. IASTED International Conference on Parallel and Distributed Computing and Networks, Singapore, Aug. 1997

16. J. Iness and B. Mukherjee, "Sparse wavelength conversion in wavelength-routed WDM networks," *Photonic Network Communications*, vol. 1, no. 3, pp. 183–205, 1999.

17. M. Ajmone Marsan, A. Bianco, E. Leonardi, F. Neri, "Topologies for wavelength-routing all-optical networks," *IEEE/ACM Trans. Networking*, vol. 1, no. 5, pp. 534-546. 1993.

18. S. Ramamurthy and B. Mukherjee, "Fixed alternate routing and wavelength converters in wavelength-routed optical networks," *IEEE GLOBECOM*, vol. 4, pp. 2295–2302, 1998.

19. P. P. Sahu and R. Pradhan, "Reduction of blocking probability in protected optical network using alternate routing and wavelength converter," *Journal of Optical Communication*, vol. 29, pp. 20–25, 2008.

20. P. P. Sahu, "Compact optical multiplexer using silicon nano-waveguide," *IEEE Journal of Selected Topics in Quantum Electronics*, vol. 15, no. 5, pp. 1537–1541, 2009.

21. R. Dutta and G. N. Rouskas, "Optical WDM networks: Principles and practice," in Design of Logical Topologies for Wavelength Routed Networks. Norwell, MA: Kluwer, pp. 79-102, 2000.

22. O. Gerstel, S. Kutten, Dynamic wavelength allocation in WDM ring networks, IBM Research Report RC 20462, 1996.

23. N. Sreenath, C. S. R. Murthy, B. H. Gurucharan, and G. Mohan, "A two-stageapproach for virtual topology reconfiguration of WDM optical networks," *Opt. Networks Mag.*, pp. 58-71, 2001.

24. J. Iness, "Efficientuse of optical components in WDM-based optical networks," PhD dissertation, University of California, Department of Computer Science, Davis, 1997.

25. I. Chlamtac, A. Ganz, and G. Karmi, "Lightpath communications: An approach to high bandwidth optical WANs," IEEE Trans. Commun., vol. 40, pp. 1171-1182, July 1992.

26. N. F. Maxemchuk, "Regular mesh topologies in local and metropolitan area networks," *AT&T Technical Journal*, vol. 64, pp. 1659–1686, 1985.

27. N. F. Maxemchuk, "Routing in the Manhattan street network," *IEEE Transactions on Communications*, vol. 35, pp. 503–512, 1987.

6 Traffic Grooming in Optical Networks

Using optical wavelength-division multiplexing (WDM), a light path in the network offers a large number of wavelength channels in physical optical fiber links between two nodes [1,2]. Since the bandwidth of a wavelength channel in an optical WDM backbone network is more (10Gbps (OC-192) [3] and grown now to 40 Gbps (OC-768)), a fraction of the customers who need the lower bandwidth of STS-11 (51.84 Mbps), OC-3, and OC-12 have to use a high-bandwidth wavelength channel for backbone applications. Since high-bandwidth wavelength channels are filled up by many low-speed traffic streams, efficiently provisioning customer connection switch such diverse bandwidth needs traffic grooming [4–26].

The formulation of the traffic-grooming problem is discussed in this chapter. For a network configuration, each edge is treated as a physical link, and there are a number of transceivers at each node, number of wavelengths on each fiber, and the capacity of each wavelength, and a set of connection requests with different bandwidth granularities, such as OC-12, OC-48 [3], etc. Here, the setting up of light paths for accommodating the connection requests is carried out by routing and wavelength assignment (RWA) with traffic grooming. Because of the sub-wavelength granularity of the connection requests, more connections are set up via multiplexing on the same light path.

6.1 REVIEW OF TRAFFIC GROOMING

The connection requests received in advance are called as static traffic, whereas those received one at a time are known as dynamic traffic. Traffic grooming with static traffic is achieved with dual optimization [4,5]. In a non-blocking scenario, the objective is to reduce the network cost and total number of wavelengths used in a WDM mesh network while establishing all the requests. In a blocking case, all connections are not accommodated due to resource restriction; the objective is to maximize the network through put. In case of dynamic traffic, where connections arrive one at a time, the objective is to minimize the network resources used for each request and to minimize the blocking probability (BP).

In traffic grooming, there are four sub-problems: (a) to designing the virtual topology that consists of light paths, (b) to find routing of the light paths over the physical topology, (c) to perform wavelength assignment to the light paths, and (d) to perform routing of the traffic on the virtual topology. The virtual-topology design and RWA are NP-hard problems, and traffic grooming is also an NP-hard problem [5].

To resolve the static traffic-grooming problem [5], the approach is to resolve four sub-problems separately – design of virtual topology, traffic grooming, carrying out of routing, and wavelength assignment. The divide-and-conquer method is used for traffic grooming, and an optimal solution is difficult to achieve. These four sub-problems are dependent on each other, and the solution to one sub-problem affects the solution of another sub-problem. Sometimes, using the optimal solution for one sub-problem may not provide the optimal solution of the whole problem. With static traffic, the traffic-grooming problem can be formulated as an integer linear program (ILP), and an optimal solution can be obtained for some relatively small networks.

Traffic grooming is an important problem for designing WDM networks with maximum utilization of each light path. The work in [5] reviews most of the recent research work on traffic grooming in WDM ring and mesh networks.

The traffic grooming has been used in synchronous optical network (SONET) and WDM ring networks [6,7]. The major cost of such a network is contributed from SONET add–drop multiplexers (ADMs). Therefore, minimizing/reducing the number of SONET/ADMs is the objective of static traffic grooming in ring networks [6].

Optical back bone networks providing nationwide communication/wide-area communication systems are based on mesh topology. Traffic grooming on WDM mesh network plays an important role in performing wavelength assignment very efficiently. The formulation of the static traffic-grooming problem is treated as an ILP, and a heuristic is used to minimize the number of transceivers [8]. The greedy and iterative greedy schemes are developed [9] in which some of the physical-topology constraints are relaxed, assuming all possible virtual topologies on the given physical topology. There are several node architectures for supporting traffic grooming in WDM mesh networks to formulate the static traffic-grooming problem as an ILP. There are many works [8–15] considering a dynamic traffic pattern in WDM mesh networks. In Ref. [12], a connection admission control scheme is proposed to ensure fairness in terms of connection blocking. A theoretical capacity correlation model is used to compute the BP for WDM networks with constrained grooming capability. There are two route computation algorithms [13] used in order to achieve good performance in a dynamic environment considering different grooming policies under different network states. These two schemes can be used to dynamically establish reliable low-speed traffic in WDM mesh networks with traffic-grooming capability.

In this chapter, the static traffic-grooming problem and its solution are discussed provided all the traffic demands are known in advance. Then, we consider the dynamic traffic-grooming problem, where connections arrive non-uniformly/randomly at a network, hold for a certain amount of time, and then serve in the network.

6.2   STATIC TRAFFIC GROOMING

Having network resources such as wavelengths and transceivers, RWA determines wavelength channel to successfully carry the connection requests (lightpaths) in an optical WDM mesh network as already discussed in Chapter 4. It is a lightpath-provisioning problem [2]. There are many RWAs reported in the optical networking based on either static traffic demands or dynamic traffic demands. In most of the previous studies, a connection request occupies bandwidth for an entire lightpath channel. In case of static traffic grooming, it is assumed that the bandwidth of the connection requests can be some fraction of the lightpath capacity. Figure 6.1 shows

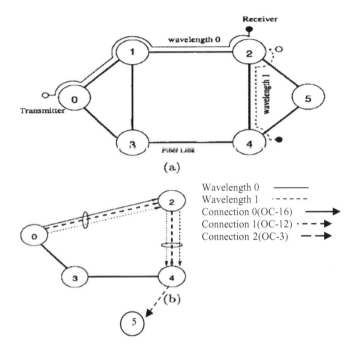

FIGURE 6.1 (a) Optical network having six nodes with connection requests and (b) the assignment of wavelength [2].

traffic grooming in a sample WDM mesh network. Figure 6.1a presents a small six-node network where each fiber has two wavelength channels. The capacity of each wavelength channel is OC-192, i.e., approximately 2.5 Gbps. There are three connection requests – (0,2) with bandwidth requirement OC-16, (2,4) with bandwidth requirement OC-12, and (0,5) with bandwidth requirement OC-3 [2]. Two lightpaths have already been setup to accommodate these three connections, as shown in Figure 6.1a.

For their source restriction where transmitter in node 0 and receiver in node 4 are busy, a lightpath directly from node 0 to node 4 cannot be set up; thus, connection 3 has to setup by the spare capacity of the two existing lightpaths, as shown in Figure 6.1b. Different connection requests between the same node pair (s,d) can be either groomed on the same lightpath, which directly joins (s,d), using various multiplexing techniques, or routed separately through different virtual paths.

The static traffic-grooming problem in a mesh network is to be optimized with the following objective functions [2]:

* to maximize the (weighted) network throughput for a given traffic matrix set and network resources
* to minimize the no of wavelength channels.

Finally, the mathematical formulation to accommodate other network optimization criteria is extended.

6.2.1 PROBLEM STATEMENT FOR TRAFFIC GROOMING

The problem of grooming low-speed traffic requests onto high-bandwidth wave-length channels in network is stated below. The following parameters are considered in the problem [2]:

1. A physical topology is represented as $G_p = (V, E_p)$ having a weighted uni-directional graph, where V is the number of network nodes, and E_p is the number of physical links connecting the network nodes. We assume the following [2]:
   - There are an equal number of fibers joining two nodes in different directions. Links are assigned weights which may correspond to physical distance between nodes.
   - All links have the same weight 1, which corresponds to the fiber hop distance.
   - A network node i is equipped with a $C_p(i) \times C_p(i)$ optical cross-connect ((OXC) based on wavelength-routing switch (WRS)), where $C_p(i)$ indicates the number of incoming fiber links to node i. For any node i, the number of incoming fiber links is equal to the number of outgoing fiber links.
2. Number of wavelength channels carried by each fiber = W. Capacity of a wavelength = C.
3. A set of $N \times N$ traffic matrices, where N = number of nodes in the network. Each traffic matrix in the traffic-matrix set indicates one particular group of low-speed connection requests between the nodes of the network. If C is OC-192, there are many traffic matrices: an OC-1, OC-3, OC-4, OC-12, OC-16, and OC-48.
4. The number of transmitters = T_i, and photodiodes (receivers) = R_i at each node i.

The goals are to determine the following [2]:

1. a virtual topology $G = (V, E)$
2. to minimize the total network cost or maximize total throughput.

Node Architecture

To accommodate connection requests in a WDM network, lightpath connections can be setup between pairs of nodes. A connection request is setup using one or more lightpaths before it reaches the destination. For the establishment of a connection request, there are two important functionalities performed by the WDM network nodes – wavelength routing and multiplexing and demultiplexing. An OXC contributes the wavelength-routing capability to the WDM network nodes, whereas an optical multiplexer/demultiplexer carries multiplexes/demultiplexes of several wavelengths on the same fiber link. Low-speed connection requests will be multiplexed on the same wavelength channel by using an electronic-domain hierarchal time-division multiplexing (TDM). There are two traffic grooming node architectures having IP over WDM and SONET over WDM in a WDM optical network [2] as shown in Figures 6.2 and 6.3.

In the figures, the node architecture consists of two components – WRS and access station. The WRS performs wavelength routing and wavelength multiplexing/ demultiplexing, whereas the access station carries out local traffic addition/dropping and low-speed traffic-grooming functionalities. WRS consists of an OXC, network control and management unit (NCM), and optical multiplexer/demultiplexer [2]. In the NCM unit, the network-to-network interfaces (NNIs) configure the OXC and exchange control messages with spare nodes on a dedicated wavelength channel (considered as wavelength 0 in Figures 6.2 and 6.3). The network-to-user interface (NUI) converses with the NNI and swaps control information with the user-to-network interface (UNI), the control component of the access station. The OXC has wavelength-switching functionality. Each fiber has three wavelengths. Wavelength 0 is used as a control channel for the NCM to exchange control messages between network nodes [2]. Other wavelengths are used to transmit data traffic.

In Figure 6.2, each access station has transmitters and receivers (transceivers) of all the wavelength channels. Traffic originating from an access station is transmitted with an optical signal on one wavelength channel by a transmitter. Traffic at an access station is converted from an optical signal to electronic data via a receiver. Both tunable transceivers and fixed transceivers are used in a WDM network. A tunable transceiver uses different wavelengths of an optical signal on any free

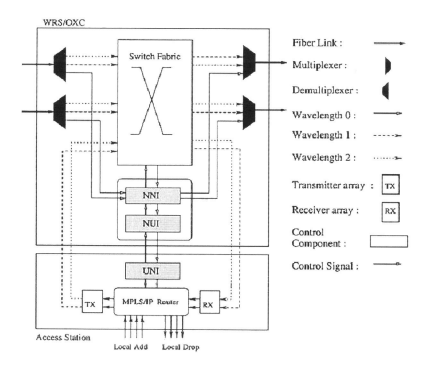

FIGURE 6.2   Node architecture with IP over WDM without traffic grooming [2].

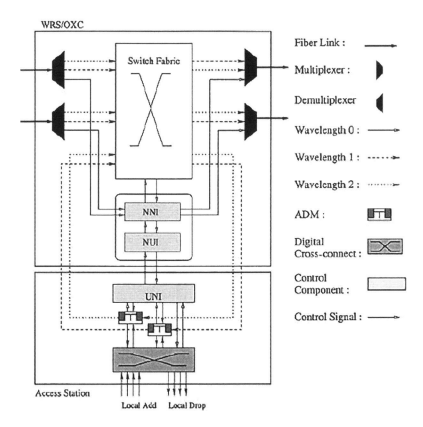

FIGURE 6.3   Node architecture with SONET over WDM without traffic grooming [2].

wavelength in its tuning range. A fixed transceiver can only get an optical signal on one wavelength. To explore all of the wavelength channels on a fiber, a set of fixed transceivers, one per wavelength, are grouped together to form a transceiver array. In Figure 6.2, the access station offers a flexible, software-based, bandwidth-provisioning capability to the network. Multiplexing low-speed connections to high-capacity light paths is done by the Multiprotocol label switching (MPLS)/IP router using a software-based queuing scheme. The MPLS/IP-over-WDM model provides flexible bandwidth granularity for the traffic requests and has less over heads than the SONET-over-WDM model. But the higher processing speed of the MPLS/IP router is a disadvantage [2].

In Figure 6.3, each access station has several ADMs with SONET [2] in which each SONET ADM has the ability to separate a high-rate SONET signal into lower-rate components, and a SONET ADM is used for dropping /adding a wavelength channel that has carried a large number of traffic. The digital cross-connect (DXC) transmits low-speed traffic streams between the access station and the ADMs. A low-speed traffic stream on one wavelength can be either dropped to the local client (IP router, ATM switch, etc.) or switched to another ADM and sent out on another wavelength [2]. Figure 6.3 presents a SONET-over-WDM node architecture where SONET components (ADM, DXC, etc.) and SONET framing schemes are to give TDM-based fast multiplexing/demultiplexing capability in comparison to the software-based scheme in Figure 6.2. Optical cross-connects (OXCs) are key elements in a carrier's WDM backbone network. There are transparent and opaque approaches to build these OXCs. The transparent approach refers to all-optical switching, whereas the opaque approach has switching with optical–electronic–optical (O–E–O) conversion [2,3].

Non-grooming [1,2] **OXC:** This type of OXC is fabricated with either transparent or opaque approach having wavelength-switching capability. In case of transparent approach, this type of OXC is able to switch traffic at higher bandwidth granularity, such as a waveband (a group of wavelengths) or fiber. There is no low-data-rate port on a non-grooming OXC.

Single-hop grooming [2] **OXC:** The OXC is employed only to switch traffic at wavelength (or higher) granularity as shown in Figure 6.4. On the other hand, it has some lower-data-rate ports/low-speed traffic streams. The traffic from these low-speed ports is multiplexed onto a wavelength channel by hierarchical TDM before it goes to the switch fabric.

Multi-hop partial-grooming OXC [2]: As shown in Figure 6.4, the switch fabric of OXC is composed of two parts: a wavelength-switch fabric (W-Fabric), which is either all-optical or electronic, and an electronic-switch fabric (G-fabric) switching low-speed traffic streams. With this hierarchical switching and multiplexing, this OXC can perform switching of low-speed traffic streams from one wavelength channel to other wavelength channels and groom them with other low- speed streams without using any extra network element. The wavelength capacity is OC-N, and the lowest input port speed of the electronic switch fabric is OC-M ($N > M$) where the ratio between N and M is called the *grooming ratio*. In this architecture, only a few of the wavelength channels can be switched to the G-fabric for switching at finer granularity. The number of ports, which connect the W-fabric and G-fabric, determines how much multi-hop grooming capability an OXC has.

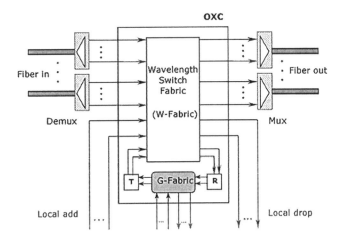

FIGURE 6.4 Node architecture with SONET over WDM with traffic grooming [2].

Multi-hop full-grooming OXC: This type of OXC shows full-grooming functionality using G-fabric. Every OC-N wavelength channel arriving at the OXC will be de-multiplexed into its constituent OC-M streams before it enters the switch fabric. The switch fabric switches these OC-M traffic streams in a non-blocking manner. Then, the switched streams are multiplexed back onto different wavelength channels. An OXC with full-grooming functionality has to be built using the opaque approach. The switching fabric of OXC is treated as a large grooming fabric.

Light-tree-based source-node grooming OXC: Optical "light-tree" supports multicast applications in optical WDM networks [16,17]. A light-tree is a wavelength tree which sets up connections between one source node and multiple destination nodes. Through a light-tree, traffic from a source node is transmitted to all destination nodes of the tree. In a light-tree, the node generating the traffic is called the "root" node, and the traffic destination nodes are called the "leaf" nodes. For multicast, an OXC needs to duplicate the traffic from one input port to multiple output ports by copying the electronic bit stream from one input port to multiple output ports.

Figure 6.5 presents a simple architecture of a multicast-capable OXC based on transparent technology. This figure indicates how the OXCs' multicast capability

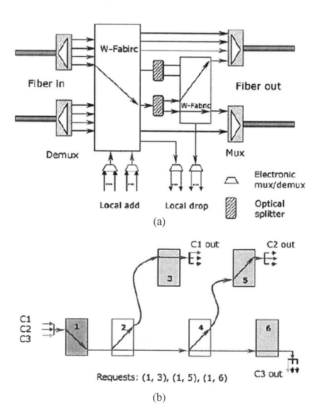

FIGURE 6.5 (a) Source grooming multicast capable OXC and (b) the operation of source node grooming [2].

is employed to carry out traffic grooming. There are three low-speed traffic steams from the source node 1 to different destination nodes 3, 5, and 6, where the aggregated bandwidth requirement is lower than the capacity of a wavelength channel in this example. At each destination node, only the appropriate traffic stream is selected and relayed to the client equipment. In this way, the low-speed traffic from the same source node is grouped to the same wavelength channel and transmitted to different destination nodes. Such an OXC can carry the light-tree-based source-node grooming scheme as well as the single-hop grooming scheme [2].

6.2.2 MATHEMATICAL (ILP) FORMULATION OF THE STATIC TRAFFIC-GROOMING PROBLEM

The static traffic-grooming problem in a mesh network is an integer linear program (ILP). The following are considered for this formulation [8]:

1. All physical links of an irregular mesh network are bidirectional between each node pair.
2. The OXC has no wavelength-conversion capability where a lightpath connection is established with the same wavelength channel if it traverses through several fibers.
3. The transceivers in a network node are tunable to any wavelength used in the fiber.
4. A connection request may have several lower-speed connections routed separately from the source to the destination.
5. The multiplexing/demultiplexing capability and time-slot interchange capability are needed in each node where the access station of a network node also has the same capability. This is used for the software-based provisioning scheme. The grooming capability of the node architecture in Figure 6.3 is restricted by the number of output ports of SONET and the size of the OXC.

Multi-Hop Traffic Grooming

In this scheme, a connection is set up through multiple lightpaths before it reaches the destination. A connection is groomed with different connections in different lightpaths. The following are the notations in our mathematical formulation [2,8]:

1. m and n represent end points of a physical fiber link.
2. i and j indicate originating and terminating nodes for a lightpath.
3. s and d indicate source and destination of an end-to-end traffic request. The end-to-end traffic may traverse through a single lightpath or multiple lightpaths.
4. y indicates the granularity of low-speed traffic requests in which traffic demands between node pairs are OC-1, OC-3, OC-12, and OC-48.
5. t indicates the index of OC-y traffic request for any given node pair (s,d).
   N = Number of nodes in the network.
   W = Number of wavelengths per fiber. We assume all of the fibers in the network carry the same number of wavelengths.

P_{mn}: Number of fibers interconnecting node m and node n.

$P_{mn} = 0$ for the node pairs which are not physically adjacent to each other.

$P_{mn} = P_{nm} = 1$ if and only if there exists a direct physical fiber link between nodes m and n.

P_{mn}^{w}: Wavelength w on fiber P_{mn}, $P_{mn}^{w} = P_{mn}$.

TR_i: Number of transmitters at node i.

RR_i: Number of receivers at node i. We consider that all the nodes have tunable transceivers, which can be tuned to any of W wavelengths.

C: Capacity of each channel (wavelength).

Λ = Traffic matrix set. $\Lambda = \{\Lambda_y\}$, where y can be any allowed low-speed streams, 1, 3, 12, etc. and $y \in \{1,3,12,48\}$. $\Lambda_{y,sd}$ is the number of OC-y connection requests between node pair (s,d).

Variables of virtual topology:

V_{ij}: Number of lightpaths from node i to node j in virtual topology in which $V_{ij} = 0$ does not imply that $V_{ji} = 0$; i.e., lightpaths may be unidirectional.

V_{ij}^{w}: Number of lightpaths from node i to node j on wavelength w in which if $V_{ij}^{w} > 1$, the lightpaths between nodes i and j on wavelength w may take different paths.

Variables of physical topology route:

$P_{mn}^{w,ij}$: Number of lightpaths between node pair (i,j) routed through fiber link (m,n) on wavelength w.

Variables of traffic route:

$\lambda_{ij,y}^{sd,t}$: The tth OC-y low-speed traffic request from node s to node d employing lightpath (i,j) as an intermediate virtual link.

$S_{sd}^{y,t}$: $S_{sd}^{y,t} = 1$ if the tth OC-y low-speed connection request from node s to node d has been successfully routed; otherwise, $S_{sd}^{y,t} = 0$.

Optimize total successfully routed low-speed traffic [2]; i.e.,

maximize

$$\sum_{y,s,d,t} y \cdot S_{sd}^{y,t} \tag{6.1}$$

Constraints:

On virtual-topology connection variables [18]:

$$\sum_{j} V_{ij} \leq TR_i \quad \forall i \tag{6.2}$$

$$\sum_i V_{ij} \le RR_i \quad \forall j \tag{6.3}$$

$$\sum_j V_{ij}^w \le V_{ij} \quad \forall i,j \tag{6.4}$$

$$\text{Int } V_{ij}^w, V_{ij} \tag{6.5}$$

On physical route variables [18]:

$$\forall i,j,w,k \quad \sum_m P_{mk}^{ij,w} = \sum_n P_{kn}^{ij,w} \quad if \ k \ne i,j \tag{6.6}$$

$$\forall i,j,w, \quad \sum_m P_{mi}^{ij,w} = 0 \tag{6.7}$$

$$\forall i,j,w, \quad \sum_n P_{jn}^{ij,w} = 0 \tag{6.8}$$

$$\forall i,j,w, \quad \sum_n P_{in}^{ij,w} = V_{ij}^w \tag{6.9}$$

$$\forall i,j,w, \quad \sum_m P_{mj}^{ij,w} = V_{ij}^w \tag{6.10}$$

$$\forall m,n,w, \quad \sum_{ij} P_{mn}^{ij,w} = P_{mn}^w \tag{6.11}$$

$$P_{mn}^{ij,w} \in \{0.1\} \tag{6.12}$$

$$\sum_i \lambda_{id,y}^{sd,t} = S_{sd}^{y,t} \quad \forall s,d,y \in \{1,3,12,48\}, t \in \{0, \Lambda_{y,sd}\} \tag{6.13}$$

On virtual-topology traffic variables [2]:

$$\sum_j \lambda_{sj,y}^{sd,t} = S_{sd}^{y,t} \quad \forall s,d,y \in \{1,3,12,48\}, t \in \{0, \Lambda_{y,sd}\} \tag{6.14}$$

$$\sum_i \lambda_{ik,y}^{sd,t} = \sum_j \lambda_{kj,n}^{sd,t} \quad \forall s,d,k,t, \ if \ k \ne s,d \tag{6.15}$$

$$\sum_i \lambda_{is,y}^{sd,t} = 0 \quad \forall s,d,y \in \{1,3,12,48\}, t \in \{0,\Lambda_{y,sd}\} \tag{6.16}$$

$$\sum_j \lambda_{dj,y}^{sd,t} = 0 \quad \forall s,d,y \in \{1,3,12,48\}, t \in \{0,\Lambda_{y,sd}\} \tag{6.17}$$

$$\sum_{y,t}\sum_{s,d} y \times \lambda_{ij,y}^{sd,t} \leq V_{ij} \times C \quad \forall i,j \tag{6.18}$$

$$S_{sd}^{y,t} \in \{0,1\} \tag{6.19}$$

The above equations follow principles of conservation of flow and resources (transceivers, wavelengths, etc.) [2,18].

- Equation (6.1) indicates the optimization of objective function.
- Equations (6.2) and (6.3) indicate that the number of lightpaths between node pair (i,j) is less than or equal to the number of transmitters at node i and the number of receivers at node j.
- Equation (6.4) shows the lightpaths between i, and j consisting of lightpaths on different wavelengths between i and j where the value of V_{ij}^w can be greater than 1.
- Equations (6.6)–(6.10) are the multi-commodity equations showing flow conservation for the routing of a lightpath from its source to its termination.
- There are two ways of formulation for flow-conservation equations [18]:
  i. disaggregate formulation
  ii. aggregate formulation.
      In the disaggregate formulation, every i–j (or s–d) pair represents a commodity, whereas in the aggregate formulation, all the traffic that originates from node i (or node s) indicates the same commodity, regardless of the traffic's destination. The disaggregate formulation is used for the flow-conservation equations since it properly describes the traffic requests between different node pairs.
- Equations (6.11) and (6.12) indicate that wavelength w on one fiber link (m,n) can only be present in at most one lightpath in the virtual topology.
- Equations (6.13)–(6.19) show that aggregate traffic flowing through lightpaths cannot exceed the overall wavelength (channel) capacity indicated for the routing of low-speed traffic requests on the virtual topology.

Single-Hop Traffic Grooming

In single-hop traffic grooming [2,18], a connection traverses a single lightpath, and this is end-to-end traffic grooming. The formulation of the single-hop traffic-grooming problem is almost the same as the formulation of the multi-hop traffic-grooming problem except for the routing of connection requests on the virtual topology.

On virtual-topology traffic variables [2]:

$$\sum_{sd} y x S_{sd}^{y,t} \leq V_{sd} \times C \quad \forall s, d \tag{6.20}$$

$$S_{sd}^{y,t} \in \{0,1\} \tag{6.21}$$

Equations (6.20) and (6.21) indicate the conservation of traffic flow for single-hop traffic grooming [18].

Formulation Extension for Fixed-Transceiver Arrays

The mathematical formulations indicate that the transceivers in a network node are tunable to any wavelength. If fixed-transceiver arrays are at every network node, M indicates the number of fixed-transceiver arrays used at each node, and the formulation is written as follows [2].

On virtual-topology connection variables:

$$\sum_{j} V_{ij}^{w} \leq M \quad \forall i, w \tag{6.22}$$

$$\sum_{i} V_{ij}^{w} \leq M \quad \forall j, w \tag{6.23}$$

$$\sum_{ij} V_{ij}^{w} \leq V_{ij} \quad \forall i, j \tag{6.24}$$

$$\text{Int } V_{ij}^{w}, V_{ij} \tag{6.25}$$

The other parts of the formulations in the previous sections have the same variables. Equations (6.22) and (6.23) indicate that the number of lightpaths between i and j on wavelength w is less than or equal to the number of transmitters at node i and the number of receivers at node j on this wavelength.

Computational Complexity

The RWA optimization problem with traffic grooming is NP-complete [2]. Each connection request has the full capacity of a lightpath, and the traffic-grooming-based routing also follows the standard RWA optimization problem. The number of variables and equations increases exponentially with size of the network. So, for large networks, heuristic approaches are employed.

6.2.3 NUMERICAL SIMULATION RESULTS
FROM ILP FORMULATIONS

The simulation results of the static traffic-grooming problem are shown in Figure 6.6 [2]. The traffic matrices are randomly generated by using the traffic demand to be any one of OC-1, OC-3, and OC-12 [2]. The traffic matrices are generated as follows:

1. The number of OC-1 connection requests between each node pair is a uniformly distributed random number between 0 and 16 and is given in the following matrix:

$$
\begin{pmatrix}
0 & 2 & 6 & 7 & 8 & 11 \\
2 & 0 & 7 & 13 & 9 & 16 \\
5 & 3 & 0 & 7 & 11 & 9 \\
4 & 7 & 14 & 0 & 8 & 7 \\
6 & 7 & 9 & 15 & 0 & 8 \\
6 & 7 & 9 & 5 & 12 & 0
\end{pmatrix}
$$

2. The number of OC-3 connection requests between each node pair is a uniformly distributed random number between 0 and 8, and the traffic matrix is given below:

$$
\begin{pmatrix}
0 & 2 & 3 & 4 & 5 & 7 \\
2 & 0 & 7 & 1 & 5 & 6 \\
5 & 3 & 0 & 3 & 3 & 7 \\
4 & 7 & 4 & 0 & 7 & 6 \\
5 & 7 & 8 & 5 & 0 & 4 \\
7 & 7 & 4 & 5 & 4 & 0
\end{pmatrix}
$$

3. The number of OC-12 connection requests between each node pair is a uniformly distributed random number between 0 and 2.

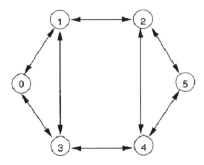

FIGURE 6.6   Example network: six-node network.

$$\begin{pmatrix} 0 & 1 & 2 & 3 & 4 & 4 \\ 2 & 0 & 3 & 1 & 3 & 4 \\ 2 & 3 & 0 & 3 & 3 & 4 \\ 4 & 3 & 4 & 0 & 2 & 4 \\ 3 & 3 & 4 & 2 & 0 & 4 \\ 3 & 3 & 4 & 2 & 2 & 0 \end{pmatrix}$$

The capacity of each wavelength (channel) is OC-48.

Table 6.1 represents the throughput estimated by using a commercial ILP solver, "CPLEX" [19] with different network resource parameters. In the single-hop case, a connection is permitted to traverse a single lightpath using end-to-end traffic grooming (multiplexing), whereas in the multi-hop case, a connection is dropped at intermediate nodes and groomed with other low-speed connections on different lightpaths before reaching its destination. Multi-hop grooming provides higher throughput than the single-hop case. When the number of tunable transceivers at each node is increased from 3 to 5, the network throughput increases significantly, both in the multi-hop case and in the single-hop case. But when the number of tunable transceivers at each node increases from 5 to 7, the network throughput remains almost same. This scenario is less likely in the single-hop case [2].

Table 6.2 shows transceiver utilization and link wavelength employment of the multi-hop case. When the number of transceivers is augmented (from 3 to 5), the overall wavelength utilization is enhanced [2]. This is due to setting up of more lightpaths to accommodate the connection requests as mentioned in traffic matrix having OC-1 connection requests only. When the links use fully the available wavelengths, increasing the number of transceivers (from 5 to 7) does not enhance network throughput but gives poor transceiver utilization, as shown in Table 6.2 ($T = 7$ and $W = 3$).

TABLE 6.1

Throughput and Number of Lightpaths Established [2] (T = Number of Transmitters and W = Number of Wavelengths)

	Multi-Hop		Single-Hop	
	Throughput	Number of Lightpaths	Throughput	Number of Lightpaths
$T = 3, W = 3$	74.7% (OC-738)	18	68.0% (OC-672)	18
$T = 4, W = 3$	93.8% (OC-927)	24	84.1% (OC-831)	24
$T = 5, W = 3$	97.9% (OC-967)	28	85.7% (OC-847)	24
$T = 7, W = 3$	97.9% (OC-967)	28	85.7% (OC-847)	24
$T = 3, W = 4$	74.7% (OC-738)	18	68.0% (OC-672)	18
$T = 4, W = 4$	94.4% (OC-933)	24	84.7% (OC-837)	24
$T = 5, W = 4$	100% (OC-988)	29	95.5% (OC-944)	28

TABLE 6.2
Results: Transceiver Utilization (Multi-Hop Case) [2]

		$T = 3, W = 3$	$T = 5, W = 3$	$T = 7, W = 3$
Node 0	Transmitter	100%	100%	71.4%
	Receiver	100%	100%	71.4%
Node 1	Transmitter	100%	100%	71.4%
	Receiver	100%	100%	71.4%
Node 2	Transmitter	100%	100%	71.4%
	Receiver	100%	100%	71.4%
Node 3	Transmitter	100%	100%	71.4%
	Receiver	100%	100%	71.4%
Node 4	Transmitter	100%	80%	57.4%
	Receiver	100%	80%	57.4%
Node 5	Transmitter	100%	80%	57.4%
	Receiver	100%	80%	57.4%

TABLE 6.3
Results: Virtual Topology and Lightpath Utilization [2] (Multi-Hop Case with $T = 5$ and $W = 3$)

	Node 0	Node 1	Node 2	Node 3	Node 4	Node 5
Node 0	0	2 (70%)	0 (100%)	1 (89%)	1 (100%)	1 (100%)
Node 1	1 (100%)	0	1 (100%)	2 (100%)	1 (100%)	0
Node 2	1 (100%)	1 (95%)	0	1 (100%)	1 (100%)	1 (70%)
Node 3	2 (100%)	1 (100%)	1 (100%)	0	0	1 (100%)
Node 4	1 (100%)	1 (100%)	0	0	0	1 (91%)
Node 5	0 (100%)	0	2 (98%)	1 (100%)	1 (100%)	0

Table 6.3 shows virtual topology and the lightpath capacity utilization for the multi-hop case of $T = 5$ and $W = 3$. In the table, most of the lightpaths provides more capacity utilization. There are some node pairs ((0,1), (1,3), etc.) having multiple lightpaths though the aggregate traffic between them can be carried by a single lightpath. The extra lightpaths are needed to accommodate multi-hop connection traffic. The results from the ILP solutions provide that if there is a lightpath set up between (s,d), the low-speed connections between (s,d) tend to be packed on this lightpath channel directly. There are two simple heuristic algorithms for resolving the traffic-grooming problem in a large network [2].

Similarly we can establish a 15-node sample network (Figure 6.7) using traffic grooming in each node through OC-48 wavelength channels.

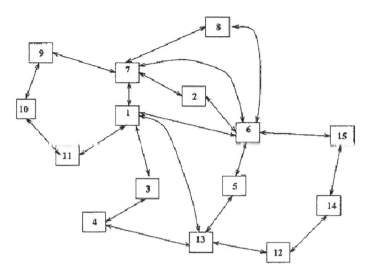

FIGURE 6.7 Example network: 15-node network.

6.2.4 HEURISTIC TECHNIQUE

The optimization problem of traffic grooming is NP-complete [2]. This NP-complete problem is divided into the following four sub-problems, which are not independent:

1. Determine a virtual topology, i.e., the number of lightpaths between any node pair.
2. Find the lightpaths over the physical topology.
3. Allot wavelengths optimally to the lightpaths.
4. Find the paths for the low-speed connection requests on the virtual topology.

Routing

There are routing schemes – fixed routing, fixed-alternate routing, and adaptive routing [20,21] which are already discussed in Chapters 4 and 5. In fixed routing, the connections are always assigned paths through a pre-defined fixed path for a given source–destination pair. The shortest-path route for each source–destination pair is calculated offline using standard shortest-path algorithms, such as Dijkstra's algorithm [22]. In fixed-alternate routing, multiple fixed routes are estimated when a connection request comes. In this approach, each node in the network needs to have a routing table with an ordered list of a number of fixed routes to each destination node. When a connection request comes, the source node attempts to establish the connection on each of the routes from the routing table in sequence, until the connection is successfully established. Since fixed-alternate routing provides simplicity of control for setting up and tearing down connections, it is also widely used in the dynamic connection-provisioning case. For certain networks, having two alternate paths gives lower blocking than having full wavelength conversion at each node with fixed routing [21]. In adaptive routing, the route from a source node to a destination node is selected dynamically, depending on the current network state. The current network state is determined by the set of all connections that are currently in progress [20]. When a connection request comes, the current shortest path between the source and the destination is estimated based on the available resources in the network; then the connection is established through the route. Heuristics is used for adaptive routing.

Wavelength Assignment

There are many wavelength-assignment approaches used in optical networks – first fit (FF), least-used wavelength assignment, most-used wavelength assignment, etc. Once the path is selected for a connection request, wavelength-continuity constraints are considered, and wavelengths are allotted to each path. Among these approaches, FF is chosen [20,23] where all wavelengths are numbered, and a lower-numbered wavelength is considered before a higher-numbered wavelength. The first available wave length is then selected as per the next lower-numbered wavelength.

Heuristics

Two heuristic algorithms are used for the traffic-grooming problem – maximizing single-hop traffic (MSHT) [1,24] and maximizing resource utilization (MRU) [24]. We consider $T(s,d)$ represents the aggregate traffic between node pair s and d, $t(s,d)$ represents one connection request between s and d, and C indicates the wavelength capacity.

MSHT: This basic heuristic was already used in Chapter 3 for the traditional virtual-topology design problem. This simple heuristic is used to set up lightpaths between source–destination pairs with the highest $T(s,d)$ values, considering the constraints on the number of transceivers at the two end nodes and the availability of a wavelength in the path connecting the two end nodes. The connection requests between s and d are allotted on the new lightpath, and every connection will traverse a single lightpath hop. The algorithm is attempted to allot the blocked connection requests using currently available spare capacity of the virtual topology. The heuristic algorithm is as follows [2].

Algorithm MSHT

1. Construct virtual topology:
   a. Arrange out all of the node pairs (s,d) according to the sum of traffic requests $T(s,d)$ not carried between (s,d) and put them into a list L in descending order.
   b. Attempt to establish a lightpath between the first node pair (s',d') in L using FF wavelength assignment and shortest-path routing, subject to the wavelength and transceiver constraints. If it fails, delete (s',d') in L; otherwise, let $T(s,d) = $ Max $[T(s,d)-C, 0]$ and go to Step 1a until L is empty.

2. Find the path of the low-speed connections on the virtual topology constructed in Step 1.

    a. Fit all of the connection requests which can be accommodated through a single lightpath hop, and revise the virtual topology network state.

    b. Find the path of the rest of connection requests based on the current virtual topology network state, in the descending order of the connections bandwidth requirement.

MRU: $H(s,d)$ represents the hop distance on physical topology between node pair (s,d). We consider $T(s,d)/H(s,d)$ as the connection resource utilization value, representing the average traffic per wavelength link. This parameter provides how efficiently the resources have been used to accommodate the connection requests. This heuristic establishes the lightpaths between the node pairs with the maximum resource utilization values. When no lightpath can be set up, the remaining blocked traffic requests will be routed on the virtual topology based on their connection resource utilization value $t(s,d)/H'(s,d)$, where $t(s,d)$ denotes blocked connection request and $H'(s,d)$ represents the hop distance between s and d on the virtual topology. The pseudo-code for this MRU heuristic has almost same steps as those of the pseudo-code of MSHT [2].

Both heuristic algorithms have two processes [2] – in the first process, they establish lightpaths as much as possible to accommodate the aggregate end-to-end connection requests. If there are enough resources in the network, every connection request can be successfully transmitted through a single-hop lightpath, and this reduces the traffic delay since the optical signals need not be converted into electronic domain.

In the second stage [2], the additional capacities of the currently established lightpath channels are allotted to accommodate the connection requests blocked in the first stage, and the algorithms provide more priority of single-hop groomable connections.

Heuristic Results and Comparison

In Table 6.4, we try to compare the throughput performance results obtained by using these heuristics based on ILP solver [20]. We consider the six-node network in Figure 6.6 for a heuristic algorithm. The heuristic approaches require

TABLE 6.4

Comparison of Throughput Results between ILP and Heuristic Algorithms (Total Traffic Demand Is OC-988) [2]

	Multi-Hop (ILP)	Single-Hop (ILP)	Heuristic (MSHT)	Heuristic (MRU)
$T = 3, W = 3$	74.7% (OC-738)	68.0% (OC-672)	71.0% (OC-701)	67.4% (OC-666)
$T = 4, W = 3$	93.8% (OC-927)	84.1% (OC-831)	89.4% (OC-883)	93.6% (OC-925)
$T = 5, W = 3$	97.9% (OC-967)	85.7% (OC-847)	94.4% (OC-933)	94.4% (OC-933)
$T = 7, W = 3$	97.9% (OC-967)	85.7% (OC-847)	94.4% (OC-933)	94.4% (OC-933)
$T = 3, W = 4$	74.7% (OC-738)	68.0% (OC-672)	71.0% (OC-701)	67.4% (OC-666)
$T = 4, W = 4$	94.4% (OC-933)	84.7% (OC-837)	93.1% (OC-920)	93.6% (OC-925)

less computation complexity than the ILP approach. The two proposed RWA algorithms are relatively simple and straight forward in comparison to adaptive routing. Here FF wavelength assignment is used to obtain heuristics to achieve better performance.

For simulations, the same traffic matrices mentioned in Section 6.2.4 are considered. The MRU heuristic [2] provides better performance than the MSHT algorithm with respect to network throughput. The number of tunable transceivers at each node is restricted to 10 in this case, and the number of wavelengths on each fiber link increases to 16 in this case. Here, increasing the number of wavelengths cannot provide more network throughput. For small values of transceivers (<7 in this case), MSHT performs better than MRU, since MRU uses wavelengths efficiently. So maximizing wavelength utilization does not improve the performance. Each node has one tunable array of 12 transceivers where the size of the transceiver array is same as the number of wavelengths supported by every fiber link.

6.2.5 MATHEMATICAL FORMULATION OF OTHER OPTIMIZATION CRITERIA

In this section, the ILP formulations are extended to handle different optimization criteria for the static traffic-grooming problem.

Network Revenue Model

As discussed earlier, the low-speed connection requests between the same node pair are inclined to be filled together on to the same lightpath channel for accommodating more traffic requests. The connections accommodated by a single lightpath channel are more than the connections which have to traverse multiple lightpaths [2]. Considering the same bandwidth requirement, the purpose of optimization is to maximize network throughput. For this problem, the priority of two connections is considered even if they have the same bandwidth requirement. Different connection requests have different end-node distance, quality-of-service requirement, etc. The priority is determined by a "weight" associated with it, and the weight is determined by the bandwidth requirement and end-node distance of the connection request. For a given network topology and traffic demand, the objective is to maximize the weighted network throughput. The following parameters are used:

W_i = the weight of connection i
D_i = the end-node distance of connection i
C_i^α = the bandwidth requirement of connection i.

The connection's weight function is written as

$$W_i = D_i \times C_i^\alpha \tag{6.26}$$

where $0 \leq \alpha \leq 1$ and D_i are measured by the shortest-path distance in the physical topology. Equation (6.26) represents the power-law cost function used to study the actual tariffs demanded by communication services, and a "quantity discount" controlled by α in that capacity cost (per unit of channel capacity) decreases as the capacity increases. Thus, the network weighted throughput becomes

$$T = \sum_{i-1}^{K} D_i \cdot C_i^{\alpha} \cdot S_i \qquad (6.27)$$

where
  $S_i = 1$, if connection request i has been satisfied
  $S_i = 0$, otherwise.

The total number of connection requests $= K$. T is also known as *network revenue*.
  Objective: Maximize network revenue [2]

$$Maximize = \sum_{i-1}^{K} D_{sd} \cdot y^{\alpha} \cdot S_{sd}^{y,t} \qquad (6.28)$$

where D_{sd} denotes the distance between node pair (s,d).
  To obtain network revenue using ILP formulation, the same network topology is used with traffic matrix set as mentioned earlier. In Equation (6.28), D_{sd} is estimated by the shortest-path hop distance between nodes s and d on the physical topology. Table 6.5 compares the results between the two optimization models in which multi-hop grooming is allowed in both models. For $T = 3$ and $W = 3$, the maximal achievable revenue is 83.7% with 72.4% of traffic requests. The maximum achievable traffic load of the network is 74.7%. In the revenue model, if there is a lightpath setup between (s,d), it carries some long multi-hop connections (with higher weight) which are transmitted with this lightpath as intermediate hops. Some connections directly between (s,d) are blocked. Packing of different connections between the same node pair with the same existing lightpath (joining the end points) is not performed in the grooming scheme anymore. Because of the quantity-discount parameter α in

---

TABLE 6.5

Results of Comparison between Revenue Model and Network Throughput Model [2]

	Optimize Revenue		Optimize Throughput
	Revenue	Throughput	Throughput
$T = 3, W = 3$	83.7%	72.4%	74.7%
$T = 5, W = 3$	98.5%	97.2%	97.9%
$T = 7, W = 3$	98.5%	97.2%	97.9%
$T = 3, W = 4$	83.7%	72.4%	74.7%
$T = 4, W = 4$	94.3%	91.7%	94.4%
$T = 5, W = 4$	100%	100%	100%

equation (13.26), lower-speed connections are more accommodated than higher-speed connection requests. Different heuristics are needed for different optimization criteria.

6.3   DYNAMIC TRAFFIC GROOMING

In this section, the dynamic traffic-grooming problem is discussed [2,25]. For dynamic grooming problem, a heterogeneous, optical WDM network with multi-granularity is taken as sample network in which connections with different bandwidth-granularity requirements arrive one at a time in the nodes, and each such connection is required to be set up through the network based on the current network state. The pieces of network equipment (NE) are from different vendors, and new equipment has to co-exist with legacy equipment.

The following are considered for the networks: (a) network nodes have OXCs that are used in different architectures and technologies; (b) all nodes may not have the capability of wavelength conversion and traffic-grooming (i.e., sparse wavelength conversion and grooming); (c) wavelength conversion and traffic grooming are performed on certain wavelength channels, and (d) different fiber links have different numbers of wavelength channels. This range of heterogeneity increases the complexity of the networks [25,26].

The problem of dynamically provisioning connections of different bandwidth granularities of mesh networks considers two levels of hierarchical switching capability – (a) wavelength switching and (b) SONET- or SDH-based low-speed circuit (time-slot) switching.

6.3.1   PROVISIONING CONNECTIONS IN HETEROGENEOUS WDM NETWORKS

Three protocols in WDM network control connections of different bandwidth granularities [25,26]: resource-discovery protocol, signaling protocol, and route-computation algorithm.

- Resource-discovery protocols estimate how the network resources are scattered and maintained in the OXCs' link-state databases for distributed control.
- Signaling protocols evaluate how the connection is reconfigured and how a network node allots its local network resources to the connection, e.g., port mapping or label assignment for a connection.
- Route-computation algorithms evaluate how the path of a low-speed connection request is estimated and selected according to the carrier's grooming policy [25].

A unified control plane for intelligent WDM mesh-based networks is made by two bodies – the International Telecommunications Union (ITU) taking care of the architecture for Automatically Switched Optical Networks (ASONs) [27,28] and the Internet Engineering Task Force (IETF) taking care of Generalized Multi-Protocol Label Switching (GMPLS) [29].

The network control plane provides an intelligent, automatic, end-to-end circuit (or virtual circuit) provisioning/signaling scheme throughout different network domains. Here the GMPLS control has been used for the network heterogeneity caused by the inter-operation of OXCs from different vendors and the co-existence of new network equipment.

Resource Discovery

There are two types of network links – physical link (optical fiber) and virtual link [2]. The physical link (fiber link) and virtual link (lightpath) are controlled on the basis of link-state information (LSI). Besides the neighbors connected by physical links, a node has other neighbors connected to the node by virtual links.

- **Fiber link:** The fiber link in a full wavelength-convertible network is represented as $f(m,n,t,w,c)$, [2] where m and n denotes the end nodes of the fiber link, t denotes fiber index (for numbering multiple fibers between the same node pair), w indicates the available (free) wavelength channels on that fiber, and c indicates the administrative link cost. In a WDM network with wavelength-continuity constraint, information regarding availability of each individual wavelength channel is required.
- **Virtual link** [2]: A lightpath is represented as $V(i,j,v,t,m_l,m_z,c)$, where i and j denote the end nodes of the lightpath; v denotes the lightpath type; t indicates the lightpath; m_l indicates the minimal reservable bandwidth on this lightpath, which is determined by the grooming ratio of the end nodes; m_z indicates the maximal reservable bandwidth on this lightpath, which is bounded by the total available (free) capacity on the lightpath; and c indicates the administrative link cost. Multiple lightpaths (of the same type) between the same node pair are also bundled as a logical link.

A lightpath is a traffic-engineering (TE) link (TE) [29] where the switching granularity of the lightpath at any intermediate node includes full wavelength-channel granularity, where low-speed connections are transmitted via multiple lightpaths. There are four types of lightpaths to be groomed in a carrier's optical WDM network, and they are as follows [2].

Multi-hop ungroomable lightpath: A multi-hop ungroomable lightpath is used if it is not connected with a finer-granularity switching element at its end nodes. This lightpath carries the traffic directly between node pair (i,j).

Source-groomable lightpath [2]: A source-groomable lightpath exists if it is connected with a finer-granularity switching element only at its source node.

Destination-groomable lightpath [2]: A destination-groomable lightpath exists if it is connected with a finer-granularity switching element only at its destination node.

Fully groomable lightpath [2]: A fully groomable lightpath exists if it is connected to finer-granularity switching elements at both end nodes. Such a lightpath carries traffic between any node pair in the network. The source-groomable, destination-groomable, and fully groomable lightpaths are called as multi-hop groomable lightpaths.

The state of a network node also needs to be sent to all the nodes. A multi-hop partial-grooming OXC carries out multi-hop grooming on a limited number of light-paths. The available multi-hop grooming capability of the OXC is also presented when the node's state changes [30].

Route Computation

In an intelligent WDM network, the path of a connection request is computed either by the source node or by the network control and management system. Following are the possibilities to find a path for the connection request having source, destination, and capacity requirements [2,30]:

> Operation 1: An existing lightpath carries a connection C (s,d,r) between nodes s and d, and r is the bandwidth requirement.
> Operation 2: Multiple existing groomable lightpaths carry a connection $C(s,d,r)$.
> Operation 3: A new lightpath (either groomable or ungroomable) carries a connection C (s,d,r) between node pair (s,d) if enough resources exist.
> Operation 4: A combination of two existing groomable lightpaths carry a connection C (s,d,r) by setting up new groomable lightpaths using available wavelength channels in fiber links and grooming resources in network nodes.

There are multiple ways for carrying a connection request. There should be an efficient technique to find a proper path from multiple candidate routes as per network operator's grooming policy. In a dynamic traffic environment, connection requests having various bandwidth requirements reach a network, stay for a certain period of time, and then leave the network after connections are set up using a groomed wavelength. A grooming policy performs better in a dynamic traffic environment than a static one. Dynamic traffic grooming adjusts the grooming policy dynamically according to the traffic pattern and current network state. Following are the four different grooming policies for dynamic traffic.

Minimize the Number of Traffic Hops on the Virtual Topology [2,30].

> We first use operation 1. When operation 1 fails, then we attempt to set up a lightpath from s to d and transmit the traffic on this lightpath (operation 3). Only when a direct lightpath is not set up, we use multi-hop grooming by either operation 2 or operation 3.

Minimize the Number of Traffic Hops on the Physical Topology (MinTHP) [2].

> We select the one with the fewest wavelength links by comparing the number of wavelength links used by all the four operations.

Minimize the Number of Lightpaths (MinLP) [2].

> The minimal number of new lightpaths is set up to transmit the traffic. Operation 1 is used first, and if it fails, the traffic is transmitted using

multiple existing lightpaths (operation 2). If operation 2 also fails, one light-path is set up with the minimal number of wavelength links by either opera-tion 3 or operation 4.

Minimize the Number of Wavelength Links (Min WL) [2].

A minimum number of extra wavelength links are used to accommodate the traffic. In this case, if both operations 1 and 2 fail, the number of wave-length links are selected by operations 3 and 4 comparing MinWL, and the network requiring fewer wavelength links is selected; the number of light-paths are estimated by operations 3 and 4 comparing MinLP, and the net-work requiring fewer lightpaths is selected, and the number of wavelength links is used for tie-breaking.

In dynamic grooming, the network status is changed as connection requests reach a node of the network non-uniformly and are transmitted from that node. The groom-ing should be adjusted according to the current network state [2].

Signaling

After the establishment of a route, every intermediate node along the route needs to be informed through appropriate signaling protocols. Two of the protocols used are resource reservation protocol (RSVP) with traffic-engineering extensions [31,32] and constraint-based routing label distribution protocol (CR-LDP) [33,34] used for signaling. In a heterogeneous WDM network, the route selection for a connection request has a sequence of intermediate nodes and link bundles. Since a link bundle consists of multiple candidate links, an intermediate node needs to select one for the connection request and to reconfigure the OXC and establish the connection.

A Graphical Model for Generic Provisioning

Due to network heterogeneity, the complexity is increased to provide services to the connection requests efficiently [2,30]. Hence, a generic bandwidth-provisioning model is employed to incorporate various network elements (NEs) and accommo-date different grooming policies to control their transmission of traffic easily and efficiently with the reduction of the overall cost (network cost and operation cost) significantly. This model is implemented and scaled by using a distributed control plane. A generic provisioning model has been developed for heterogeneous opti-cal WDM networks. Figure 6.8 shows a provisioning model having three nodes 0, 1, 2 in which the node 0 uses a multi-hop partial-grooming OXC and the nodes 1 and 2 have single-hop grooming OXCs. Each link in Figure 6.8a indicates a free wavelength channel between a node pair, whereas each link in Figure 6.8b has an established lightpath. The lightpath (0,2) is a source-groomable lightpath, the light-path (1,0) is a destination-groomable lightpath, and the lightpath (2,1) is a multi-hop ungroomable lightpath. A low-speed connection request from node 1 to node 2 is transmitted by lightpaths (1,0) and (0,2). On the other hand, a request from node 2 to node 0 cannot be transmitted by lightpaths (2,1) and (1,0) since node 1 has no multi-hop-grooming capability. Figure 6.8c is a graphical representation of the network

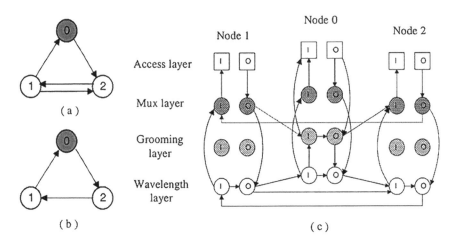

FIGURE 6.8   (a–c) Network states of three-node provisioning model [2].

state of a three-node model having four layers – access layer, multiplexer (mux) layer, grooming layer, and wavelength layer. The access layer represents the access point of a connection request where a customer's connection starts and terminates. The access layer has an IP router, an ATM switch, or any other client equipment. The mux layer represents the OXC ports from which low-speed traffic streams are directly multiplexed (demultiplexed) onto (from) wavelength channels before being transmitted to the grooming fabric. The grooming layer represents the grooming component of the network node.

A network node has two vertices at each layer representing the input and output ports of the network node at that layer. The links in this graphical model are named as follows.

Grooming-switching link has a connection between the input port and the output port of the grooming layer at a given node i, when node i has multi-hop traffic-grooming capability [2].

Wavelength-switching link has a connection between the input port of the wavelength layer and the output port of the same layer at a given node i. It represents the wavelength-switching capability of the network node [2].

Mux link has a connection between the output port of the access layer and the output port of the mux layer at a given node i, representing the traffic starting from node i and getting transmitted to another network node without going through any grooming fabric [2].

Demux link has a connection between the input port of the mux layer and the input port of the access layer at a given node i, representing the traffic on a wavelength channel to be demultiplexed and terminated at this node without going through any grooming fabric [2].

Mux to wavelength transmitting link [2] has a connection between the output port of the mux layer and the output port of the wavelength layer at a given node i.

Wavelength to mux receiving link [2] has a connection between the input port of the wavelength layer and the input port of the mux layer at a given node i.

Grooming link has a connection between the output port of the access layer and the output port of the grooming layer at a given node *i*, when node *i* has multi-hop-grooming capability (i.e., free outgoing grooming ports). In this link, the traffic starting from node *i* is groomed with other traffic streams (originating from node *i* or from other network nodes) to the same wavelength channel and transmitted to the next network node together.

De-grooming link [2] has a connection between the input port of the grooming layer and the input port of the access layer at a given node *i*, when node *i* has multi-hop-grooming capability (i.e., free incoming grooming ports), representing the traffic streams on a wavelength channel to be demultiplexed and terminated at node *i* or switched to other lightpaths.

Grooming to wavelength transmitting link [2] has a connection between the output port of the grooming layer and the output port of the wavelength layer at a given node *i*, having multi-hop-grooming capability (i.e., free incoming grooming ports). A multi-hop groomable lightpath (i.e., either a source-groomable lightpath or a multi-hop fully groomable lightpath) originates at node *i*.

Wavelength to grooming receiving link [2] has a connection between the input port of the wavelength layer and the input port of the grooming layer at a given node *i*, having multi-hop grooming capability (i.e., free incoming grooming ports). A multi-hop groomable lightpath (i.e., either a destination-groomable lightpath or a multi-hop fully groomable lightpath) terminates at node *i*.

Wavelength link [2] has a connection between the output port of the wavelength layer at node *i* and the input port of the wavelength layer at node *j* set up as per the availability of the wavelength channels between the node pair (*i*,*j*).

Lightpath link [2] has a connection at the output port of the mux layer (grooming layer) at node *i* that terminates at the input port of the mux layer (grooming layer) at node *j*.

Each link in Figure 6.8c represents the availability of the corresponding network resource. A link is removed if the corresponding network resource is not available, and a link is again added if the corresponding network resource becomes available from the state of being unavailable (e.g., addition of a lightpath link). After properly adjusting the administrative link costs, suitable paths are determined according to different grooming policies for a request by applying shortest-route-computation algorithms. Figure 6.9 presents the corresponding graph of network nodes having three different traffic-groomable OXC architectures – single-hop grooming, multi-hop partial grooming, and multi-hop full grooming.

Engineering Network Traffic Using the Proposed Graphical Model

This technique exhibits a platform for network operators for the realization of different grooming policies for improving the provisioning flexibility and network resource efficiency. Figure 6.10 presents the routes of connection request (1,2) with different TE objectives through different grooming policies by using auxiliary graphical model. A new traffic request is transmitted from node 1 to node 2, and two possible paths (thick links) are used for this connection request in which the path shown in Figure 6.10a has two existing lightpath links and the path shown in Figure 6.10b use two new wavelength channels. The connection requires full wavelength-channel

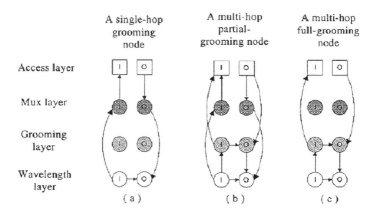

FIGURE 6.9   (a–c) Different grooming OXCs and their representations in the auxiliary graph [2].

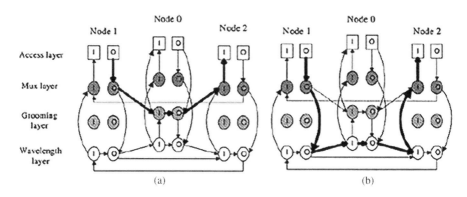

FIGURE 6.10   (a and b) Examples of routes for connection requests [2].

capacity, or the overall bandwidth requirement of the future traffic demands between the nodes in a pair is estimated to be close to full wavelength-channel capacity; the path in Figure 6.10b is considered if the wavelength channels are fully utilized and no grooming is required at node 0; otherwise, the path in Figure 6.10a is selected if enough free capacity is available on the existing lightpaths.

6.3.2   ILLUSTRATIVE NUMERICAL EXAMPLES

Comparison of Grooming Policies

The performance of different grooming policies are compared by analyzing a mesh network in Figure 6.11 having 19 nodes and 36 links. We consider the following:

- All the nodes with grooming capability do not have wavelength-conversion capability, and each link is bidirectional with 16 wavelengths in each direction.

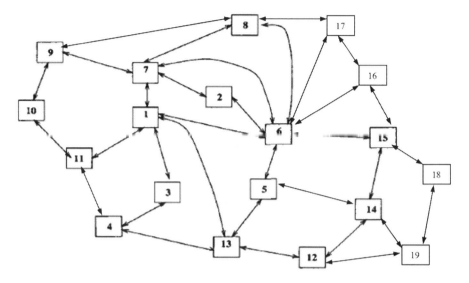

FIGURE 6.11   A mesh network having 19 nodes and 36 physical links [2].

- Each wavelength transmits maximum number of requests of bandwidth 10 Gbps (capacity ~ OC-192).
- The traffic arrival follows a Poisson process, and the connection holding time is exponentially distributed (whose average value is normalized to unity in our studies reported here).
- The traffic is uniformly distributed among all node pairs. The four types of connection requests are OC-3, OC-12, OC-48, and OC-192, and the ratio of the number of these connections is 6:6:6:1.

100,000 connection requests are considered for simulations to obtain the network performance under a certain scenario and a grooming policy. Table 6.6 shows the utilization of wavelength links (U_w) and transceivers (U_T) for the network load L of 300 Erlangs. For 16 transceivers in each node, the utilization of transceivers is very high, but these are more constrained. For 32 transceivers at each node, the utilization

TABLE 6.6

Performance of Different Traffic Grooming Policies in Terms of U_W and U_{Tx} [2]

	$T_x = 16$		$T_x = 32$		$T_x = 40$	
	U_w	U_{Tx}	U_w	U_{Tx}	U_w	U_{Tx}
MinTHV	0.7819	0.9858	0.8878	0.7264	0.8905	0.5884
MinTHP	0.5674	0.9901	0.7354	0.8165	0.7361	0.6910
MinLP	0.7403	0.9807	0.8890	0.8007	0.8918	0.6651
MinVVL	0.6201	0.9859	0.8133	0.8683	0.8120	0.7825

of both transceivers and wavelengths is relatively high. The percentage of blocked traffic also increases with the increase of the network load, but different grooming policies have different BPs.

6.4 ADAPTIVE GROOMING (AG)

The MinTHV [2] performs better than the others. When transceivers have more constrained resources, MinTHP has lower blocking. When wavelength links have shortage of resources, these two grooming policies are combined together. The combined algorithm is an Adaptive Grooming Policy (AGP) where, for each connection request, switches of MinTHV and MinTHP [2] are active on the basis of the current network state. If the ratio is larger than A_1, MinTHV is used for avoiding the setting up of lightpaths since transceivers have a scarcity of resources at this time. If the ratio is less than A_2, MinTHP is used to try to save wavelength links as much as possible. If the ratio is in between, the policy will not be changed. We found that the choices of values $A_1 = 1.2$ and $A_2 = 1.5$ result in the best performance for the network topology in Figure 6.11.

6.4.1 PERFORMANCE IN TERMS OF DIFFERENT PARAMETERS [2]

We analyzed the performance in a dynamic traffic environment with the following considerations:

- The traffic arrival process follows the Poisson distribution, whereas the connection-holding time follows a negative exponential distribution.
- The capacity of each wavelength channel is OC-192 which is 10 Gbps.
- The bandwidth requirement of the connection requests may be OC-3, OC-12, OC-48, or OC-192 [2].
- Connection requests are uniformly distributed among all node pairs.
- Average connection-holding time is normalized to unity.
- The cost of a fiber link is considered to be unity, the load (in Erlang) is taken as the arrival times of connection requests, and the average connection-holding time indicates the average number of connections the network carries at any instant.

From the connections' bandwidth distribution and Erlang load, the network's input load is estimated in the unit of OC-1. Three metrics are used to evaluate the network performance – traffic blocking ratio (TBR), connection blocking probability (CBP), and resource efficiency ratio (RER).

- *TBR* [2] is the ratio of the amount of blocked traffic to the amount of bandwidth required by all traffic requests during the entire simulation period.
- *CBP* [2] is the ratio of the total number of blocked connection requests to the total number of traffic requests during the entire simulation period.
- *RER* [2] indicates the number of connections routed and groomed efficiently.

The bit error rate (BER) is computed as [2]

$$BER = \frac{\sum_i \rho_i \times t_i}{\sum_i \gamma_i \times t_i}$$

where t_i = time period between the ith event (connection arrival or departure) and $(i + 1)$th event, ρ_i = load carried by the network during the time period t_i, and γ_i = the total number of wavelength links that do not change during the time period t_i. If in the network every connection is assigned full wavelength-channel capacity (i.e., no capacity is wasted) and is assigned the shortest path (in hop distance), the RER is equal to unity. A connection considers the following policies:

- The connection is accommodated through the least-cost path. The cost of a lightpath link is equal to the overall cost of the fiber links it traverses.
- If there are multiple least-cost paths and the connection does not require full wavelength-channel capacity, the required path is chosen as per the minimal number of free wavelength links.
- If there are multiple least-cost paths and the bandwidth requirement of the connection and full wavelength-channel capacity, the route is chosen as per the minimal number of electronic grooming fabrics.

There are five different configurations of network: in configuration 1, all nodes have single-hop grooming OXCs; in configurations 2, 3, and 4, few nodes have multi-hop partial-grooming OXCs [2]; and in configuration 5, some nodes have multi-hop full-grooming OXCs. As the load increases, the TBR increases in all five configurations. When the network has more grooming capability, the TBR is reduced. Multi-hop grooming capability enhances the RER and decreases the TBR. Multi-hop grooming capability does not increase the probability of connection blocking due to the lack of a free wavelength and low wavelength-capacity utilization due to underutilization of the established lightpaths. The connections needing a full wavelength-channel capacity k have more chances of blocking. There is a fairness concern. When low-speed connections are carried by the network, the resources tend to be fragmented. Therefore, high-speed connections are blocked more if they are not fitted into the available capacities [11].

6.5 HIERARCHICAL SWITCHING AND WAVEBAND GROOMING

Day by day, the need for higher capacity of optical networks is skyrocketing, and there is also for the development of a new switching paradigm to find the path and provision connections in such networks [35,36]. Optical networks require the switching of lower-speed TDM circuits (e.g., STS-1, 51.84 Mbps) within the switch fabric. Besides the switch fabric, expensive optical-to-electrical and electrical-to-optical (OEO) conversions are needed for the switch to operate. The limitations of all-optical switches are the lack of traffic grooming, arbitrary wavelength conversion, and the lack of multicast capabilities for a flexible network. The concept of using both waveband-level and wavelength-level switching in a hierarchical manner is receiving increasing attention. The authors in Ref. [37] proposed a destination-based lightpath-grouping heuristic algorithm to take advantage of waveband switching. The heuristic approaches are used for designing two-layer (waveband and wavelength) networks.

A hybrid (OEO) hierarchical switch architecture is incorporated into all-optical waveband switching, and OEO sub-wavelength grooming is also considered.

6.5.1 HYBRID NODE ARCHITECTURE

The hybrid architecture contains all-optical waveband switching and OEO grooming switching. Figure 6.12 represents hybrid node architectures consisting of WDM transmission equipment and switch equipment. The connection between the WDM transmission equipment and the switch equipment is transmitted by very-short-reach (VSR) optical interfaces. In Figure 6.12a, the WDM signals arrive at an input fiber; a WDM demultiplexer (DEMUX) separates different wavelengths. Each wavelength comes to a receiver where the bits are converted to electrical signals and then converted back to optical signals on a wavelength. At the other end of the VSR interface, an O/E converter converts the signals back to the electrical form and sends them into the switch fabric. For N number inputs/outputs ports, the electrical switch fabric consists of $4N$ O/E and E/O transponders. The architecture in Figure 6.12b is almost the same as that of the electrical fabric, where $4N$ OEO transponders are required. In another configuration (Figure 6.12c), the OEO transponders are not needed around the switch fabric and switch directly the optical signal from the VSR optical interface. All-optical switching/networking needs high-quality and long-reach transmission. In Figure 6.12, the electrical switch fabric is basically a non-blocking switch architecture having single switch chips with a small number of ports. A commonly used switch architecture is a

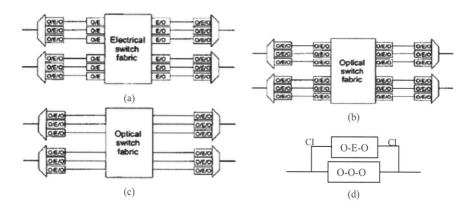

FIGURE 6.12   (a–d) Different hybrid architectures having Optical to Electrical Conversion (OEC) and optical switch fabric [2].

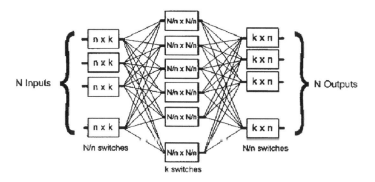

FIGURE 6.13 Clos switch having *N* input/output ports [2].

non-blocking three-stage Clos switch shown in Figure 6.13. Besides the required number of switch elements in a Clos switch, another factor indicating the size of a switch fabric is the maximum port data rate of a single switch chip. The highest data rate of commercial clos-switch chips is approximately 3.25 Gbps per port, which is more than that of OC-48. The input data streams need to be inverse-multiplexed into lower-speed streams (e.g., one OC-192 channel to four OC-48 channels), and multiple switch ports are used in parallel. The port has a data-rate limit, and the Clos architecture makes electrical switch fabrics very difficult to scale with the growing number of wavelengths per fiber and increasing data rate per wavelength. Other limitations of electrical switch fabrics include space and power consumption. Moreover, the signal traces on a single backplane can only travel a limited distance; thus, a large-scale switch with too many switching elements provides a large footprint in terms of power and size. Apart from these limitations, the currently deployed switches use mainly OEO switches [2] which are commercially available as mature switch chips. However, they can provide valuable traffic grooming, wavelength conversion, and multicast functions. The switching architectures shown in Figure 6.12a–c use OEO conversions, whereas the architecture in Figure 6.12d shows optical-optical-optical (OOO) conversion [2] with OEO conversion used for control information transmission waveband switching and OEO traffic-grooming switching. The node based on an all-optical switch has a fiber panel, an all-optical waveband switch, an OEO grooming switch, traffic aggregation equipment, and waveband and wavelength multiplexers/demultiplexers. The fiber patch panel is used for configuring the fiber topology whenever it is required. In Figure 6.14, the fibers that terminate at the node first pass through the waveband demultiplexers and are then sent into the waveband switch after separation.

The waveband switch utilizes the entire waveband as its switching granularity. A particular waveband needs to be switched to a corresponding output fiber. A wavelength or a sub-wavelength TDM circuit needs to be switched to a different waveband or a different output fiber; the parent waveband is switched to the OEO grooming switch. The grooming switch has an electrical grooming switch fabric and is capable of performing sub-wavelength TDM circuit switching (traffic grooming) and arbitrary wavelength conversion. The input and output ports of the grooming switch operate at a wavelength channel's data rate. Both the waveband switch and grooming switch have add and drop ports for locally generated and terminated traffic. The traffic aggregation/deaggregation equipment performs TDM multiplexing/demultiplexing for local lower-speed connections. If the local connections between a source–destination pair has enough load to fill a whole waveband; these connections will be aggregated and directly sent to the waveband add port. A waveband path accommodates an end-to-end waveband connection that originates and terminates at a waveband add/drop port or a grooming switch. The wavelength multiplexers/demultiplexers are responsible for separating/merging wavelengths in a waveband [2].

An optical switch architecture in Figure 6.14 has both an all-optical switch matrix and an electrical grooming switch fabric. Since an all-optical waveband switch does not provide wavelength-conversion capability, it can be constructed in a waveband-layered approach, shown in Figure 6.14. The B wavebands accommodate B number of individual all-optical switch fabrics operating for a specific waveband. The number of input/output ports of a single switch fabric is determined by the number of input/output fibers and the number of waveband add/drop ports.

An OEO grooming switch is made with grooming switch chips whose size is evaluated by the number of waveband ports from/to the waveband switch, and a number of local add/drop ports are needed for its design. The all-optical waveband switching grooms a large amount of traffic into wavebands and avoids switching in

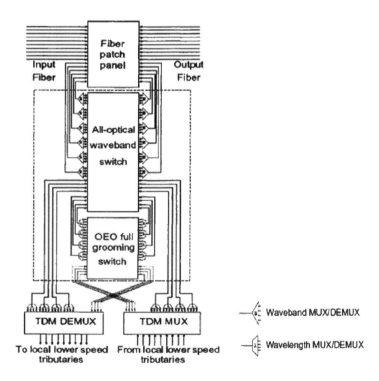

FIGURE 6.14   An architecture consisting of OEO grooming switching and optical wide-band switching [2].

the electrical domain. Its traffic is uniformly distributed among all nodes; on average, at each node, only a fraction $1/H$ of the total incoming and outgoing traffic is added or dropped. (H is a network's average hop distance.) By making a group of pass-through traffic, in the electrical grooming switch, the number of OEO transponders are reduced [2].

The performance differences between a hybrid hierarchical switch and an OEO grooming switch are shown in Figure 6.15. For F = number of fibers = 6 in the node, the number of wavebands in a fiber B is ~4. B_A and B_D indicate the number of waveband add and drop ports, whereas B_W denotes the number of ports on the waveband switch leading to the grooming switch, and W_B indicates the number of ports on the waveband switch received from the grooming switch. W_A and W_D indicate the number of wavelength add/drop ports, respectively. The same procedure is followed for the waveband and wavelength add ports. B_W and W_B are selected on the basis of how much traffic-grooming capability the node has. In the case of maximum traffic grooming, B_W and W_B are both equal to the total number of input wavebands for the node having the capability to groom all the incoming traffic.

In the node architecture in Figure 6.15b, with OEO grooming switch, all incoming fibers separate directly into wavelengths through demultiplexing, and the wavelengths are then sent to the OEO grooming switch. The number of add/drop ports is estimated by the amount of add/drop traffic. For wavelengths operating at data rates beyond OC-48, demultiplexing occurs, and extra ports are required.

6.5.2 Issues and Problems

Challenging issues for hierarchical switching are the design of network, path provisioning, and wavelength provisioning. The main aim in design is to achieve the minimization of the network cost by planning wave paths for a set of known

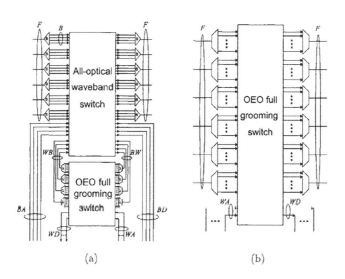

(a)   (b)

FIGURE 6.15   (a) A hybrid architecture and (b) a fully OEO architecture [2].

client connections over the wave paths. The main issue is virtual-topology design which was discussed earlier and traffic-grooming virtual-topology problem which has also been discussed in this chapter. The wavebands per fiber and wavelengths per waveband are found to be shared among connections. The main purpose of service provisioning is to minimize the BP. The impairment-aware wavelength routing in a hybrid network is another issue of both network design and provisioning.

6.6 VIRTUAL CONCATENATION

Optical networks using SONET/SDH have become the efficient high-speed optical backbone to successfully support a significant amount of data and voice traffic in this backbone. With the development of WDM switch having opaque (OEO) switching [2], a multi-service optical WDM network requires intelligent optical cross-connects (OXCs). In this direction, SONET/SDH is made to be a very important component as a framing layer for more efficient and intelligent network operation, administration, maintenance, and protection in an irregular mesh-based or interconnected-ring-based network. For sudden increase of data traffic, the inefficiency of transporting packet data with a SONET/SDH frame becomes an issue where network operators are required to optimize the usage of their current bandwidth to accommodate various types of new services and applications, based on IP, frame relay, Ethernet, Fiber Channel, etc. From a single network node, traffic streams from different client network equipment are to be discretely mapped into different tiers of SONET/SDH bandwidth trunks (data containers), which produce a huge amount of waste. Accommodating a Gigabit Ethernet connection, a concatenated OC-48 (approx. 2.5 Gbps) provides 60% bandwidth wastage. From a network perspective, the time-slot requirement of a SONET/SDH concatenated channel provides a constraint for traffic provisioning and degrades network performance in a dynamic traffic environment.

6.6.1 VIRTUAL CONCATENATION ARCHITECTURE

Virtual concatenation (VC) architecture is employed in a SONET- or SDH-based optical network to accommodate data traffic in a finer granularity to enhance link efficiency. In VC, the use of small data content makes a bigger data content that provides more data per second. Considering a network's switching granularity, VC is used for small basket size of VT-1.5 up to STS-3c. Figure 6.16 indicates a VC architecture to accommodate multiple services with a single OC-48 channel through VC, and an OC-48 channel is used to accommodate two Gigabit Ethernet traffic streams and six STS-1 TDM voice traffic streams. The traffic is obtained in different OC-12 pipes, and these OC-12 pipes are transmitted through the network over various paths balancing benefit.

When multiple traffic streams from one client are transmitted over different paths, the VC mappers at the destination nodes require compensation for the differential delay between the bifurcated streams constructing the information at the destination node. The maximum 50 ms (±25 ms) delay is obtained in a commercially available device with external RAM over a 10,000 km distance transmission. In general, a virtually concatenated SONETISDH channel has $N \times$ STS-1 and is transmitted as

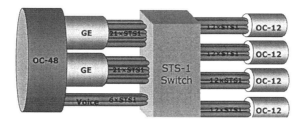

FIGURE 6.16 VC architecture [2].

individual STS-1s across the network, and at the receiver, the individual STS-1s are realigned and sorted to recreate the original payload. The advantages of VC are

- relaxing of time-slot alignment and continuity constraints
- more efficient utilization of channel capacity for multiple types of data and voice services
- bifurcation of traffic streams to balance network load.

With VC, a high-speed traffic stream is split into multiple low-speed streams, and paths are assigned for them separately in the network. The traffic is distributed across the network more evenly. Hence, the network's blocking performance is also enhanced. The advantage of SONET/SDH VC is the operation of an optical WDM mesh network under a dynamic traffic environment. The assumptions are as follows [2]:

- Connections with different bandwidth granularities are considered to arrive at and depart from the network, one at a time, following a Poisson arrival process and negative-exponential-distribution holding time.
- There are two types of traffic pattern considered – pattern I and pattern II.
  In traffic pattern I, there are five classes of services with data service rates of approximately 51 Mbps, 153 Mbps, 622 Mbps, 2.5 Gbps, and 10 Gbps, which are perfectly groomed into the tiered SONET/SDH containers and their corresponding optical carriers, i.e., OC-1, OC-3c, OC-12c, OC-48c, and OC-192c.
  In pattern-II, there are ten service classes with service rates of 50 Mbps, 100 Mbps, 150 Mbps, 200 Mbps, 400 Mbps, 600 Mbps, 1 Gbps, 2.5 Gbps, 5 Gbps, and10 Gbps corresponding to SONET/SDH containers with VC denoted by STS-1, STS-2, STS-3, STS-4, STS-8, STS-12, STS-21, STS-48, STS-96, and STS-192, respectively, and without VC, these containers are denoted by STS-1c, STS-2c, STS-3c, STS-4c, STS-8c, STS-12c, STS-21c, STS-48c, STS-96c, and STS-192c, respectively.
- Here the capacity of each wavelength channel is OC-192.

When traffic bifurcation is needed, a simple routing and traffic-bifurcation heuristic is applied to a connection request, and the heuristic is as follows:

Step 1: Determine the shortest path based on the network administrative cost between the node pair.

Step 2: Update the available capacity of the links along the path for decreasing the available capacity by the minimal capacity as per the estimation of bandwidth of the path considering bandwidth of the route constrained by the link along the route, which has minimal free capacity.

Step 3: Eliminate the link without free capacity and repeat steps 1 and 2 until the connection is accommodated by the set of paths evaluated or no more paths exist for the connection.

After t routes' verification, two types of network configuration are found as per all nodes equipped with STS-1 full-grooming switches and nodes having partial-grooming switches. In a partial-grooming switch architecture, only a limited number of wavelength channels are used for a grooming switch (or grooming fabric within an OXC) to perform traffic grooming. The 5%–10% improvement in network performance is obtained through VC. In traffic pattern I, every service class is grouped into one of the tiered SONET/SDH containers, and we consider no traffic bifurcation. The performance improvement is due to the capability of eliminating the time-slot alignment and contiguity constraints provided by VC. Bandwidth blocking probability (BBP) is significantly reduced by using VC, and more improvement is obtained by using a simple traffic-bifurcation scheme in which a connection is bifurcated if and only if no single path with enough capacity exists. More advanced network-load-balancing and traffic-bifurcation approaches further maximize network throughput. SONET/SDH VC also provides the benefits of network compatibility, network resiliency, network management and control, etc.

Other technologies related to traffic grooming are link-capacity adjustment scheme (LCAS) [2] and generic framing protocol (GFP) [2]. LCAS based on VC is a two-way signaling protocol operating continuously between the source and destination to adjust the pipe capacity while it is in use. It provides on-demand traffic provisioning and online traffic grooming/regrooming. GFP is a traffic-adaptation protocol for broadband transport applications, providing a standard mapping of a physical layer or logical link layer signal into a byte-synchronous channel having SONET/SDH links or wavelength channels in an optical transport network (OTN). There are two approaches used for mapping protocols onto GFP: frame-mapped GFP [2] and transparency-mapped GFP [2]. Frame-mapped GFP is used for a packet-switched environment for point-to-point protocol (PPP), IP, and Ethernet traffic, whereas transparency-mapped GFP works in the transport mode and is intended for delay-sensitive storage-area network (SAN) applications in the case of FC and Fiber Connection (FICON) technologies [2].

6.7   RWA OF TRAFFIC GROOMING CONNECTIONS

There are four processes–link state information collection, working path selection, traffic grooming in working paths, and traffic grooming in backup paths for the protection of connections. We have collected local state information in a link-by-link manner. The working paths of the source–destination pairs are developed on the basis of shortest paths using Dijkstra's algorithm. As discussed in Section 6.2, here, we use both dedicated wavelength grooming (DWG) and shared wavelength grooming (SWG). Here, for the purpose of traffic grooming, we have considered two types of algorithms – destination shared wavelength grooming (DES_SWG) and source shared wavelength grooming (SOURCE_SWG) [38]. In the case of DES_SWG, we consider the traffic grooming of the connections of $s-d$ pairs of same destination along same path whereas for SOURCE_SWG, we consider the connections of $s-d$ pairs of same source along same path. As indicated in the example shown in Figure 6.2, traffic grooming for DES_SWG is performed between the connections of 1-6 and 2-6 $s-d$ pairs on the path <1-2-6>. In the case of SOURCE_SWG, it is performed between the connections of 1-2 and 1-6 pairs on the same path.

The backup paths are formed for the survivability of the traffic grooming network. The problem for backup path formation of the connections is how to protect low-speed sub-wavelength granularity connections separately if the working path of the connection fails. In the case of SOURCE_SWG, [38] the backup paths for 1-2 and 1-6 are <1-3-2> and <1-3-5-6>, respectively, as indicted by the dashed line in Figure 6.17. In case of DES_SWG, the backup paths for 1-6 and 2-6 are <1-3-5-6> and <2-3-5-6>, respectively.

6.7.1 SOURCE _ SWG Algorithm

The SOURCE_SWG algorithm [38] is also based on alternate path routing. Here, each node maintains a fixed alternate path routing table, which has K routes (R_i) (in descending order of the shortest path, where $i = 1, 2, 3…k$ correspond to the first, second,… kth shortest time delay path) for reaching each destination node. Initially, the connections are groomed on first shortest path. If it is groomed to the first shortest path, then the algorithm is attempted for other alternate routes in descending order. It is assumed that a connection can occupy more than one connection slot according to its bandwidth requirement. The algorithm procedure is given below [38].

1. Start a network with T connection requests from all $s-d$ pairs of the network
2. Construct the grouping of same source and different destination along the first shortest path obtained from the routing table.

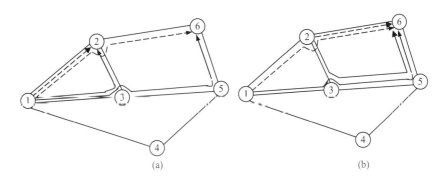

(a)   (b)

FIGURE 6.17 A network having five nodes: (a) SOURCE_SWG and (b) DES_SWG [41].

3. Construct shared protection paths for each s–d pair.
4. Choose a wavelength along the first shortest path using parallel reservation schemes and groom the connections of s–d pairs of the same group on the wavelength using the G-switch till all available wavelengths are filled up in a connection slot.
5. If the connections are not assigned to the lightpath due to unavailability of wavelengths, then the connections are assigned wavelengths by using alternate routing paths in descending order.
6. Repeat step-4 and step-5 till all the connections of all s–d pairs are assigned.

6.7.2   DES_SWG Algorithm

Like the SOURCE_SWG algorithm, the DES_SWG algorithm is based on alternate path routing. The difference is in the grouping of the s–d pairs. Here we consider grouping using same destination and different sources, as discussed earlier. The rest of the algorithm is same as that of SOURCE_SWG algorithm.

6.7.3   Problem Formulation [41]

Following are the objectives of optimization:

1. For a given number of traffic grooming nodes, the number of wavelength channels per fiber is to be minimized.
2. For a given number of wavelength channels and traffic grooming nodes, the total number of connections is to be maximized.

Each link indicates the geographical distance between nodes. The following notations are considered for the formulation:

- i, j indicates the end points (i.e., nodes) of a physical link that might occur in the route of a connection.
- w indicates an index for the wavelength number of link e, where $w = 1,2,3,$ ... W.
- l = an index of a s–d pair/connection where $l = 1, 2, 3, L$; L is the total number of possible s–d pairs.
- c = an index of a connection where $c = 1, 2, 3, $ ... t; t is the total number of connections of an s–d pair.
- e = an index for link number where $e = 1,2,3, $ ... E.

We require the following integers for the solution of the problem.

- $SW_{c,w,d}^{R}$ is an integer that takes a value 1 if the connection c to destination node d in a groomed path R is assigned a wavelength w; otherwise it is zero, where d is the last node of the path R.
- $SW_{c,w,s}^{R}$ is an integer that takes a value 1 if the connection c from source node s in a groomed path R is assigned a wavelength w; otherwise it is zero, where s is the first node of the path R.

- $DW_{c,w,s,d}^{R}$ is an integer that takes a value 1 if the connection c from source s to destination d is assigned a wavelength w; otherwise it is zero, where d is the last destination node of groomed path R and s is the first node of the groomed path R.
- $P_{ij}^{l,w,b}$ indicates an integer having a value 1 if the link ij of the route b is assigned a wavelength w for shared protection of the lth s–d pair/connection; otherwise it is zero.

Objective for traffic grooming:

$$\text{Minimize } W. \tag{6.29}$$

Grooming constraint:

For a groomed path R in DES_SWG algorithm,

$$\sum_{c} SW_{c,w,d}^{R} + \sum_{c} DW_{c,w,s,d}^{R} \leq g \quad \forall\, 1 \leq w \leq W \tag{6.30}$$

$$\sum_{w=1}^{W} SW_{c,w,s}^{R} = 1 \quad \forall\, 1 \leq c \leq t, \quad s \in V \tag{6.31}$$

$$\sum_{w=1}^{W} DW_{c,w,s,d}^{R} = 1 \quad \forall\, 1 \leq c \leq t, \quad s,d \in V \tag{6.32}$$

Protection constraints:

$$\sum_{j:(ij \in E)} P_{ij}^{l,w,b} = \sum_{j:(ji \in E)} P_{ji}^{l,w,b}, i \neq s,d, \forall (w \in W) \text{ and } (i,j \in V) \tag{6.33}$$

$$\sum_{w \in W} \sum_{j:(ij \in E)} P_{ij}^{l,w,b} - \sum_{w \in W} \sum_{j:(ji \in E)} P_{ji}^{l,w,b} = 1, i = s \tag{6.34}$$

$$\sum_{w \in W} \sum_{j:(ij \in E)} P_{ij}^{l,w,b} - \sum_{w \in W} \sum_{j:(ji \in E)} P_{ji}^{l,w,b} = -1, \quad i = d \tag{6.35}$$

For a groomed path R in SOURCE_SWG algorithm,

$$\sum_{c} SW_{c,w,s}^{R} + \sum_{c} DW_{c,w,s,d}^{R} \leq g \quad \forall\, 1 \leq w \leq W \tag{6.36}$$

$$\sum_{w=1}^{W} SW_{c,w,d}^{R} = 1 \quad \forall\, 1 \leq c \leq t, d \in V \tag{6.37}$$

Equation (6.29) represents the objective, which minimizes the number of wave-lengths per fiber. Equations (6.33)–(6.35) show the flow balance constraints of restricted shared backup paths. Equation (6.30) ensures that at the most g connections are groomed into a wavelength channel, and equation (6.32) indicates that only one wavelength is assigned to a connection in DES_SWG algorithm. Similarly, equations (6.36) and (6.37) are used for grooming constraints of SOURCE_SWG algorithm. Equation (6.35) shows that only one wavelength is assigned to a connection between a source node s and a destination node d in the groomed path R, and it can be used for both the algorithms.

SUMMARY

In this chapter, we have discussed the need for traffic grooming and the principle of traffic grooming. We have discussed different schemes used in static traffic grooming and compared them. We have mentioned the basics of dynamic traffic grooming and its advantages over static traffic grooming. Adaptive traffic grooming is also explained. We have described a practical hierarchical optical switch architecture integrating all-optical wave band switching and OEO traffic grooming. This architecture has been demonstrated to achieve large capacity, scalability, high flexibility, and a significantly lower cost. A part from hierarchical optical switch architecture, the traffic grooming architecture can be designed entirely with mature components, providing a very practical solution. VC and LCAS support a SONET- or SDH-based optical network to evolve toward a data-centric intelligent automatically switched optical network. The following benefits are obtained: (a) the time-slot continuity and alignment constraints of traditional SONET/SDH concatenation are relaxed, (b) bandwidth efficiency of a wavelength channel is improved, (c) inverse-multiplexing (i.e., traffic bifurcation) and load balancing are enabled, and (d) service-related performance is improved. Moreover, VC is also discussed where network operator adjusts the pipe capacity while it is in use. This increases the possibility for on-demand traffic provisioning, and it makes SONET- or SDH-based optical WDM networks more data friendly.

PROBLEMS

6.1. How many OC-192 links are needed to carry all the traffic if we use packing techniques separately for each of the seven OC-3, five OC-12, and six OC-48 connections and pack all of the above connections together.

6.2. Design an architecture for traffic grooming in airing topology having a smaller number of ADMs. Design a traffic-grooming architecture to show that, for a ring topology, minimizing the number of ADMs may need more wavelengths than the minimal.

6.3. Consider the following network (Figure P6.1) having four nodes. The corresponding link utilization matrix can be seen below (with 0.3 signifying 30% utilization of a wavelength):

$$
\begin{array}{cccc}
0.0 & 0.4 & 0.2 & 0.5 \\
0.4 & 0.0 & 0.3 & 0.7 \\
0.3 & 0.1 & 0.0 & 0.4 \\
0.6 & 0.4 & 0.7 & 0.0
\end{array}
$$

1. How many wavelengths are saved with and without traffic grooming.
2. What is the average utilization for each wavelength in the non-grooming mode? What is the utilization in traffic grooming mode?

6.4. Consider the network in Figure P6.2. The traffic demands between each pair of nodes are as shown in the figure (next to each link). The capacity of each link is OC-192.

1. Without traffic grooming, how many lightpaths are required, and how many wavelengths need to be allocated? Show all the lightpaths and the traffic carried by them.
2. With traffic grooming, how many lightpaths are required, and how many wavelengths need to be allocated? Show all the lightpaths and the traffic carried by them.

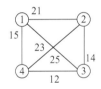

6.5. The network topology is shown in Figure P6.2, with capacities on links.
Given the following traffic matrix (the traffic is assumed to be symmetrical in both directions), try to accommodate the traffic optimally in terms of network through put by using traffic grooming (with multiple paths). Show the aggregation of traffic into the lightpaths.

$$
\begin{array}{cccc}
0 & 4 & 2 & 5 \\
4 & 0 & 3 & 7 \\
3 & 1 & 0 & 4 \\
6 & 4 & 7 & 0
\end{array}
$$

6.6. Consider the network topology in Figure P6.3. The capacity of each wavelength is OC-192. Given the following traffic matrix in Gbps, what is the number of wavelengths required with traffic grooming and without traffic grooming?

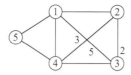

6.7. Consider the network in Figure P6.4. The capacity of each wavelength channel is OC-192. Assume that there are no limits on the number of wavelengths on each link and on the number of transceivers at each node. The following three lightpaths are already established (including their routes and capacities used):

1. from 8 to 12 at OC-96 (8-7-9-12)
2. from 12 to 21 at OC-48 (12-16-22-21)
3. from 10 to 16 at OC-144 (10-9-12-16)
4. from 9 to 17 at OC-144 (9-12-13-17)
5. from 11 to 16 at OC-144 (11-9-12-16)
6. from 7 to 16 at OC-144 (7-9-12-16).

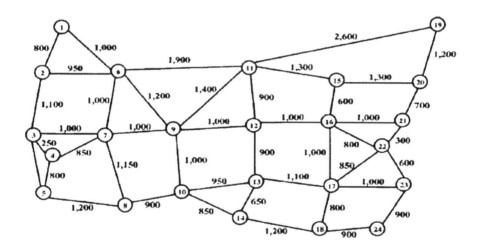

When a new low-speed connection request arrives, one of the following two methods can be used to establish the connection.

Method 1: Groom the new connection onto an existing lightpath with sufficient spare capacity.

Method 2: Groom the new connection onto existing lightpaths, thereby establishing a multi-hop connection.

Method 3: Create a new lightpath for the new connection.

We need to establish the following connections. Find the routes – if possible – for each of the following connections, indicating the use of existing lightpaths, wherever applicable, and new lightpaths. If a connection cannot be established using a specified method, please justify why it cannot established.

  a. Node 5 to 20 at OC-48 using method 1
  b. Node 5 to 20 at OC-48 using method 2
  c. Node 8 to 20 at OC-96 using method 3
  d. Node 8 to 21 at OC-48 using method 2
  e. Node 8 to 21 at OC-192 using method 3
  f. Node 8 to 21 at OC-96 using method 1
  g. Node 10 to 21 at OC-144 using method 2
  h. Node 10 to 21 at OC-96 using method 1
  i. Node 10 to 21 at OC-48 using method 3

6.8. A waveband path can originate from a waveband add port (one of the BA ports) or from an OEO-switch or waveband-switch port (one of the WB ports) (see Figure 6.4). Similarly, it can terminate at one of the BD ports or BW ports. How many types of waveband paths are there based on how such a path originates and terminates? What constraints does each type of waveband path impose on routing connection requests?

6.9. It is not desirable to route a connection to a large number of paths from the point of view of network operation and management. How can you modify the heuristic (traffic bifurcation and route) in Section 6.6.2 so that a connection will be routed onto no more than K paths, where K is constant?

6.10. We considered a generic graph model for traffic grooming. Can this model also handle the hierarchical optical switches shown in Figure 6.14? Justify your answer.

6.11. Another hierarchical optical switch architecture might contain a separate wavelength switch between the waveband switch and the OEO-grooming switch. Does the architecture proposed in this chapter have wavelength switching capability? What are the advantages and disadvantages of each architecture?

6.12. In a static network design case, all traffic requests are known and do not change. Which of the VC benefits listed in Section 6.6 may not be obtained? In dynamic network design, mention a list of benefits obtained by VC.

6.13. Discuss challenging issues for hierarchical switching in a netwvork with and without traffic grooming.

6.14. Discuss problems of static traffic grooming and how dynamic traffic grooming removes these problems.

REFERENCES

1. R. Ramaswami and K. Sivarajan, *Optical Networks: A Practical Perspective*, 2nd ed., Morgan Kaufmann Publishers, 2001.
2. B. Mukherjee, *Optical WDM Networks*, Springer-Verlag, 2006.
3. G. Keiser, *Optical Fiber Communication*, McGraw Hill, 2002.
4. E. Modiano and P. Lin, "Traffic grooming in WDM networks," *IEEE Communications Magazine*, vol. 39, no. 7, pp. 124–129, July 2001.
5. K. Zhu and B. Mukherjee, "A review of traffic grooming in WDM optical networks: Architectures and challenges," *SPIE Optical Networks Magazine*, vol. 4, no. 2, pp. 55–64, March/April 2003.

6. X. Zhang and C. Qiao, "An effective and comprehensive approach for traffic grooming and wavelength assignment in SONET/WDM rings," *IEEE/ACM Transactions on Networking*, vol. 8, no. 5, pp. 608–617, October 2000.

7. W. Cho, J. Wang, and B. Mukherjee, "Improved approaches for cost-effective traffic grooming in WDM ring networks: Uniform-traffic case," *Photonic Network Communications*, vol. 3, no. 3, pp. 245–254, July 2001.

8. V. R. Konda and T. Y. Chow, "Algorithm for traffic grooming in optical networks to minimize the number of transceivers," *Proceedings, IEEE Workshop on High Performance Switching and Routing 2001*, Dallas, TX, pp. 218–221, May 2001.

9. M. Brunato and R. Battiti, "A multistart randomized greedy algorithm for traffic grooming on mesh logical topologies," *Proceedings, Workshop on Optical Network Design and Modelling (ONDM) '02*, Torino, Italy, pp. 417–430, February 2002.

10. R. Srinivasan and A. K. Somani, "A generalized framework for analyzing time-space switched optical networks," *IEEE Journal on Selected Areas in Communications*, vol. 20, no. 1, pp. 202–215, January 2002.

11. S. Thiagarajan and A. K. Somani, "Capacity fairness of WDM networks with grooming capabilities," *SPIE Optical Networks Magazine*, vol. 2, no. 3, pp. 24–31, May/June 2001.

12. S. Thiagarajan and A. K. Somani, "Traffic grooming for survivable WDM mesh networks," *Proceedings, OptiComm '01*, Denver, CO, pp. 54–65, August 2001.

13. K. Zhu and B. Mukherjee, "On-line approaches for provisioning connections of different bandwidth granularities in WDM mesh networks," *Proceedings, OFC '02*, Anaheim, CA, pp. 549–551, March 2002.

14. L. A. Cox and J. Sanchez, "Cost saving from optimized packing and grooming of optical circuits: mesh versus ring comparisons," *SPIE Optical Networks Magazine*, vol. 2, no. 3, pp. 72–90, May/June 2001.

15. A. Lardies, R. Gupta, and R. A. Patterson, "Traffic grooming in a multi-layer network," *SPIE Optical Networks Magazine*, vol. 2, no. 3, pp. 91–99, May/June 2001.

16. L. H. Sahasrabuddhe and B. Mukherjee, "Light-trees: optical multicasting for improved performance in wavelength-routed networks," *IEEE Communications Magazine*, vol. 37, no. 2, pp. 67–73, February 1999.

17. N. K. Singhal and B. Mukherjee, "Protecting multicast sessions in WDM optical mesh networks," *IEEE/OSA Journal of Lightwave Technology*, vol. 21, no. 4, pp. 884–892, April 2003.

18. R. Ramaswami and K. N. Sivarajan, "Design of logical topologies for wavelength-routed optical networks," *IEEE Journal on Selected Areas in Communications*, vol. 14, no. 5, pp. 840–851, June 1996.

19. CPLEX, http://www.ilog.com.

20. H. Zang, J. P. Jue, and B. Mukherjee, "A review of routing and wavelength assignment approaches for wavelength-routed optical WDM networks," *SPIE Optical Networks Magazine*, vol. 1, no. 1, pp. 47–60, January 2000.

21. R. Ramamurthy and B. Mukherjee, "Fixed-alternate routing and wavelength conversion in wavelength-routed optical networks," *IEEE/ACM Transactions on Networking*, vol. 10, no. 3, pp. 351–367, June 2002.

22. W. Stalling, *Data Communication and Networks*, PrenticeHall of India, 1999.

23. B. C. Chatterjee, N. Sarma, P. P. Sahu, "Review and Performance Analysis on Routing and Wavelength Assignment Approaches for Optical Networks," *IETE Technical Review*, vol. 30, no. 1, 2013, 12–23.

24. J. Wang, W. Cho, V. R. Vemuri, and B. Mukherjee, "Improved approaches for cost-effective traffic grooming in WDM ring networks: ILP formulations and single-hop and multihopconnections," *IEEE/OSA Journal of Lightwave Technology*, vol. 19, no. 11, pp. 1645–1653, November 2001.

25. K. Zhu, H. Zhu and B. Mukherjee, *Traffic Grooming in Optical WDM Mesh Networks*, Optical Network Series, Springer, 2005.

26. K. Zhu and B. Mukherjee, "Traffic grooming in an optical WDM mesh network," *IEEE Journal on Selected Areas in Communications*, vol. 20, no. 1, pp. 122–133, January 2002.

27. ITU-T, Recommendation G.707, "Network node interface for the synchronous digital hierarchy (SDH)," April 2002.

28. ITU, Recommendation G.694.1, "Spectral grids for WDM applications," June 2002.

29. E. Mannie, et al., "Generalized multi-protocol label switching (GMPLS) architecture," Internet Draft, Work in Progress, draft-ietf-ccamp-gmpls-architecture-02.txt, March 2002.

30. H. Zhu, H. Zang, K. Zhu, and B. Mukherjee, "A novel generic graph model for traffic grooming in heterogeneous WDM mesh networks," *IEEE/A CM Transactions on Networking*, vol. 11, no. 2, pp. 285–299, April 2003.

31. R. Braden, L. Zhang, S. Berson, S. Herzog, and S. Jamin, "Resource reservation protocol (RSVP) - version 1, functional specification," RFC 2205, September 1997.

32. Awduche, et al., "RSVP- TE: Extensions to RSVP for LSP tunnels," RFC 3.209, December 2001.

33. L. Andersson, P. Doolan, N. Feldman, A. Fredette, and B. Thomas, "LDP specification," RFC 3036, January 2001.

34. B. Jamoussi, et al., "Constrained-based LSPsetup using LDP," RFC 3212, January 2002.

35. O. Gerstel, R. Ramaswami, and W. K. Wang, "Making use of a two-stage multiplexing scheme in a WDM network," *Proceedings, OFC '00*, Baltimore, MD, pp. 44–46, March 2000.

36. O. Gerstel and R. Ramaswami, "Optical layer survivability-an implementation perspective," *IEEE Journal on Selected Areas in Communications*, vol. 18, pp. 1885–1899, October 2000.

37. M. Lee, J. Yu, Y. Kim, C. Kang, and J. Park, "Design of hierarchical crossconnect WDM networks employing a two-stage multiplexing scheme of waveband and wavelength," *IEEE Journal on Selected Areas in Communications*, vol. 20, pp. 166–171, January 2002.

38. P. P. Sahu, "New traffic grooming approaches in optical networks under restricted shared protection," *Photonics Communication Networks*, vol. 16, 2008, 223–238.

7 Survivability of Optical Networks

Due to failures that occurred in a network, the traffic transportations are disrupted [1,2]. Different types of failures that occurred in an optical wavelength division multiplexing (WDM) network [2–4] are shown in Figure 7.1. In the network, "duct" is a bidirectional physical cylindrical enclosure between two nodes in which fibers are kept aligned into bundles under the ground [2]. A fiber cut is usually found due to a duct cut during construction or destructive natural calamities, such as earthquakes and flood.

A central office (CO) can also be disrupted where optical cross-connects (OXCs) are located, usually because of disasters such as fire or flooding [2,5]. This is known as node failure. Although node failures are uncommon, the disruption will damage heavily in optical network. Besides node and link failures, there is channel failure in optical WDM networks [5]. A channel failure usually occurs due to the failure of transmitting and/or receiving equipment operating on that channel. The failure in time (FIT) represents the average number of failures in 10^9 hours; Tx indicates the optical transmitter systems, whereas Rx denotes the optical receiver systems; and $MTTR$ is the mean time to repair. Single-fiber failures are dominant over other failures in communication networks [2].

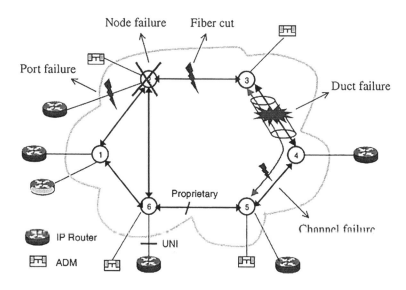

FIGURE 7.1 Different types of failures in an optical WDM network [2].

Due to fiber cut, the huge traffic loss occurs, and network survivability has become one among issues in network design and its real-time operation. It is needed to devise effective methods to recover from failures of links and nodes. Apart from the failures occurred in the infrastructure of an optical network, other failures can also affect other network layers, e.g., the port failure at an IP router, as shown in Figure 7.1. Normally, these failures are dealt with the fault-management schemes in the network layers. The link failures in an IP network will be handled by IP rerouting. Most of the development survivability in WDM networks focuses on the recovery from a single-link or node failure where one failure is repaired. This is called as single failure scenario. There is also possibility of multiple simultaneous failures in a realistic network (Figure 7.1), and appropriate recovery methods can also be designed [6,7].

Survivability can be implemented in many layers such as ATM, IP, and SONET/SDH [2]. The fault-management schemes in each layer deal with their different character. In an IP network, the Internet is seen as a collection of autonomous systems (ASS) having a set of routers. Routers employ an interior gateway protocol (IGP for ASS, which is as, for example, open shortest-path first (OSPF) protocol [2]). The routing table at each router/node within the domain is simplified in a distributed manner. It can take few seconds after the failure is identified before the routing tables have consistent routing information. The issue for fault management in IP network is the long failure-detection time. As soon as there is a loss of signal on an optical link or the bit-error rate (BER) on a link exceeds a threshold, the node identifies the failures in milliseconds.

7.1 PARAMETERS FOR SURVIVAL SCHEMES

The following some terminologies are used for survival schemes of the network [2–5].

Restoration time (*RT*) is the exact disruption-holding time and should be minimized as much as possible.

Restoration success rate (*RSR*) denotes the ratio between the number of successfully restored traffic streams and the number of disrupted traffic streams after a network failure occurs.

Service restorability is usually a network-wide parameter representing the capability that a network can survive a specific failure scenario.

Availability is defined *as* the asymptotic probability that a system (a connection in the case of this chapter) will be found in the operating state at a random time in the future. Availability of a system can be computed statistically, based on the failure frequency and failure repair rate of the underlying network components.

Reliability is the probability that a system will operate without any disruption for a predefined period of time. Service reliability can be represented by the number of "hits" or disruptions in a unit of time. Availability and reliability are different measures of service quality.

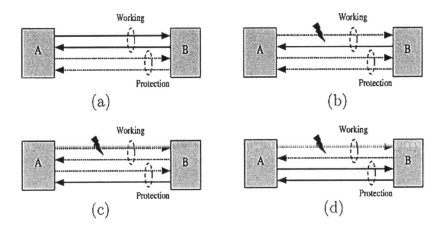

FIGURE 7.2   An APS protocol: (a) working path and protection path having signal in both directions, (b) failure occurs on the working path from A to B, (c) B detects failure and switches traffic, (d) A switches traffic to protection path.

7.2 FAULT MANAGEMENT

SONET and SDH networks give high accessibility through protection techniques [2,8]. In networks with protection schemes, two paths are provided: a connection-working path to carry traffic under normal operation (also called as a primary path) and the protection path (also called as a backup path) to transmit traffic while failures occur. There are two types of connection-unidirectional traffic transmitted simultaneously on the working and protection paths and bidirectional, traffic in both direction needed to be switched from the working path to the protection path even if the failure is only on the working path of one direction. This process needs a signaling protocol known as an automatic protection-switching (APS) protocol [2]. In Figure 7.2, if a failure has an effect on the working path A to B without affecting the working path B to A, as per APS protocol, then the receiver in B identifies the failure on the working path, and node switches off its transmitter on the working path and then transmits the traffic to the protection path. The receiver in A detects the signal loss and switches the traffic to the protection path as well. If traffic is transmitted in both directions over a single fiber, a fiber cut can be detected by both ends. An APS protocol is not used here but it needs to handle the equipment failures as shown in Figure 7.2.

7.2.1 FAULT MANAGEMENT IN RING TOPOLOGY

SONET/SDH rings having multiple nodes are used for signal transmission between two nodes. A ring uses a two connected topology having two disjoint paths between any pair of nodes that have no node or link in common except the source and destination nodes. The *self-healing* (SH) [2] process is used for failures automatically detected and traffics are rerouted from failed links or nodes using the other half of the ring. There are three ring architectures: two-fiber unidirectional path-switched ring (UPSR [2]), two-fiber bi-directional line-switched rings (BLSR/2) [2], and four-fiber bi-directional link-switched rings (BLSR/4) [2].

7.2.1.1   Unidirectional Path-Switched Ring (UPSR)

In a UPSR, there are two adjacent ring connections: adjacent nodes are connected [28] in the clockwise direction as the working fiber and adjacent nodes are connected in the counter clockwise direction as the protection fiber. For each SONET connection, traffic is sent to both the working fiber and the protection fiber. The destination controls both the working and protection fibers and selects the better signal between the two.

The destination gets signal from the working fiber under normal conditions. If there is a link or node failure on the working path, the destination will switch over to the protection fiber and continue to receive the data. This scheme is referred to as 1 + 1 (or *dedicated*) *path protection* scheme [2] as protection resource is dedicated to one working connection at one time. Figure 7.3a presents a UPSR having four nodes. The protection scheme in a UPSR needs no communications between the nodes providing fast failure recovery. But this architecture is not efficient as half of the bandwidth/capacity is for protection purposes. There is no sharing of the protection bandwidth between connections. This is the main disadvantage with the UPSR [2].

The UPSR has two data speeds: OC-3 ad OC-12. There is no restriction in the number of nodes in the UPSR. But the clockwise (working) and counterclockwise (protection) paths taken by a signal have different delays with them. The ring length is limited with the requirement of the RT.

7.2.1.2   Bidirectional Line-Switched Ring (BLSR)

Figure 7.3b shows a BLSR/4 where two adjacent nodes in the ring are joined by four fibers: two for working traffic and two for protection traffic [2,8]. Usually, the traffic between two nodes is sent via the shortest path. The traffic for node A to node B is transmitted clockwise, whereas the traffic from node B to node **A** is transmitted counterclockwise. When a working fiber fails, the traffic is transmitted with the protection fiber between the two nodes on the same link. This scheme is known as span switching [2]. Both working fibers and protection fibers between two nodes fail simultaneously.

BLSR/2: The nodes usually transmit jointly, and if there is a node failure in the network, the traffic between the two nodes is transmitted around the ring on the protection fiber. This is known as *ring switching* [2].

There is a BLSR/2. Two adjacent nodes in the ring are joined by two fibers in different direction. The half of the bandwidth of fiber is allotted for protection purpose. In case of a link failure, the traffic on the failed link is transmitted along the other part of the ring using the protection bandwidth. This is known as 1:1 protection [2].

ABLSR permits protection bandwidth to be shared between spatially separated connections. Figure 7.3c shows such a case where protection capacity from node A to node D (from node D to node C) on the fiber in the counterclockwise direction is shared by connections 1 and 2. Furthermore, the protection bandwidth is used to carry low-priority traffic during normal operation. This traffic is attempted for transmission when the bandwidth is required for protection [2].

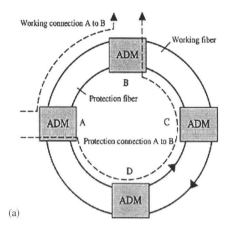

(a)

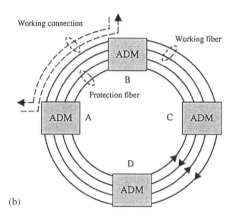

(b)

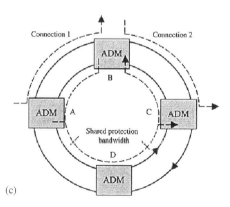

(c)

FIGURE 7.3   (a) A unidirectional path- switched ring (UPSR) [2]. (b) A bidirectional path switch ring with four fibers (BPSR/4) [2]. (c) A two fiber BSLR/2

ABLSR accommodates maximum 16 nodes due to having a 4 bit addressing field used for the node identifier. The ring length is controlled by the requirement on the RT. The BLSRs are employed in long-haul and inter-office networks. These rings use OC-12, OC-48, and OC-192 rates [8].

7.2.2 FAULT MANAGEMENT IN WDM MESH NETWORKS

The mesh network has been extensively used because of the poor scalability of interconnected rings and the excessive resource redundancy used in ring-based fault-management schemes. The design and operation of a survivable WDM mesh network have received increasing attention [9–17]. Here, abroad overview of the fault-management issues involved in designing an optical mesh network employing OXCs, and also their real-time network operation, including dynamic connection provisioning, is discussed. The basic mechanisms in fault-management schemes and protection cycles are mentioned. The various challenges in fault-management schemes, e.g., primary (working) and backup (protection) route computations, maximizing shareability for the shared-protection schemes, and different considerations in dynamic restoration, are mentioned a long with the appropriate techniques to resolve fault-management problems.

7.3 FAULT-RECOVERY MECHANISM

Two types of fault-recovery mechanisms for both node failure and link failure are protection and restoration [2]. Figure 7.4 shows the classification of different protection and restoration schemes [18]. The backup routes with wavelength assignments (WAs) are estimated in advance for a protection scheme, whereas in restoration after a failure occurrence, another route and a free wavelength have to be selected dynamically for each interrupted connection. The dynamic restoration schemes have more utilization of network bandwidth because of not having allocation of spare capacity in advance,

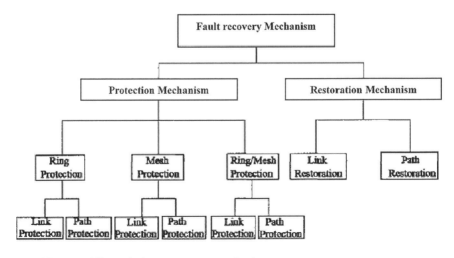

FIGURE 7.4   Different fault-management mechanisms.

and also providing, whereas protection schemes have faster recovery time and provide guaranteed recovery from disrupted traffic. Different protection schemes are discussed in this chapter, whereas different restoration schemes are discussed in Chapter 18.

Protection schemes can be classified as ring protection and mesh protection. Ring-protection schemes include APS [2] and self-healing rings (SHRs) [2], which are already discussed in previous section. Further, both ring protection and mesh protection can be made in two ways: path protection and link protection.

7.3.1 PATH AND LINK PROTECTION

In path protection [2], the traffic is transmitted through a backup path (or protection path) after a link failure occurs on its working/primary path. The primary and backup paths for a connection must have different link so that no single-link failure can affect both of these paths. In link protection, the traffic is transmitted only around the failed link. The path protection gives more utilization of backup resources and has lower end-to-end propagation delay for the recovered route, whereas link protection gives faster protection-switching time. UPSR uses path protection in a ring, whereas BLSR follows link protection implemented in a ring. Figure 7.5 shows the operation of link protection and path protection. The path protection schemes are made in two ways: dedicated (1 + 1 and 1:1) and shared (M:N), whereas link protection scheme is also made by dedicated-link and shared-link protection.

Recently, sub-path protection is introduced in a mesh network with partitioning of a primary path into a sequence of segments and protecting each segment separately where the whole network is divided into different domains. In this case, a light path segment in one domain must be backed up by the resources of the same domain [19–21]. In comparison with path protection, sub-path protection provides high scalability and fast recovery time. Protection schemes are usually designed to protect against a single-link failure, whereas node failures are dealt with the estimation of node-disjoint routes. There are mainly two types of protection schemes: dedicated or shared. In dedicated protection scheme, sharing is not allowed between backup bandwidth, while in shared-protection scheme, backup bandwidth can be shared on some links as long as their protected segments (links, sub-paths, paths) are mutually diverse or not in the same SRGs. But the recovery time and the resource efficiency

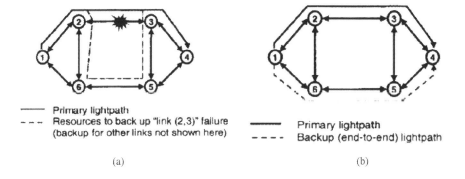

(a)   (b)

FIGURE 7.5   Operation of protection schemes: (a) link protection and (b) path protection [2].

in shared-protection scheme are more than those of dedicated protection scheme. Figure 7.4 provides an overview of some related work on dynamic shared-path-protected connection provisioning on WDM mesh networks.

7.3.2 DEDICATED PROTECTION (1:1 AND 1 + 1) AND M:N SHARED PROTECTION

In dedicated protection [22], the traffic is allotted simultaneously on both primary and backup paths from the source node to the destination node. If one is cut, the destination simply switches over to the other path and continues to receive the data. This form of protection is usually referred to as $1 + 1$ protection, which provides very fast recovery and requires no signaling protocol between two end nodes. If traffic is only transmitted on the primary path, the source and destination nodes both switch over to the backup path when the primary path is cut. In 1:1 protection, the backup bandwidth is also used to carry low-priority perceptible traffic during normal operation apart from high-priority traffic. Shared-protection scheme is also referred to as M:N protection where M primary paths share N backup paths. Figure 7.6 shows $1 + 1$, 1:1, and M:N path protection schemes and the comparison between them [2].

Protection schemes are either reverting or non-reverting. In reverting protection scheme, the traffic is switched back to its primary path after recovery of the failure on the primary path. In non-reverting protection scheme, the traffic is transmitted through the backup path for the remaining service time even after repair of the

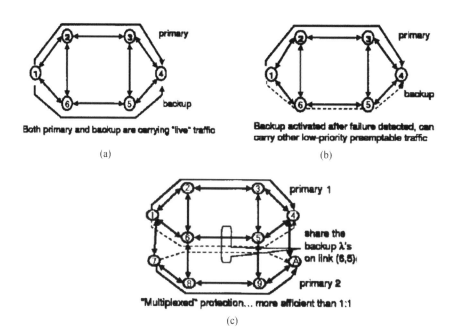

Both primary and backup are carrying "live" traffic

(a)

Backup activated after failure detected, can carry other low-priority preemptable traffic

(b)

"Multiplexed" protection... more efficient than 1:1

(c)

FIGURE 7.6 Examples of protection: (a) 1 +1 protection, (b) 1:1 protection, and (c) M:N protection [2].

failure. Normally, reverting may be applied for a shared-protection scheme, but dedicated protection schemes can be either reverting or non-reverting. Since multiple connections are sharing the common backup bandwidth, the backup bandwidth must be freed up immediately after the failure will be repaired, so that it can be used to protect other connections when another failure occurs [2].

Present day's backbone networks are more densely connected in a ring, and traffic is becoming more and more distributed. The mesh protection schemes offer more bandwidth-efficient solution than rings.

The various issues of WDM mesh network survivability are discussed for designing a survivable WDM mesh network. Network traffic can be static, dynamic, or incremental; these techniques can be applied to different provisioning scenarios according to different network characteristics.

7.4 PROTECTION ISSUES RELATED TO RING COVER, STACKED RINGS

Optical networks using single ring have very fast restoration since a simple switching mechanism makes restoration within 50–60 ms. SHR techniques are used in digital cross-connect (DXC)-based networks [2]. To enhance the restoration speed, an irregular mesh-based network topology is made by using multiple logical rings [9,23,24]. This scheme is known *as ring cover* scheme. These logical rings work as physical SHRs, which are only used for protection purpose. Both UPSR and BLSR are employed. There are two types of traffic used in mesh ring networks: intra-ring traffic and inter-ring traffic. The intra-ring traffic is regular traffic in a single SHR, whereas the inter-ring traffic requires extra switching and add/drop operations at the node crossed over by two or more rings. Figure 7.7 shows ten-node mesh network replaced by five logical BLSRs [2]. Node A and node B are not covered by the same ring so the traffic from node A to node B needs to traverse multiple rings. In the figure, traffic from node **A** to node B can be routed via rings 5 and 2. It can also take other routes through other rings, e.g., using rings 5 and 4. This type of interconnected logical rings is usually called *stacked rings*. Since the logical rings are taken for protection purpose, a set of directed cycles are needed for covering all the links in the network in which any failure link has been recovered for the traffic transmission on the failed link, using APS [2] in a mesh network.

Minimization problem of the total (working and protection) bandwidth required in a given network topology is divided into three sub problems [2]:

- For every traffic demand, routing of the working light path in the mesh topology.
- Identifying the ring(s) covering the link and protecting the traffic.
- Provisioning the spare bandwidths that are necessary to protect the working light path.

There are design constraints such as limitation of the size of each ring (number of nodes) in order to achieve fast restoration [25]. ILP approach is used to resolve this problem. Protection cycle (p-cycle) is used in ring cover techniques [25]. In a p-cycle,

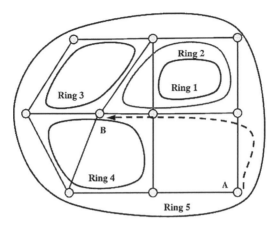

FIGURE 7.7   Five logical BLSRs covering mesh optical network having ten nodes [2].

not only could these on-cycle links be protected in an SHR manner, but also the spanning links could be protected.

7.5   SURVIVABLE ROUTING AND WAVELENGTH ASSIGNMENT (S-RWA)

In a WDM mesh network having end-to-end path protection, there is a problem of estimating a link-disjoint primary and backup path pair and assigning a proper wavelength channel to each path under survivability, which is presented as the survivable routing and wavelength assignment (S-RWA) problem [26].

The cost of a dedicated path-protected connection includes both the costs of the primary and backup light paths. In case of shared-path protection, the cost of a connection also includes sum of the cost of primary light path and the additional backup links. The path pair is selected from a set of preplanned alternate routes or dynamically estimated on the basis of current network state.

Depending on different traffic-engineering considerations for dynamic traffic, different cost functions are used for network links, i.e., to minimize hop distance, the length of the links, the fraction of the available capacity on the links (to balance network load) and the network cost (total equipment cost plus operational cost) on the links (to minimize cost). For wavelength assignment (WA) problem, different WA heuristics have been reported in the literature [26]. The problem of computing a pair of link-disjoint paths in a WDM network is considered as a NP-complete [26].

7.5.1   ALGORITHMS FOR COMPUTING LINK-DISJOINT PATHS

Two-step algorithm is used to compute link-disjoint routes [2]: in the first step, the primary path is determined using a shortest-path algorithm, and in the second step, the backup path is determined by first removing the edges along the primary path and then finding the shortest path in the reduced graph potential weaknesses in that its computation is sequential. A pair of link-disjoint paths is found between the source and destination nodes.

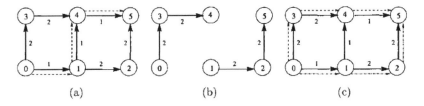

FIGURE 7.8 Two-step algorithm having six number nodes for survivability of network [2] (a) route 0-1-4-5 is determined (b) node 0 and node 5 disconnected (c) link disjoint paths existed.

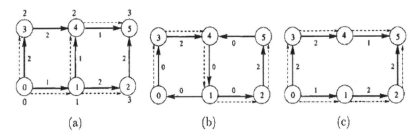

FIGURE 7.9 A sample network in which one-step algorithm is operated [2] (a) link disjoint paths determined (b)shortest path node 0 and node 5 determined (c) link-disjoint paths (0,3,4,5) and (0,1,2,5).

Figure 7.8 represents the diagram where the two-step algorithm is unsuccessful in finding a pair of link-disjoint paths from node 0 to node 5, where such a pair of paths is present. The link weights are mentioned besides the link. In the first step, route 0-1-4-5 is determined as a (least-cost) primary path indicated by dashed lines in Figure 7.8a. After link failures along the primary path, node 0 and node 5 become disconnected, as shown in Figure 7.8b. A pair of link-disjoint paths does exist as route (0-1-2-5) and route (0-3-4-5), as shown in Figure 7.8c.

To reduce time complexity (O (N^2), where N = number of nodes), one-step algorithm is used to compute simultaneously two link-disjoint paths of a SD pair [27]. The total cost of the two paths estimated by this algorithm is reduced to minimum. Figure 7.9 represents a sample network to describe how the one-step algorithm operates. A pair of link-disjoint paths from node 0 to node 5 needs to be computed. The link costs are mentioned besides the links in Figure 7.9a. The shortest-path distance between the nodes source 0 and destination u (i.e., $d(0,u)$). The dashed line indicates the shortest-path tree starting at node 0, from which we can determine

Algorithm: One-Step Algorithm

We consider graph G having source node s and destination node d. The primary and protection paths between nodes s and d are calculated **as** follows:

1. Estimate the shortest-path tree rooted at node s using Dijkstra's shortest-path algorithm. The $d(s, u)$ represents the shortest-path distance from node s to node u.

2. Change the original graph G to an auxiliary graph GA **as** follows.
   Nodes and links are kept untouched.
   - The cost of each link (u, v), $c'(u,v) = cI(u,v) + d(s,u) - d(s,v)$, where $ci(u,v)$ denotes the cost of link (u,v) in graph G_I and $c(u, v)$ denotes the cost of link (u, v) in graph G.
   - Reverse the directions of the links along the shortest path from node s to node d.
3. Estimate the shortest path from node s to node d in graph G_I.
4. The shortest path between nodes s and d in G (GA) is represented as T (T_I). After removing the links appearing in both T and T (in opposite direction), all the other link sin T and T' form a cycle when ignoring their directions.

There is another one-step algorithm developed [2] where the graph transformation reverses the directions of the links along the shortest path from the source to the destination node in graph G, by using a modified Dijkstra's algorithm where the shortest-path problem can be resolved when a graph has some links with negative cost.

Figure 7.9 presents sample network to describe how the one-step algorithm works, in which a pair of link-disjoint paths from node 0 to node 5 is required to be determined. In Figure 7.9a, link cost is mentioned beside and the number near to a node u represents the shortest-path distance between the nodes 0 and u (i.e., d (0, u)). Two link-disjoint paths between nodes s and d can be found from the cycle that the shortest path from node 0 to node 5 is $T = (0,1,4,5)$. After graph transformation, the shortest path between nodes 0 and 5 is $T' = (0,3,4,1,2,5)$, as shown in Figure 7.9b. After removing the interlacing link (1,4), Figure 7.9c shows the two link-disjoint paths (0,3,4,5) and (0,1,2,5). The dashed line provides the shortest-path tree rooted at node 0, from which the shortest path from node 0 to node 5 is determined as $T = (0,1,4,5)$. The overall computational complexity for the one-step algorithm is $O(N^2)$ (where N = number of nodes).

7.5.2   ILP OF S-RWA FOR STATIC TRAFFIC DEMANDS

Like different S-RWA heuristics [26], linear program-based approaches deal with the problem formulation where a set of candidate routes are estimated to obtain a pair of primary backup paths on the basis of current network state by optimizing the total amount of bandwidth used on all the links for a given set of static traffic demands in an off-line manner [2]. Different approximation schemes have been taken for ILP-based approaches suited for use in a practical network with a reasonable volume of traffic demands [28].

Here, the ILP formulations of path protection schemes are made to protect against single-link failures. The demand matrix has the number of connections to be established between each node pair, and the set of alternate routes up to three ($k = 3$) that are used to satisfy any demand between each node pair are pre-estimated.

The minimization of the total number of wavelengths used on all the links is made in the network considering both the primary paths and backup paths. The ILP solution also determines the routing and wavelength assignment (RWA) of the primary and backup paths.

We consider the following:

- The network topology considered to be a directed graph G.
- A demand matrix, i.e., the number of light path requests between node pairs.
- Alternate routing tables at each node.

The parameters are as follows:

N: Number of nodes in the network (numbered 1 through N). There are node pairs numbered 1 through $N \times (N - 1)/2$.

E: Number of links in the network (numbered 1 through E).

W: Number of wavelengths on a link.

R_i: Set of alternate routes for node pair i.

$M_i = |R_i|$: Number of alternate routes between node pair i, and M is the maximum number of alternate routes between any node pair, i.e., $M = \max_i M_i$.

R_j^i: Set of eligible alternate routes between node pair i after link j fails.

End nodes (j): Set of alternate routes between the node pair adjacent to link j.

d_i: Demand for node pair i, in terms of number of connection requests. (Each connection requires the bandwidth of a full-wavelength channel.)

Apart from the above parameters, the following variables have been considered for ILP formulation [2]:

w_j: Number of wavelengths used by primary light paths on link j.

s_j: Number of *spare* wavelengths used on link j.

$\gamma_w^{i,r}$ takes on the value of 1 if the *r*th route between node pair i utilizes wavelength w before any link failure; 0 otherwise. These variables are employed in both ILPs.

$\alpha_{w,p}^{i,b}$ takes on the value of 1 if the dedicated backup route b on wavelength w is employed for protecting a primary route p between node pair i; 0 otherwise. These variables are employed only in ILP formulation.

$\delta_{w,p}^{i,b}$ takes on the value of 1 if the shared backup route b on wavelength w is employed for protecting a primary route p between node pair i; 0 otherwise. These variables are employed only in ILP2.

m_w^j takes on the value of 1 if wavelength w is utilized by some backup route r that traverses link j; 0 otherwise. These variables are employed only in ILP2.

7.5.2.1 ILP1: Dedicated Path Protection [2]

Minimize the total capacity used:

$$\text{Minimize: } \sum_{j=1}^{E} \left(w_j + s_j \right) \tag{7.1}$$

Number of light paths on each link is bounded:

$$\left(w_j + s_j \right) \leq W \quad 1 \leq j \leq E \tag{7.2}$$

Demand between each node pair i is satisfied:

$$d_i = \sum_{r=1}^{M_i} \sum_{w=1}^{W} \gamma_w^{i,r} \quad 1 \le i \le N(N-1) \tag{7.3}$$

Number of primary light paths traversing link j:

$$w_j = \sum_{i=1}^{N(N-1)} \sum_{r \in R_i, j \in r} \sum_{w=1}^{W} \gamma_w^{i,r} \quad 1 \le j \le E \tag{7.4}$$

Number of spare channels utilized for link j:

$$s_j = \sum_{i=1}^{N(N-1)} \sum_{b \in R_i, j \in b} \sum_{p \in R_i, p \ne w} \sum_{w=1}^{W} \alpha_{w,p}^{i,b} \quad 1 \le j \le E \tag{7.5}$$

Wavelength-continuity constraint, i.e., only one primary or backup light path can use wavelength w on link j:

$$\sum_{i=1}^{N(N-1)} \sum_{r \in R_i, j \in r} \gamma_w^{i,r} + \sum_{i=1}^{N(N-1)} \sum_{b \in R_i, j \in b} \sum_{p \in R_i, p \ne b} \alpha_{w,p}^{i,b} \le 1 \tag{7.6}$$

where $1 \le j \le E, 1 \le w \le W$

Due to a link failure, if route p fails between node pair i, then the demand between node pair i should still be satisfied:

$$\sum_{w=1}^{W} \gamma_w^{i,r} = \sum_{b \in R_i, p \ne b} \sum_{w=1}^{W} \alpha_{w,b}^{i,b} \tag{7.7}$$

where $1 \le i \le N(N-1), p \in R_i$

7.5.2.2   ILP2: Shared-Path Protection [2]

Minimize the total capacity used:

$$\text{Minimize:} \sum_{j=1}^{E} \left(w_j + s_j \right) \tag{7.8}$$

Number of light paths on each link is bounded:

$$\left(w_j + s_j \right) \le W \quad 1 \le j \le E \tag{7.9}$$

Demand between each node pair i is as follows:

$$d_i = \sum_{r=1}^{M_i} \sum_{w=1}^{W} \gamma_w^{i,r} \quad 1 \le i \le N(N-1) \tag{7.10}$$

Numbers of primary light paths traversing link j are

$$w_j = \sum_{i=1}^{N(N-1)} \sum_{r \in R_i, j \in r} \sum_{w=1}^{W} \gamma_w^{i,r} \qquad 1 \le j \le E \qquad (7.11)$$

The spare channels for link j:

$$s_k = \sum_{w=1}^{W} m_w^k \quad 1 \le k \le E \qquad (7.12)$$

Constraints whether wavelength w is reserved for some restoration path on link k:

$$m_k^w \le \sum_{i=1}^{N(N-1)} \sum_{p,b \in R_i, k \in b} \delta_{w,p}^{i,b}, \quad \text{where,} 1 \le j \le E, 1 \le w \le W$$

$$(7.13)$$

$$N(N-1) \times E \times M \times m_k^w \ge \sum_{i=1}^{N(N-1)} \sum_{p,b \in R_i, k \in b} \delta_{w,p}^{i,b}, \quad 1 \le j \le E, 1 \le w \le W$$

Wavelength-continuity constraint, i.e., only one primary or backup light path can use wavelength w on link j, is written as:

$$\sum_{i=1}^{N(N-1)} \sum_{p,b \in R_i, k \in b} \gamma_w^{i,r} + m_w^j \le 1, \quad 1 \le j \le E, 1 \le w \le W \qquad (7.14)$$

Constraints that two backup light paths can share wavelength w on the link k if corresponding primary paths are fiber disjoint

$$\sum_{i=1}^{N(N-1)} \sum_{p,\in R_i, f \in p} \sum_{b \in R_i, k \in b} \delta_{w,p}^{i,b} \le 1, \quad 1 \le f \le E, 1 \le j \le E, 1 \le w \le W \qquad (7.15)$$

Constraints ensure that every primary light path is protected by a backup light path:

$$\sum_{w=1}^{W} \gamma_w^{i,r} = \sum_{b \in R_i, b \neq p} \sum_{w=1}^{W} \delta_{w,p}^{i,b}$$

$$(7.16)$$

$$\text{where } 1 \le i \le N(N-1), \forall p \in R_i, 1 \le w \le W$$

7.5.3 Maximizing Share Ability for Shared-Protection Schemes

WDM mesh networks are more well organized than SONET-based interconnected ring networks in terms of protection as it supports differentiated protection scheme. In a shared-protection scheme, the backup path is shared between different connections. One connection traffic transmits from the primary path to the backup path when a network failure occurs. To enhance e resource use, there are reports on shared-protection scheme in WDM mesh networks in order to optimize network resource efficiency [29]. Recently, many approaches are proposed to handle multiple link failures [25,30,31].

7.5.3.1   Backup Route Optimization

Planning of primary path and their backup paths with sharing is an approach for protected optical backbone [2]. An alternative is to fix the primary path on the basis of current network state first before obtaining the backup route for a connection request. This is implemented by adjusting link costs based on current resource usage information of network links. p represents the number of wavelength channels allotted for the primary paths (connections) on link j; b_j represents the number of wavelength channels allotted on link j to protect failure of link i ($1 \le i \le N$, and N is the number of links in the network) when link i fails, and b is the total number of allocated spare wavelength channels for protection purposes. Under the assumptions that there is a single-link failure at any time and shared-path protection is employed in the network, b is equal to the maximal value of b_i. The primary path of a connection transmits the signal through links $l_{m1}, l_{m2}, ..., l_{mn}$. The link cost of l_j is written as [2]:

$$\text{Cost}\left(l_j\right) = \infty \text{ if } l_j \text{ is on the primary path,}$$

$$= 0 \text{ if } b_{m1} \le b,... \text{ and } b_{mn} \le b, \qquad (7.16)$$

$$= 1 \text{ otherwise.}$$

Using this link-cost adjustment function, the link cost is zero if none w wavelength channel is required to be allocated; otherwise, the link cost is set to 1. After assignment of link cost to each network link, the backup path is determined by using any shortest-path algorithm such as Dijkstra's algorithm.

Figure 7.10 represents how to estimate a backup path optimizing resources h are ability after the primary path is known. Figure 7.10a represents the state of a part of a network where only a few network nodes and links are shown. There is a connection request between node pair (s,d), whose primary path traverse's links l_1, l_2, and l_3 as the solid. The candidate backup path is found for the connection request that either traverses links 14 and 15 or link 16. Considering enough wavelength channels on each link, Figure 7.10b shows the link vector for each network link. Figure 7.10c shows the network state with adjusted link cost from which are source–shareability–optimized back up route [2] is estimated for the given primary path using a standard shortest-path algorithm. Figure 7.10d illustrates the updated link information after getting both primary and backup paths for the connection request.

7.5.3.2   Physical Constraint on Backup Route Optimization

The backup path selection is an issue. There is a backup path having long (hop) distances of connections even though the length of primary path is short. The long backup path exhibits a signal-quality degradation at the destination, especially in an all-optical WDM network. An optical signal traversing a long distance provides more BER at the destination node, which is not tolerable for services at upper network layers. When a network failure occurs, a preplanned backup path restores the

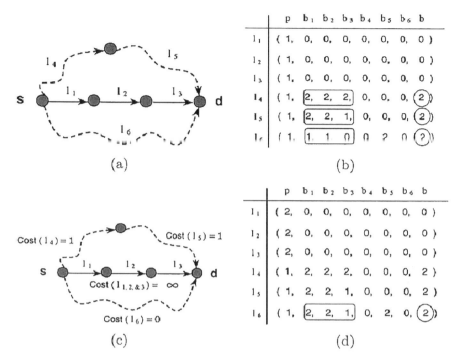

FIGURE 7.10 An example of backup path optimization [2] (a) state of a part of a network (b) link vector (c) network state with adjusted link (d) updated link information.

affected connections in which the unexpected long backup paths can potentially degrade signal quality. The authors in Ref. [31] have proposed an ILP-based model to jointly calculate the shared-protected primary–backup path pair for dynamic traffic.

7.6 DYNAMIC RESTORATION

The traffic restoration schemes are used for survivability of network as an important area of research interest. The following performance evaluates a restoration scheme [2]:

RSR is the ratio between the number of successfully restored connections and the number of affected connections after a network failure happens.
RT is the average time needed to successfully restore a disrupted connection request.

Different considerations for restoration schemes in optical WDM mesh networks are given below [2].

- **Distributed Control:** In a distributed control, the source node of each interrupted connection restores the service following either a pre-estimated route or a dynamically estimated route. As the connections are restored in a distributed manner, resource contention may occur at some network link.

- **Centralized Control:** In a centralized control system, failure-affected connections will be restored one by one so that resource contention is avoided but this scheme may degrade the RT performance of some connections. A centralized controlled restoration scheme may get higher RSR than that for distributed control, as it can carry out global optimization of network resource usage.
- **Preplanned Restoration Routes vs. Online Dynamically Computed Restoration Routes:** In a distributed control system, the restoration routes can be preplanned or dynamically computed. In a preplanned scheme, a candidate restoration route set is evaluated for each connection. When a connection fails, the restoration path is selected without online computation. RT is increased for this performance. The route set may be periodically updated in order to improve the successful restoration.
- **Path vs Sub-path vs Link Restoration:** The restoration schemes are of different types: path-based, sub-path-based, and link-based schemes, according to which anode initializes the alternate path and how the new path is routed to bypass the failed link. The work [2] shows the performance trade-off of these different restoration mechanisms under a distributed control and signaling system.

7.7   OTHER NETWORK SURVIVABILITY ISSUES

More attention is set to a service view point [32–35]. A certain quality of service (QoS) per a customer's needs and service quality provisioning are critical issues in optical backbone. AWDM mesh network is used as an optical backbone to most of the network. The QoS for the services such as online trading, military applications, and banking services along with stringent reliability is diverse due to their diverse requirements. Service quality can be obtained in different ways such as signal quality, service availability, service reliability, RT, and service restorability. Signal quality is basically measured by optical signal-to-noise ratio (OSNR), BER, etc. in optical network and is affected during transmission.

7.7.1   SERVICE AVAILABILITY

A protection scheme provides a connection's availability after traffic on the failed primary path, as it is transferred to the backup path where a path-protected connection is 100% available in the presence of any single failure.

A path-protected connection may be unsuccessful in some multiple failure scenarios where two concurrent failures occur: one on the backup path and the other on the primary path. The connection availability again depends on the details of the failures (locations, repair times, etc.) and reserved backup resources. We need to have a relatively accurate estimation of the connection availability.

Availability is dependent on a Service-Level Agreement (SLA) [36]. The SLA is a contract between the network operator and a customer. ASLA violation provides penalty to the network operator, according to the contract which may be provisioning of free services for one additional period of time. A cost-effective, availability-aware, connection-provisioning scheme is very much desirable for each customer's service request (static or dynamic) [2].

A network component's availability is a static quantity depending on the component's failure rate and average time to fix a failure [2]. Then, in online provisioning, one of the routes will carry a new connection as long as its availability is larger than that required by the customer.

A connection availability is a dynamic quantity as and when the connection (t) is sharing some wavelengths on its backup path with others. Normally, we consider S_t containing all the connections that share some backup wavelength on some link with t. The various backup sharing-related operational decisions also change the connection availability. The connection t's traffic can be transferred to its primary path after the failure of the primary path is repaired and known as *reverting*; however, the traffic can stay on the backup path for the remaining service time, which is known as *non-reverting*.

7.7.2 Availability Study

For the improvement of QoSs, an availability analysis is needed for a connection with different protection schemes in a WDM mesh network [2]. For analysis, the following typical assumptions are considered:

- A system is either functional or nonfunctional while failure occurs.
- Different network components fail independently.
- The "up" time (i.e., time to failure) and the "down" time (i.e., time to repair) of components are exponentially distributed stochastic variables represented as mean values: mean time to failure (MTTF) and mean time to repair (MTTR), respectively.

The availability of a system is a fraction of time that the system is "up" or functional during the entire service time. If a connection t is carried by a single path, its availability (A_t) is the path availability and t is dedicated or shared protected. A_t is estimated by both the primary and backup paths.

7.7.2.1 Network Component Availability

The component's availability is determined by considering its failure characteristics. The failed component is needed to be repaired /restored. This procedure is called as an alternating renewal process. The availability of a network component j (denoted as a_j) is formulated as [37]

$$a_j = \frac{MTTF}{MTTF + MTTR} \tag{7.17}$$

Where *MTTF* of a fiber link is derived according to measured fiber-cut statistics. *MTTR* is the mean time to repair, which is normally considered to be 2 hours for equipment and 12 hours for cable cut.

7.7.2.2   End-to-End Path Availability

The availability of ith path (A_i) is estimated by considering the known availabilities of the network components on this path. Path i is only used when all the network components along its route are free. a_j is indicated as the availability of network component j. G_i is represented as a set of network components used by path i. Then, A_i is written as:

$$A_i = \prod_{j \in G_i} a_j \qquad (7.18)$$

7.7.2.3   Availability of Dedicated Path-Protected Connection

The connection t is set up by one primary path p and provisioned by one backup path b that is link disjoint with primary path p disrupted. The path of the traffic is changed to backup path, and the receiver receives the signal from path p to path b as long as b is accessible; otherwise, the connection becomes occupied until the failed component is repaired/restored [38]. A_t is estimated by using the following formula:

$$A_t = 1 - (1 - A_p)(1 - A_b) = A_p + (1 - A_p)A_b \qquad (7.19)$$

Where A_p and A_b represent the availabilities of primary path p and backup path b. The connection uses multiple backup paths to increase its availability. In sub-path protection [21,39], the primary path is usually divided into non overlapping segments, and each segment is protected by a link-disjoint backup route.

7.7.2.4   Availability in Backup Sharing

There are various problems or restrictions in backup sharing affecting the availability of a shared-path-protected connection [2,40].

7.7.2.4.1   Reverting vs. Non-reverting

In shared-path protection, connection t is set up by the primary path p and the protection is provisioned by a link-disjoint backup path b, and the protected wavelength on each link of b is also shared by other connections where SRLG constraints are satisfied. We take S_t (sharing group of t) having all the connections that share some backup wavelength on some link with connection. There are two approaches for recovery of failure: reverting and non-reverting. In reverting case, connection t's traffic will be accommodated by backup path b when a failure occurs on p, and after the failure is repaired, connection t's traffic can be accommodated back by the primary path p. In case of non-reverting, after repairing of failures, it can stay on b for the remaining service time (or till b fails). There are advantages and disadvantages of the reverting and non-reverting strategies. The traffic flow will be troubled twice at least in the reverting strategy, which may be undesirable for some services such as online trading/banking. The non-reverting may provide un preferred service degradation as same backup path is shared by many users and connections sharing same backup path may find disturbed in backup re computation and backup path resource processes.

7.7.2.4.2 *Active Recovery vs. Lazy Recovery*

In the reverting model, after reverting back to primary path, the backup resources on backup path *b* are accessible to recover the other failed traffics in S_t. This process is active recovery [2]. If the backup resources stay to be activated when the next failure arrives, these currently failed connections cannot be recovered even though their backup path is accessible. This process is lazy recovery [2].

In active recovery, the backup resources on released path *b* recover more than one connection. When a connection makes the backup resource available after repairing and for more failed connections waiting for the backup resources, problem is which connection should be chosen to recover next.

The availability of connection $t(A_t)$ is a function of the size of S_t and the availabilities of the connections in S_t. When one or more connections in S_t are having connected at same time with *t*, either *t* or some of the failing connections in S_t acquire the backup wavelengths. The conditional probability is needed to compute and represented as ∂_k that *t* obtains the backup wavelengths when *k* connections in S_t be unsuccessful to communicate simultaneously with *t*. A continuous-time Markov chain is used to compute ∂_k, as shown in Figure 7.11a, which is basically the corresponding state-transition diagram (when $k = 1$) for the Markov chain having employment of an active recovery. t_1 represents the other connections sharing backup resources with *t*. The label for each state in Figure 7.11 represents three tuples (x,y,z), where *x* and *y* indicate the state of the primary paths of connections *t* and t_1, respectively, and *z* shows which connection the backup path is recovering. Tuples *x* and *y* could be "Up" (U) or "Down" (D), and *z* could be "None" (0), "*t*", or "t_1"; we obtain $MTTF = 1/\lambda$ and $MTTR = 1/\mu (MTTF1 = 1/\lambda_1$ and $MTTR1 = 1/\mu_1)$ as the mean failure parameters for connection $t(t_1)$.

$$\partial_k = \frac{\mu_1}{\mu_1 + \mu} \tag{7.20}$$

The value of ∂_k depends on repair rates of the concurrently failed connections (but not their failure rates)! If all the connections have the same repair rate, then

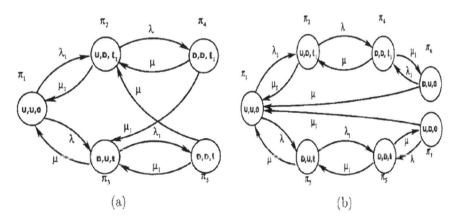

(a)   (b)

FIGURE 7.11 Markov chain representing state transition of the network system [2] (a) state transition diagram (b) Markov chain to compute lazy-recovery policy.

we get $\partial_k = 1$ from the above equation. Similarly, we can compute the conditional probability for one connection to acquire the backup wavelengths when k connections in S_t are failed and the statistics to repair a failed link will be independent of the link; i.e., it will be the same network-wide. The same approach is used to derive ∂_k for lazy recovery even though the Markov chain is different. Figure 7.11b represents the corresponding Markov chain to compute ∂_k with the lazy-recovery policy. With the value of ∂_k, the availability of a shared-path-protected connection t is estimated. Connection t will be available if:

1. Path p is accessible.
2. p is engaged, b is obtainable, and p can get the backup wavelengths when other connections in the sharing group S_t have also failed.

Therefore, A_t can be derived as

$$A_t = 1 - \left(1 - A_p\right)\left(1 - A_b\right) = A_p + \left(1 - A_p\right)A_b$$

where A_p and A_b denote the availability of p and b, respectively, and N is the size of $S_t . \partial_k$ is a probability that the connection t can get the backup resources when both p and the other primary paths in S_t fail and p_k is a probability that exactly primary paths in S_t are unavailable. All possible k connection failures are taken to estimate p_k in which the probability of k simultaneous failures decreases drastically as k increases.

7.8 DYNAMIC ROUTING AND WAVELENGTH ASSIGNMENT UNDER PROTECTION

Dynamic RWA under protection is very difficult task. There are different strategies reported by different authors. Very simple strategy is hybrid type, which is as follows: RWA is dynamic one and protection made is static one. In this section, we have discussed about such strategy with alternate path routing, the use of wavelength converter, and traffic grooming [2].

7.8.1 PROTECTION SCHEMES IN ALTERNATE PATH ROUTING AND WAVELENGTH ASSIGNMENT

There are two protection schemes used: shared protection and restricted shared protection. These protections are employed on NSFNET $T1$ backbone having 14 nodes and 21 physical links, as shown in Figure 7.12. To decrease the cost of network, the use of wavelength converter in the node is avoided. Here, we have discussed both the shared-protection and restricted shared-protection schemes [41–43].

7.8.1.1 Shared protection

Figure 7.13a represents the protection path tree based on existing shared protection for source node (2) to other destination nodes. The wavelength-1 is assigned for source (2) to other destination [43]. Paths 2-1-8-9-12-14, 2-1-8-9-12-11, 2-1-8-9-10,

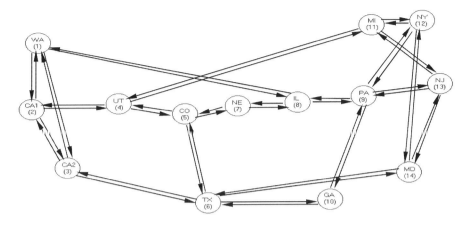

FIGURE 7.12   NSFNET *T1* backbone having 14 nodes and 21 physical links [2].

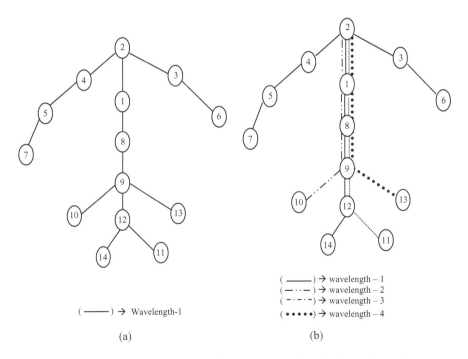

FIGURE 7.13   Protection tree having node 2 as source node and other nodes as destination node in NSFNET: (a) shared protection and (b) restricted shared protection [43].

and 2-1-8-9-13 share the same wavelength-1. Paths 2-4-5-7, 4-5-7, 2-4-5, and 4-5 also share thewavelength-1. Similarly, for other source nodes to their destinations, shared-protection path trees are made. The protection is not reliable against occurrence of simultaneous failures in sub-branching of the tree (2-1-8-9-12-14, 2-1-8-9-12-11, 2-1-8-9-10, and 2-1-8-9-13).

7.8.1.2 Restricted Shared Protection

For more reliability against simultaneous failures, we restrict sharing of the sub-branching paths of the tree [43]. Figure 2b represents the restricted sharing protection tree of source node 2 and other nodes as destinations. It is seen from the figure that 2-1-8-9-12-14, 2-1-8-9-12-11, 2-1-8-9-10, and 2-1-8-9-13 routes are not shared by same wavelength and so are assigned by wavelength-1, wavelength-2, wavelength-3, and wavelength-4, respectively. Like Figure 2a, same wavelength-1 is shared by 2-4-5-7, 4-5-7, 2-4-5, and 4-5 routes. So shared protection requires single wavelength, as shown in Figure 7.13a, whereas in case of restricted shared protection, four wavelengths are required for backup paths of node 2 as a source and other nodes as destination, as shown in Figure 7.13b. Similarly, we have made restricted protection trees for other sources and their destinations. Table 7.1 shows the protection trees

---

TABLE 7.1 [43]

Protection Trees for Other Source Nodes (N_R = Number of Wavelengths Used in Restricted Shared Protection)

Source	N_R	Protection Tree
1	4	2-4-5-7 13 11 / 1-8-9-12-5-7 / 3-6 10 14
3	4	6 / 13 14 / 3-2-1-8-9-12 / 4-5-7 10 11
4	5	3-6 / 13 14 / 4-2-1-8-9-12 / 5-7 10 11
5	5	3-6 / 13 14 / 5-4-2-1-8-9-12 / 7 10 11
6	5	4-5-7 13 14 / 6-3-2-1-8-9-12 / 10 11
7	5	13 14 / 7-5-4-2-1-8-9-12 / 3-6 10 11
8	4	13 14 / 8-9-12 11 / 10 7-6 / 1-2-4-5-7
9	2	13 14 / 9-12 11 / 10 3-6 / 8-1-2-4-5-7

(Continued)

TABLE 7.1 [43] (*Continued*)
Protection Trees for Other Source Nodes (N_R = Number of Wavelengths Used in Restricted Shared Protection)

Source	N_R	Protection Tree
10	5	13 14 / 10 –9 –12 +1 / 8 –1– 2 – 4 –5 – 7 / 3– 6
11	5	14 13 / 11–12 –9 +0 / 8 -1– 2 – 4 –5 – 7 / 3– 6
12	4	14 13 / 12 –9 +0 / 1 / 8 –1– 2 – 4 –5 – 7 / 3– 6
13	5	12 – 14 / 13–9 +0 1 / 1 / 8 -1– 2 – 4 –5 – 7 / 3– 6
14	5	11 / 14–12–9 – 10 / 13 / 8 -1– 2 – 4 –5 – 7 / 3– 6

that are developed by using $k+1$ alternate path routing ($k = 2$). These protection trees for different nodes are used for shared and restricted shared-protection paths. In case of restricted sharing of all nodes, different branches of shared-protection path are restricted and assigned with different wavelengths as discussed for node 2 protection trees.

When paths are established dynamically, WA decisions must be taken on the basis of link state information, which is updated periodically through the link state routing protocol. To formulate the problem, physical topology of the network has been modeled as a unidirectional graph $G(V, E)$, where V is the set of nodes and E is the set of links between nodes in network. As mentioned earlier, we consider each link consisting of two unidirectional fibers (two fibers can carry same wavelengths in reverse direction). The formulation of RWA related to objective function, traffic flow constraint, and wavelength constraint is already discussed in Chapter 12.

After finding a route from the routing table (which is updated periodically), a wavelength reservation mechanism is required to assign a wavelength for the route. The previous studies show that there are two types of wavelength reservation mechanism: parallel reservation [6] and hop-by-hop basis reservation [6]. It is reported [6] that the set time for parallel reservation is less than that for hop-by-hop reservation. So, we have used parallel reservation. In this case, after selection of the route, the source attempts to reserve the wavelength by sending a reservation request to each node of the route. The nodes of the route send positive and negative acknowledgments of reservation back to the source. After receiving the positive acknowledgment, the light path is established for the entire session.

The algorithm is based on alternate path routing (already discussed in Chapter 11) in which k number of paths are computed for the entire session of a connection request. Each node in the network has traffic and network information, which is updated on the basis of link state information periodically. The algorithm procedure is given below [43]:

1. Initialize with network configuration, $G(V, E, W)$.
2. If a connection request $c(s, d, t_H)$ comes, compute k number routes (R_i) on the basis of link state information in descending order of time delay T_d (where $I = 1, 2, 3,...k$ correspond to 1st, 2nd,... kth shortest time delay path).
3. For $i = 1$ to k,
   Try to assign the wavelength to the connection in ith shortest path within holding time t_H using the constraints from (17) to (20). If it is assigned, go to step 4; otherwise, it is treated as a blocked connection.
4. Set up backup path using protection trees (Figure 7.13 and Table 7.1) for the assigned connection using $(k + 1)$th alternate path.

7.8.1.2.1   Computational Complexity

The time complexity of the algorithm is written as $O(E \cdot W + n^{3.5})$, where $n = K \cdot W$ and $K =$ number of distinct paths, and N c total number of nodes in the network. The complexities of steps 1 and 2 are $(E \cdot W)$ and $O(1)$, respectively. The complexities of steps 3 and 4 are written as $O(n^{3.5})$ and $O(W^{3.5})$, respectively [18]. The reliability of both restricted shared and shared protection has been discussed in next section of this chapter.

7.8.2   ROUTING AND WAVELENGTH ASSIGNMENT BASED ON WAVELENGTH CONVERTER UNDER PROTECTION

The use of wavelength converter in each node network is costly [2,41]. Moreover, multiple conversion on a light path in the network exhibits signal degradation. The partial placement of wavelength converter is used for the improvement of performance. The problem of partial placement of wavelength converter is resolved to reduce the blocking probability. The network with wavelength converters corresponds to a directed graph, $G(V, E, C)$, where $V =$ set of nodes, $E =$ set of links in the network, and $C =$ set of wavelength converters placed in the network. The nodes of the network are numbered as $1, 2, ...N$.

- $e_{i,j} =$ link from node i to node j.
- $W =$ total number of wavelengths per fiber link.
- $W_{i,j} =$ number of free wavelengths available after protection of connection in a link $e_{i,j}$.
- $\lambda_{i,j} =$ the amount of traffic per second passing through link $e_{i,j}$.

- $\rho_{i,j}$ = the probability that a given wavelength in a link $e_{i,j}$ is occupied.
- X = the status vector for the placement of converter in the nodes and represented by

$\{x_1, x_2, \ldots x_i, \ldots x_N\}$, where $x_1, x_2, \ldots x_i \ldots x_N$ correspond to node 1, node 2,... node i ... node N, respectively. In general, x_i is written as:

$$x_i = 1, \quad \text{if converter is used at node,} i.$$

$$= 0, \text{otherwise}$$

The variable x_i is estimated from the placement of converters in small fraction of nodes in such a way that the overall blocking probability is minimized. The partial placement of wavelength converter in the network is represented by status vector:

$$\{x_i\} = \{x_1, x_2, \ldots x_N\} \tag{7.21}$$

where $x_1, x_2, \ldots X_n$ correspond to wavelength-conversion status for node 1, node 2 .... node, respectively. The status of node i is defined as

$$x_i = 1, \quad \text{if wavelength converter is placed at node} i.$$

$$= 0, \text{otherwise}$$

Objective:

$$\text{Maximize: } \sum_{i=1}^{N} U_i \cdot x_i \tag{7.22}$$

Where U_i is the utilization factor that is defined as the number of times converter at node i used for the connection of the SD pairs.

    Constraint:

$$\sum_{i=1}^{N} x_i = C \tag{7.23}$$

Where C = set of converters placed in the network

$$C \leq N \tag{7.24}$$

Each connection request is specified by source and destination information and its bandwidth. Here, working light path is established (i.e., the path carries the traffic during normal operation) for a connection between source and destination using Dijkstra's algorithm and also backup light path for protection. Other source and destination (SD) pairs may also share the backup paths if the links of backup paths are common.

We require the following integers for the solution of the protection problem.

- $W_{e_{i,j}}$ = the number of wavelengths used in a link $e_{i,j}$ for backup paths of the connections.
- $P_{ij}^{l,w,b}$ = integer that takes a value 1 if the link ij of the route b is assigned a wavelength and w for restricted shared protection of the connection; otherwise, it is zero.

The RWA problem of restricted shared-protected optical network is formulated as follows [41]:

Objective:

$$\text{Minimize:} \sum W_{e_{i,j}} \tag{7.25}$$

Constraints:

$$\left(W_{e_{i,j}} + W_{i,j}\right) \le W \tag{7.26}$$

$L = N(N-1)$, where N = total number of nodes and

$$\sum_{j:(ij \in E)} P_{ij}^{l,w,b} = \sum_{j:(ji \in E)} P_{ji}^{l,w,b}, \quad i \ne s,d, \forall (w \in W) \text{ and } (i,j \in V). \tag{7.27}$$

$$\sum_{w \in W} \sum_{j:(ij \in E)} P_{ij}^{l,w,b} - \sum_{w \in W} \sum_{j:(ji \in E)} P_{ji}^{l,w,b} = 1, \quad i = s \tag{7.28}$$

$$\sum_{w \in W} \sum_{j:(ij \in E)} P_{ij}^{l,w,b} - \sum_{w \in W} \sum_{j:(ji \in E)} P_{ji}^{l,w,b} = 1, \quad i = d \tag{7.29}$$

Equation (7.26) shows the wavelength-continuity constraint, whereas equations (7.27)–(7.29) represent the flow balance equation of shared backup paths. The other constraints such as traffic flow and wavelength constraints are already discussed in Chapter 12.

For an efficient utilization of network resources, the RWA algorithm needs to serve maximum connection requests. For this purpose, we have proposed the algorithm-based alternate path routing with optimal placement of converters. In this algorithm, each node in the network is required to keep record of link state information, which is updated periodically. On the basis of link state information, ordered lists of m number of routes are selected for connection request of an SD pair for entire session. The proposed algorithm is described below:

When a connection request is generated, the following steps are used [44]:

1. Estimate m routes (R_i) (on the basis of link state information) in descending order of time delay (where, $i = 1,2,3,...m$ correspond to 1st, 2nd,... m^{th} shortest time delay path).

2. Choose first shortest time delay path ($i = 1$) and allot the wavelength using parallel reservation scheme [5]. If the wavelength is not accessible, CONV_PLACE algorithm is employed for the establishment of connection of the SD pair. If it is assigned, set up protection path, or else go to the next step.

3. Assign wavelength for next alternative path in descending order of time delay. If it is not allotted to wavelength, then we place converters in same positions as that in the step 2 for assigning a wavelength to the connection. If it is assigned, set up protection path using protection tree and equations (3)–(5), or else go to the next.

4. If it is not allotted wavelength, repeat step 4 up to ($m - 2$) times, or else go to next step.

5. Connection is blocked, and backup path for the connection is rejected.

CONV_PLACE algorithm:

We put the converter properly in such a way that the blocking probability should be reduced. The procedure of CONV_PLACE algorithm is given below [44]:

1. Fix the number of converter (C).
2. Select a particular converter placement $\left\{ x^k{}_i \right\} = \left\{ x_1^k, x_2^k, \ldots x_N^k \right\}$ and determine $\sum_{i=1}^{N} U_i^k \cdot x_i^k$, where U_i^k is the utilization factor of ith node in kth placement.

3. Repeat step 2 till maximizing $\sum_{i=1}^{N} U_i^k \cdot x_i^k$.
4. End.

7.8.3 Traffic Grooming-Based RWA under Protection Tree

Our proposed approaches are used to reduce the number of wavelengths needed to accommodate the maximum number of connection requests. It has three phases: working path selection, traffic grooming in working paths, and restricted shared backup paths [10] for protection of connections. The working paths of the SD pairs are developed using K-shortest-path approach. As discussed in Section 6.7, here, we have used both dedicated wavelength grooming (DWG) and shared wavelength grooming (SWG) together. For traffic grooming, we have proposed two types of approaches: destination shared wavelength grooming (DES_SWG) and source shared wavelength grooming (SOURCE_SWG) [42]. In case of DES_SWG, the connections of same destination and different sources are groomed along a routing path, whereas for SOURCE_SWG, the connections of same source and different destinations are groomed along a routing path. As for an example shown in Figure 7.14a, traffic grooming of SOURCE_SWG is made between the connections of 1-2 and 1-6 pairs on the same path <1-2-6> as indicted by the dashed line. The backup paths are formed for survivability of traffic grooming network. In SOURCE_SWG, the backup paths for 1-2 and 1-6 are <1-3-2> and<1-3-5-6>, respectively, which are indicated by the solid line in Figure 7.14a.

In case of DES_SWG, traffic grooming is made between the connections of 1-6 and 2-6 SD pairs on the path <1-2-6>, as shown in Figure 7.14b. Like SOURCE_SWG,

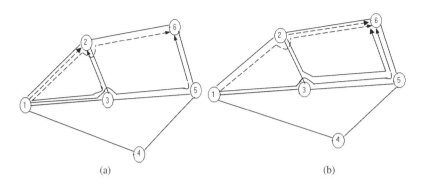

(a)   (b)

FIGURE 7.14 Example network showing (a) SOURCE_SWG and (b) DES_SWG algorithm [42].

in DES_SWG, the backup paths are required for each groomed connection. The backup paths for 1-6 and 2-6 are <1-3-5-6> and <2-3-5-6>, respectively, as indicated by the solid line in Figure 7.14b.

7.8.3.1   Problem Formulation

The physical topology of WDM network with traffic grooming is written as an undirected graph, $G_T(V, E, W)$, where V is the set of nodes and E is the set of links between the nodes in the network. We consider that each link carries the same number of wavelength (W) and each wavelength channel has g connection slot, where each connection slot is represented as minimum bandwidth of STS-4 and g is also called as the grooming factor. Each connection is specified by the source node (s), the destination node (d), bandwidth requirement (B), and holding time (t_H), and in short form, it is written as $c(s, d, B, t_H)$. To solve the shared-protected traffic-groomed problem in the network carrying all traffic requests of all possible SD pairs of the protected network simultaneously, the number of wavelength channels per fiber is minimized [42].

The following notations of the parameters are used in formulation.

- i, j represents the end points (i.e., nodes) of physical link that is along the route of a connection.
- w represents the index for the wavelength number of link e, where $w = 1,2,3, \dots W$.
- l represents the index of a SD pair/connection where $l = 1, 2, 3, \dots L$; L represents the total number of possible SD pair.
- c is an index of a connection where $c = 1, 2, 3, \dots t$; t represents the total number of connections of a SD pair.
- e is an index for link number where $e = 1,2,3, \dots E$.

We require the following integers for the solution of the problem.

- $SW_{c,w,d}^R$ is an integer that takes a value 1 if the connection c to destination node d in a groomed path R is assigned by a wavelength w; otherwise, it is zero, where d is the last node of the path R.

- $SW_{c,w,s}^{R}$ is an integer that takes a value 1 if the connection c from source node s in a groomed path R is assigned by a wavelength w; otherwise, it is zero, where s is the first node of the path R.
- $DW_{c,w,s,d}^{R}$ is an integer that takes a value 1 if the connection c from source s to destination d is assigned by the wavelength w in groomed path R; otherwise, it is zero.
- $P_{ij}^{l,w,b}$ is an integer that takes a value 1 if the link ij of the shared-protected route b is assigned a wavelength w for the lth SD pair/connection; otherwise, it is zero.

Objective for traffic grooming:

$$\text{Minimize } W \tag{7.30}$$

Protection constraints:

$$\sum_{j:(ij\in E)} P_{ij}^{l,w,b} = \sum_{j:(ji\in E)} P_{ji}^{l,w,b}, \quad i \neq s,d, \forall (w \in W) \text{ and } (i,j \in V). \tag{7.31}$$

$$\sum_{w\in W}\sum_{j:(ij\in E)} P_{ij}^{l,w,b} - \sum_{w\in W}\sum_{j:(ji\in E)} P_{ji}^{l,w,b} = 1, \quad i = s \tag{7.32}$$

$$\sum_{w\in W}\sum_{j:(ij\in E)} P_{ij}^{l,w,b} - \sum_{w\in W}\sum_{j:(ji\in E)} P_{ji}^{l,w,b} = -1, \quad i = d \tag{7.33}$$

Grooming constraint:
For a groomed path, R in DES_SWG algorithm

$$\sum_{c} SW_{c,w,d}^{R} + \sum_{c} DW_{c,w,s,d}^{R} \leq g \qquad \forall 1 \leq w \leq W \tag{7.34}$$

$$\sum_{w=1}^{W} SW_{c,w,s}^{R} = 1 \qquad \forall 1 \leq c \leq t, \qquad s \in V \tag{7.35}$$

$$\sum_{w=1}^{W} DW_{c,w,s,d}^{R} = 1 \qquad \forall 1 \leq c \leq t, \qquad s,d \in V \tag{7.36}$$

For a groomed path, R in SOURCE_SWG algorithm [42]

$$\sum_{c} SW_{c,w,s}^{R} + \sum_{c} DW_{c,w,s,d}^{R} \leq g \quad \forall 1 \leq w \leq W \tag{7.37}$$

$$\sum_{w-1}^{W} SW_{c,w,d}^{R} = 1 \qquad \forall 1 \leq c \leq t, \qquad d \in V \tag{7.38}$$

Equation (7.30) is objective, which minimizes the number of wavelengths per fiber. Equations (7.31)–(7.33) show the flow balance constraints of restricted shared backup paths. Equation (7.34) ensures that the most g connections are groomed into a wavelength channel, and equation (7.35) indicates that only one wavelength is assigned to a connection in DES_SWG algorithm. Similarly, equations (7.37) and (7.38) are used for grooming constraints of SOURCE_SWG algorithm. Equation (7.36) shows that only one wavelength is assigned to a connection between a source node s and a destination node d in the groomed path R, and it can be used for both the algorithms.

7.8.3.2   SOURCE_SWG

SOURCE_SWG approach is based on alternate path routing [42]. Here, each node maintains a fixed alternate path routing table, which has K number routes (R_i) (in descending order of shortest path, where $i = 1,2,3,... K$ correspond to 1st, 2nd, ... Kth shortest path) for all its destinations. The routes of the SD pairs of same source and different destinations of the network are grouped. If the connections are not accommodated on the first shortest path of its corresponding group or groom, then it is attempted for other alternate routes of their corresponding group in descending order. It is assumed that a connection can occupy more than one connection slot according to its bandwidth requirement. The algorithm procedure is given below.

   Input: Network configuration and set of connection requests.

   1. Initialize SOURCE_SWG with network configuration $G(V, E, W)$.
   2. Select a connection request $c(s, d, B, t_H)$ from the set of connection requests.
   3. For $i = 1$ to K,
         Try to accommodate the connection in ith shortest path of its corresponding group within holding time t_H (using constraints (7)–(9)). If it is accommodated, go to step 4; otherwise, it is treated as a blocked connection.
   4. Set up restricted shared backup path for the groomed connection using $(K + 1)$th alternate path satisfying the constraint in equations (2)–(4).
   5. Repeat steps 3 and 4 until all the connections are accommodated using objective: minimize W.

7.8.3.3   DES_SWG Algorithm

Like SOURCE_SWG approach, DES_SWG approach is also based on alternate path routing. The difference is in grouping formation of the SD pairs. Here, we consider grouping using different sources and same destination, as discussed earlier. We have minimized W using constraints in equations (5)–(7). The algorithm procedure is same as that of SOURCE_SWG approach.

7.8.3.4   Analytical Model for Blocking Probability Analysis under Protection Tree

The analytical model exhibits basically blocking analysis considering alternate path routing under shared-protected optical network [44–46]. The analysis estimates the overall blocking probability of the network with and without converter. The end-to-end success probability is taken for first shortest path of any source and destination (SD) pair. Then, we estimate the blocking probability for all alternate paths for the same SD pair. Finally, we take overall blocking probability of the network. The following parameters are taken:

- $e_{i,j}$ = directed link from node i to node j.
- W = the total number of wavelengths per fiber link.
- $W_{i,j}$ = the number of free wavelengths available after protection of connection in a link $e_{i,j}$.
- $e_{i,j}$ the amount of traffic per second passing through link $e_{i,j}$.
- $e_{i,j}$ the probability that a given wavelength in a link $e_{i,j}$ is occupied.
- X = the status vector for the placement of converter in the nodes and represented by $\{x_1, x_2, \ldots x_i, \ldots x_N\}$, where $x_1, x_2, \ldots x_i, \ldots x_N$ correspond to node 1, node 2, ... node i... node N, respectively. In general, x_i is defined as

$$x_i = 1, \text{if converter is placed at node } i.$$

$$= 0, \text{otherwise}$$

The success probability $S\left(R_{sd}^k\right)$ for a connection is written as follows [41]:

For one hop path ($H = 1$),

$$S\left(R_{sd}^k\right) = \left(1 - \rho_{0,1}^{W \ 0,1}{}^{k}\right) \tag{7.39}$$

For two hop paths ($H = 2$),

$$S\left(R_{sd}^k\right) = \left(1 - \rho_{0,1}^{W \ 0,1}{}^{k}\right)^{x_1} \left(1 - \rho_{0,2}^{W \ 1,2}{}^{k}\right)^{x_1} \cdot \left(1 - \left\{(1 - \rho_{0,1})(1 - \rho_{1,2})\right\}^{W \ 0,2}{}^{k}\right)^{(1-x_1)} \tag{7.40}$$

For more than two hop paths ($H > 2$),

$$S\left(R_{sd}^k\right) = \left(1 - \rho_{0,1}^{W \ 0,1}{}^{k}\right)^{x_1} \left(1 - \rho_{1,2}^{W \ 1,2}{}^{k}\right)^{x_1 x_2}$$

$$\left(1 - \rho_{H-2,H-1}^{W \ H-2,H-1}{}^{k}\right)^{x_{H-2} x_{H-1}} \left(1 - \rho_{H-1,H}^{W \ H-1,H}{}^{k}\right)^{x_{H-1}}$$

$$\left(1 - \left\{1 - (1 - \rho_{0,1})(1 - \rho_{1,2})\right\}^{W \ 0,2}{}^{k}\right)^{(1-x_1)x_2}$$

$$\left(1 - \left\{1 - (1 - \rho_{H-3,H-2})(1 - \rho_{H-2,H-1})\right\}^{W \ H-3,H-2}{}^{k}\right)^{x_{H-3}(1 \ x_{H-2})x_{H-1}}$$

$$\left(1 - \left\{1 - (1 - \rho_{H-2,H-1})(1 - \rho_{H-1,H})\right\}^{W \ H-2,H}{}^{k}\right)^{x_{H-2}(1-x_{H-1})}$$

$$\left(1-\left\{1-\prod_{i=0}^{H-1}\left(1-\rho_{i,i+1}\right)\right\}^{W_{0,H}^{k}}\right)^{\prod_{i=1}^{H-1}(1-x_i)} \qquad (7.41)$$

where R_{sd}^k represents kth alternate route for the SD pair. The subscript ρ denotes the index of the nodes in the route. The subscript of status vector x denotes the index of the nodes in the route R_{sd}^k. $W_{i,j}^k$ denotes the widespread available wavelengths of the links between ith and jth nodes of the route R_{sd}^k.

Blocking probability is the probability that a route R_{sd}^k for SD pair is blocked, and formulated as

$$P\left(R_{sd}^k\right) = \left[1 - S\left(R_{sd}^k\right)\right] \qquad (7.42)$$

The blocking probability of connection considering alternate paths of the SD pair is written as [41]:

$$P_{sd} = \prod_{k=1}^{K}\left[\left(1 - S\left(R_{sd}^k\right)\right)\right] \qquad (7.43)$$

and K represents the maximum number of alternate path for the establishment of light path for the SD pair.

Considering all SD pairs, the overall blocking probability of the network is obtained using equation (7.40):

$$P = \frac{\displaystyle\sum_{\forall s,d}\lambda_{sd}P_{sd}}{\displaystyle\sum_{\forall s,d}\lambda_{sd}} \qquad (7.44)$$

Equation (7.44) exhibits the overall blocking probability analysis considering alternate path routing and partial placement of converters in the network.

7.9 SERVICE RELIABILITY AND RESTORABILITY

The service reliability disruption rate, failure rate, RT, and service restorability parameters are required for the analysis of service reliability.

7.9.1 SERVICE RELIABILITY DISRUPTION RATE

Service disruption rate is a function of the failure rate and operation policies. The traffic is disrupted twice in the reverting strategy, whereas in then on-reverting strategy, the backup paths for the connections share network resources (such as physical links, wavelength channels, and node hardware) with connection t are reorganized as some resources on parts of their backup paths are used by t after t is switched to its backup path.

7.9.2 RESTORATION TIME

Service RT is mainly a function of protection-switching time of different fault-management scheme in which traffic is restored through predesigned protection resources. In dedicated (link, path, or sub-path) protection case, OXCs on the backup paths are preconfigured very fast when the connection is established. For no OXC configuration case for a failure, this type of recovery can be very fast. In shared-path protection, the OXCs on the backup paths cannot be used until the failure occurs. The protection-switching time is longer in this scheme. For dynamic restoration, the service RT includes the time for route computation and resource discovery besides failure detection, notification, OXC reconfiguration, and propagation delay.

Network partitioning has been proposed to achieve fast fault RT. In this scheme, a large network is partitioned into several smaller domains, and then, it protects each connection such that an intra-domain light path does not use resources of other domains and the primary and backup paths of an inter domain light path exit a domain (and enter another domain) through a common egress (or ingress) domain-border node [OZSM04]. If a failure occurs, only the domain having failure will activate its protection *sub-path*. This results in computation only in that domain (not in whole network). Hence, the RT is reduced.

7.9.3 SERVICE RESTORABILITY

Service restorability is also a network-wide parameter showing the capability that a network repairs a specific failure. The restorability $R_f(i)$ of a network for a specific f-order ($f \geq 1$) failure scenario (i) is written as the fraction of failed working capacity restored by a specified mechanism within the spare capacity in a network.

7.9.4 ESTIMATION OF RELIABILITY OF PROTECTION IN NSFNET T1 BACKBONE

There are two types of path for a source–destination pair: primary routing path and backup path. For the calculation of reliability, we consider that each primary routing path has same failure rate of f_p, and each backup routing path has same failure rate of f_s. Figure 7.15a shows the reliability model of shared protection of node pair 2-9 (2 represents the source node and 9 the destination node) using single wavelength. The reliability as function failure rate of node pair 2-9 is written as [43]:

$$R_{2,9}^{sp} = e^{-f_p t} + \frac{f_p}{f_p - f_s}\left[e^{-f_s t} - e^{-f_p t} \right]$$

(7.45)

Similarly, for other branches,

$$R_{2,7}^{sp} = R_{2,6}^{sp} = R_{9,13}^{sp} = R_{12,11}^{sp} = R_{12,14}^{sp} = R_{9,10}^{sp} = e^{-f_p t} + \frac{f_p}{f_p - f_s}\left[e^{-f_s t} - e^{-f_p t} \right]$$

(7.46)

So, the reliability at time t for the protection tree of node 2 is written as:

$$R_{2,}^{sp} = \left[R_{2,9}^{SP} \right]^6$$

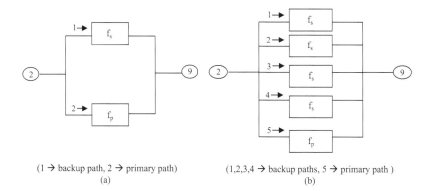

(1 → backup path, 2 → primary path)   (1,2,3,4 → backup paths, 5 → primary path)
(a)   (b)

FIGURE 7.15   Reliability model of 2–9 (SD) node pair: (a) shared and (b) restricted shared protection.

Figure 7.15b shows the reliability model of restricted shared protection of node pair 2-9 (2 represents the source node and 9 the destination node) using four wavelengths. The reliability as function failure rate of node pair 2-9 is written as [43]:

$$R_{2,9}^{RSP} = e^{-f_p t} + \frac{f_p}{f_p - f_{rsp}} \left[e^{-f_s t} - e^{-f_p t} \right] \tag{7.47}$$

where f_{rsp} is the failure rate of restricted shared protection having four wavelengths.

Similarly, for other branches [43],

$$R_{2,7}^{RSP} = R_{2,6}^{RSP} = R_{9,13}^{RSP} = R_{12,11}^{RSP} = R_{12,14}^{RSP} = R_{9,10}^{RSP} = R_{2,9}^{SP} = e^{-f_p t}$$

$$+ \frac{f_p}{f_p - f_s} \left[e^{-f_s t} - e^{-f_p t} \right] \tag{7.48}$$

So, the reliability at time t for the protection tree of node 2 is written as:

$$R_{2}^{RSP} = R_{2,6}^{RSP} \cdot R_{2,7}^{RSP} \cdot R_{12,14}^{RSP} \cdot R_{2,9}^{SP} \cdot R_{9,10}^{SP} \cdot R_{12,11}^{SP} \cdot R_{9,13}^{SP} = R_{2,9}^{RSP} \cdot \left[R_{2,9}^{SP} \right]^6 \tag{7.49}$$

The overall probability of all SD pairs of shared-protection network for NSFNET is given by

$$R^{SP} = \prod_{i=1}^{N} R_{i}^{SP} \tag{7.50}$$

Where R_{i}^{SP} is the reliability for shared-protection tree of node i as a source node and $i = 1, 2, 3, \ldots N$.

The overall probability of all SD pairs of restricted shared-protection network for NSFNET is given by [43]

$$R^{RSP} = \prod_{i=1}^{N} R_{i}^{RSP} \tag{7.51}$$

where R_i^{RSP} is the reliability for restricted shared-protection tree of node i as a source node and $i = 1, 2, 3, \dots N$.

Considering failure rates f_s of 5×10^{-3} hours^{-1}, f_{rsp} of 5×10^{-3} hours^{-1}, and f_p of 7×10^{-2} hours^{-1}, we have determined the reliability of *SP* and *RSP* NSFNET network for 24 hours as $R^{SP} = 0.4673$ and $R^{RSP} = 0.906$, respectively. As the failure rate increases, reliability decreases in both cases, but reliability becomes more and more in case of restricted sharing.

7.10 MULTICAST TREES FOR PROTECTION OF WDM MESH NETWORK

In this section, we mention architectures and approaches for establishing multicast connections in a WDM mesh network based on "light-trees" where protections are made by using multicast trees in a mesh network. Applications are bandwidth-intensive multicast networks for HDTV, interactive distance learning, live auctions, distributed games, and movie broadcasts from studios [46–48]. These applications need point-to-multipoint connections from a source node to the destination nodes in a network. The logical (or virtual) connectivity of the network is enhanced, and the hop distance is reduced. A multicast request is served by using light paths or a hybrid of light paths and light-trees [49]. The section mainly focuses on light-tree multicasting. Since a light-tree is a generalization of a light path, the set of light-tree-based virtual topologies is a *superset* of the set of light path-based virtual topologies.

7.10.1 LIGHT-TREE FOR UNICAST TRAFFIC

We consider NSFNET backbone topology for making light-tree with unicast traffic. Figure 7.16 represents a light-tree (thick lines) connecting source node *UT* to destination nodes *TX*, *NE*, and *IL* [2]. The considerations are as follows:

- The *BR* (bit rate) of each light path is normalized to1 unit, and the node *UT* has packet traffics to transmit to nodes *NE*, *IL*, and *TX*.

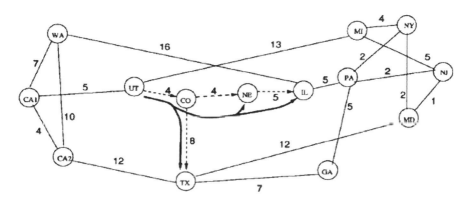

FIGURE 7.16   Light-tree-based structure of NSFNET [2].

- Node *UT* has 0.25 units of traffic to node *NE*, 0.25 units of traffic to node *IL*, and 0.5 units of traffic to node *TX*. Only one free wavelength is taken for the links *UT–CO*, *CO–NE*, *NE–IL*, and *CO–TX*.
- A path has the following four light paths: (i) from *UT* to *CO*, (ii) from *CO* to *NE*, (iii) from *CO* to *TX*, and (iv) from *NE* to *IL*. Thus, the light path-based solution requires an electronic switch at node *CO* and node *NE*, and a total of 8 transceivers (one transmitter and one receiver per light path).
- One the other hand, a light-tree (shown in Figure 7.16 by thick lines) needs 4 transceivers (one transmitter at *UT* and one receiver per node at *NE*, *IL*, and *TX*), and it does not require the electronic switch at node *CO* or at node *NE*.

The light-tree configuration needs multicast-capable wavelength-routing switches (MWRS) at every node (or at least at some nodes) and more optical amplifiers and wavelength converters in the network because a typical light-tree uses more fiber links than the number of fiber links used by a typical light path [49].

7.10.1.1 Layered-Graph Model

The light-trees not only give an improved performance for unicast traffic but also accommodate broadcast traffic and multicast traffic because of their inherent point-to-multipoint nature. Figure 7.17a shows the wavelength-routed optical network represented as a layered graph [50,51], whereas Figure 7.17b gives its layered-graph representation. Now, switching state of each wavelength-routing switch (WRS) is controlled by a controller that communicates with one another using a control network, either in-band out-band, or in-fiber, out-band. In in-fiber, out-band signaling (which we advocate for a WDM mesh network), a wavelength layer is dedicated for the control network. The wavelength λ_0 layer is used for the control network, and the controller uses multiple light-trees for fast information distribution among controllers.

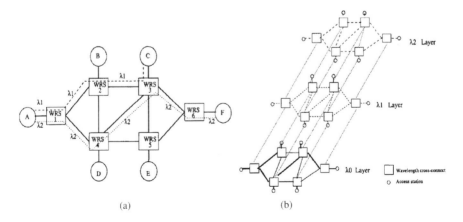

FIGURE 7.17 (a) WRS-based network having three wavelength channels and (b) its layer graph model having three wavelength channels, and wavelength λ_0 layer is a control layer [2].

7.10.2   STEINER TREES

We have already mentioned static protection tree for survivability of network traffic. This problem is minimizing the cost of setting up a multicast tree in a network and has been resolved with and without QoS guarantees such as bandwidth, reliability, delay, and jitter [2,51–53]. This minimum-cost multicast tree is modeled using a minimum Steiner tree [54], and the problem of finding a Steiner Minimum Tree (SMT) in a graph is NP-complete [55]. Two heuristics are employed to resolve SMT problems: Pruned Prim's Heuristic (PPH) and Minimum-cost Path Heuristic (MPH). In PPH, a minimum spanning tree (MST) is formed by using Prim's MST algorithm and then clipped by eliminating unwanted arcs. In MPH, the closest destination nodes are selected one by one and added to a partially built tree. Alight-tree is based on a directed Steiner tree, and the problem of finding a Directed Steiner Minimum Tree (DSMT) is NP-complete. In the absence of wavelength converters in a network, this light-tree-based multicast session reveals the wavelength-continuity constraint [56].

7.10.2.1   General Problem Statement of Light-Trees for Unicast Traffic

Here, we formulate the problem of light-tree for unicast traffic [2]. The following considerations are taken for the problem.

1. A physical topology G_p (V has a weighted undirected graph representing light-tree, V is the set of network nodes, and E_p is the set of links connecting the nodes and each link in the physical topology is bidirectional [2]). The network nodes mainly have packet switches and links corresponding to the fibers between nodes. Since links are undirected, each link has two fibers, or two channels multiplexed (using any suitable mechanism) on the same fiber. Links are assigned to the weights as per physical distances between the nodes. **A** network node i assumed to be equipped with a $D_p(i) \times D_p(i)$ WRS, where $D_p(i)$, called the physical degree of node i, equals the number of physical fiber links emanating out.
2. The number of wavelength channels in each fiber = W.
3. An $N \times N$ traffic matrix, where N is the number of network nodes, and the (i,j)th element shows the average rate of traffic flow from node i to node j. Note that the traffic flows in two reverse directions may be different.
4. The number of wavelength-tunable lasers (transmitters) and that of wavelength-tunable filters (receivers) at node i are T_i and R_i, respectively.

Our goal is to determine the following.

A virtual topology $G_v(V,F_v)$ is a graph where the out-degree of anode is the number of transmitters at that node and the in-degree of a node is the number of receivers at that node.

7.10.2.2   Formulation of the Optimization Problem: Unicast Traffic

The problem of finding an optimum light-tree-based virtual topology is considered as an optimization problem, using principles of multi commodity flow for routing of light-trees on the physical topology and for routing of packets on the virtual topology, and using the following notation:

1. s and d are the subscript and superscript indicating source and destination of a packet, respectively.
2. i and j represent the originating node (or root) and the terminating node (or leaf) in a light-tree, respectively.
3. m and n are the end points of a physical link occurred in a light-tree.
   Number of nodes in the network $= N$.

Maximum number of wavelengths per fiber $= W$.

- **Physical Topology:** The variable P_{mn}, where $P_{mn} = P_{nm} = f =$ number of fiber link indicates that there are f direct physical fiber links between nodes m and n, where $f = 0,1,2,\ldots$ and $m,n = 1,2,3,\ldots,N$. If there is no fiber link between nodes m and n, then $P_{nm} = P_{nm} = 0$.
  - Traffic matrix Λ_{sd}, which denotes the average rate of traffic flow from node s to node d, for $s,d = 1, 2,\ldots, N$.
  - Maximum loading per channel $= \beta$, $0 < \beta < 1$. B restricts the queueing delay on a light path from getting unbounded by avoiding excessive link congestion. The queueing delays are negligible compared to propagation delays for a large network. But it cannot be neglected under extremely heavy loading.
  - Capacity of each channel $= C$ (normally expressed in bits/second but converted to units of packets/second by knowing the mean packet length).
  - The variable $V_i = 0,1,2,\ldots,T_i$ is the number of transmitters in use at node i.
  - The variable $s_{ij} > 0$ if a transmitter at node i and a receiver at node j are members of the same light-tree in the virtual topology;
       $= 0$ otherwise.
    where s_{ij} is a real-valued variable, which can range from 0 to V_i, where V_i is the number of "busy" transmitters at node i. For example, if $s_{ij} = 0.5$, then it implies that the virtual link $i - j$ can use only 0.5 times the capacity of a light-tree channel.
- **Traffic Routing:** The variable λ_{ij}^{sd} denotes the traffic flowing from node s to node d and employing s_{ij} as an intermediate virtual link. Note that traffic from node s to node d may be "bifurcated" with different components taking different sets of light-trees.
- **Physical Topology Route:** The variable
  $p_{mn}^{i} > 0$ if the fiber link P_{mn} is present in a light-tree rooted at node i;
  $= 0$ otherwise.
  Where p_{mn}^{i} is a real-valued variable, which can range from 0 to V_i, where V_i is the number of "busy" transmitters at node i. A fractional value of $p_{mn}^{i} = 0.5$ means the following: from the total amount of traffic transmitted on a light-tree rooted at node i, the physical fiber link $m - n$ can

carry an amount of traffic, which is less than 0.5 times the capacity of a light-tree channel.

Integer variable: Variable $Q^i_{mn} > p^i_{mn}$ is used to take the "ceiling" of variable p^i_{mn}

Optimize:

Minimize one of two objective functions given below.

1. Average packet hop distance minimization [2]:

$$\text{Minimize:} \frac{1}{\displaystyle\sum_{s,d} \Lambda_{sd}} \sum_{i,j} \sum_{s,d} \lambda^{s,d}_{i,j} \tag{7.52}$$

This is an objective function where *the sum* is a linear sum of variables.

2. Total number of transceivers in the network:

$$\text{Minimize} \sum_{i} V_i + \sum_{i,j} V_{i,j} \tag{7.53}$$

Where $\sum_{i} V_i$ is the total number of transmitters in the network and $\sum_{i,j} V_{i,j}$ is the total number of receivers in the network. Thus, the objective function minimizes the total number of transceivers in the network.

Constraints:

On virtual-topology connection variables [2]:

$$V_{i,} \leq T \tag{7.54}$$

$$s_{ij} \leq V_{ij} \tag{7.55a}$$

$$V_{ij} = s_{ij} + 1 \tag{7.55b}$$

$$\sum_{ij} V_{ij} \leq V_i \tag{7.55c}$$

$$\sum_{ij} V_{ij} \leq R_i \tag{7.56}$$

Where V_i, and V_{ij} are the integers.

On physical route variables

$$\sum_{n} p^i_{in} = V_i \tag{7.57}$$

$$\sum_{j} s_{ij} = V_i \tag{7.58}$$

$$\sum_m p^i_{mk} = \sum_n p^i_{nk} + s_{ik}, \quad k \neq i \tag{7.59}$$

$$p^i_{mn} \leq Q^i_{mn} \tag{7.60}$$

$$Q^i_{mn} \leq V_i \tag{7.61}$$

$$\sum_{ij} Q^i_{mn} \leq W \times p_{mn} \tag{7.62}$$

Where Q^i_{mn} is an integer.

On virtual-topology traffic variables

$$\lambda^{sd}_{ij} \geq 0 \tag{7.63}$$

$$\sum_j \lambda^{sd}_{sj} = \Lambda_{sd} \tag{7.64}$$

$$\sum_i \lambda^{sd}_{id} = \Lambda_{sd} \tag{7.65}$$

$$\sum_i \lambda^{sd}_{ik} = \sum_j \lambda^{sd}_{kj} \quad \text{if } k \neq s,d \tag{7.66}$$

$$\sum_{sd} \lambda^{sd}_{ij} \leq s_{ij} \times \beta \times C \tag{7.67}$$

where Q^i_{mn} is an integer.

Optional constraints: Physical topology as a subset of virtual topology:

$$p_{mn} = 1, V_{mn} \geq 1, p^m_{mn} \geq 1, Q^m_{mn} \geq 1 \tag{7.68}$$

The above equations are based on the conservation of flows and resources (transceivers, wavelengths, etc.). Equation (7.54) indicates the number of busy transmitters (i.e., transmitters from which a light-tree originates) at node i, which is less than the total number of transmitters at node i. Equations (7.55) and (7.56) state that the integer variable is always equal to the "ceiling" of s_{ij}. Equation (7.57) recovers the convergence of the mixed integer linear program (MILP). Equation (7.58) shows a smaller number of light-trees terminating at node j than the number of receivers at node j. Equations (7.59) and (7.60) are the flow-conservation equations that account for the routing of a light-tree from its origin to its termination. Equation (7.61) indicates that the number of light-streams (a path within a light-tree) originating at node i is the same as the number of busy transmitters at node i. Equation (7.58) states that the sum of all "light-streams" at the terminating nodes is equal to the number of busy transmitters at the originating node (i). The s_{ij} values are integers. Equation (7.59) indicates that the difference between the sum of incoming "light-streams" to node *I* and the sum of outgoing "light-streams" from node *k* is equal to the amount of "light-streams" sinking at node *k*. Equation (7.60) represents that one wavelength be allocated to each "light-stream" passing through link *P*. Equation (7.62) represents that the number of wavelengths used on any physical link is less than the maximum number of wavelengths per fiber; however, this equation does not ensure the wavelength-continuity constraint. Equations (7.63)–(7.66) are used for the routing of packet traffic on the virtual topology, and the combined traffic flowing through a channel cannot exceed the channel capacity. Equation (7.67) embeds the physical topology as a subset of the virtual topology [2], i.e., every link in the physical topology is also a light path in the virtual topology.

The solutions of these problems are obtained by MILP solver such as CPLEX package [57] for different networks. The traffics are randomly distributed between [0, *C/a*], where *C* is the light-tree channel capacity, which is the same as the rate at which a transmitter can transmit, and *a* is an arbitrary integer.

7.11   LIGHT-TREES FOR BROADCAST TRAFFIC

There are broadcast traffics flowing to all nodes treated as destination nodes. The problem statement with formulation in light-trees of broadcast traffic is different from that in unicast traffic [2].

7.11.1   General Problem Statement

In case of broad cast traffic, we find a set of light-trees which is treated as a "broadcast layer" over a single wavelength.

A virtual topology $G_v = (V, E_v)$ is a graph where the out-degree of anode is the number of transmitters at that node and the in-degree is the number of receivers at that node.

7.11.2   Formulation of the Optimization Problem: Broadcast Traffic

The formulation of the problem for broadcast traffic is similar to the formulation of the problem for unicast traffic except one difference: i.e., the formulation for unicast traffic is in the routing of traffic on the virtual topology. Two types of equations are used for routing broadcast traffic: one set of the equations for constructing a spanning tree on the virtual topology for each node in the network; thus, the spanning tree rooted at node *s* will route the broadcast traffic originating at nodes, and other set of the equations for the capacity constraint of the virtual links. The formulation is given below [2].

Variables:

- Some of the variables are the same as those of unicast traffic as mentioned in previous section
- $0 \leq \lambda_{ij}^s \leq N$, determines the spanning tree rooted at node *s*. $\lambda_{ij}^s > 0$ if virtual link V_{ij} belongs to the spanning tree rooted at node *s*. λ_{ij}^s is used only to "construct" a spanning tree on the virtual topology where the broadcast

traffic flows on the spanning tree. The traffic on each link of a spanning tree rooted at node s is equal to the amount of broadcast traffic that is sourced at node s (denoted by Λ^s), where $0 \leq \Lambda^s_{ij} \leq 1$ is a 0–1integer variable.

- Λ^s_{ij} =1if virtual link $V_{,j}$ belongs to the spanning tree rooted at node s.
     = 0, otherwise.
- Broadcast traffic Λ^s originating from node s denotes its average rate, where $s = 1,2,3, \ldots, N$.

Constraints:
- On virtual-topology connection variables: same as in the previous section.
- On physical route variables: same as in the previous section.
- On "spanning tree" variables $\lambda^k_{ij}, \Lambda^k_{ij}$

$$\sum_j \lambda^s_{sj} = N - 1 \tag{7.69}$$

$$\sum_m \lambda^s_{mk} = \sum \lambda^s_{kn} + 1, \quad k \neq s \tag{7.70}$$

$$\Lambda^s_{ij} \leq V_{ij} \tag{7.71}$$

$$\lambda^s_{ij} \leq (N - 1) \times \Lambda^s_{ij} \tag{7.72}$$

$$\sum_{i,j} \Lambda^k_{sj} = N - 1 \tag{7.73}$$

$$\sum_{kj} \Lambda^k_{sj} \times \Lambda^k = s_{ij} \times \beta \times C \tag{7.74}$$

Optimize:

$$\text{Minimize} \sum_i V_i + \sum_{i,j} V_{i,j} \tag{7.75}$$

Equations (7.69)–(7.73) show a spanning tree rooted at node s on the virtual topology, whereas equation (7.74) indicates the light-tree capacity constraints. Equation (7.69) represents a flow of "commodity" from node s. Equation (7.70) shows the multi-commodity flow-conservation equation stating that one unit of flow is "sunk" at every node in the network, except the node where the flow originated. Equation (7.71) indicates that if link i–j belongs to the spanning tree, then it must necessarily exist in the virtual topology. Equation (7.72) states that variable $\Lambda^s_{sj} = 1$ for all virtual link ss_{ij} which belong to the spanning tree rooted at node s. Equation (7.73) states that the number of virtual links in the spanning tree is equal to $N-1$ (this constraint is required to avoid spanning-tree solutions that contain a cycle). Equation (7.74) indicates the capacity constraint of virtual link sij.

7.12 LIGHT-TREES FOR MULTICAST TRAFFIC

7.12.1 GENERAL PROBLEM STATEMENT

The problem of accommodating a group of multicast sessions with over all minimum cost, using a light-tree-based approach on a given physical topology (fiber network), is formally stated below. In the virtual-topology design problem, we are now considering circuit traffic where a set of multicast sessions are given for setting up each session requiring full capacity of a wavelength channel initially [2]. The following inputs are considered for the multicast traffic problem [58]:

- A physical topology G_p (V, E_p) has a weighted undirected graph, where V is a set of network nodes and E_p is the set of links connecting the nodes. Each link in the physical topology corresponds to two fibers connected between nodes in bidirectional. Each link is allocated a weight corresponding to the length of fiber between node pairs. A network node j is made to be equipped with a $D_p(j) \times D_p(j)$MWRS, where $Dp(j)$, called the physical degree of node j, is equal to the number of physical fiber links emanating out of and terminating at node j.
- Number of wavelength channels c by each fiber = W.

Our goal is to establish all the k multicast sessions on the given physical topology minimizing the overall cost. Each multicast session makes a light-tree with the source as root and the destination nodes as leaves. The problem of establishing several directed multicast trees at a minimum aggregate cost is a generalized version of the DSMT problem, which is NP-complete [58].

7.12.2 PROBLEM FORMULATION FOR A NETWORK WITH CONVERTERS

Here, the problem of establishing a group of multicast sessions for a network equipped with such "opaque" switches is made [58,59]. Following are a few notations which we will be using:

- s and d indicate the source node and destination node, respectively, in a multicast session.
- m and n indicate the *endpoints of* a *physical link* that might occur in a light-tree.
- i is taken as an index for multicast session number, where i = 1,2, ..., k.
- Number of nodes in the network = N.
- Maximum number of wavelengths per fiber − W (a system wide parameter).
- Physical topology P_{mn}, where $P_{mn} = P_{nm} = f$ (i.e., fiber links are assumed to be bidirectional), indicates that there are f direct physical fiber links between nodes m and n, where f = 0,1,2,... and m, n = 1,2,3,...,N. For no fiber link between nodes m and n, $P_{mn} = P_{nm} = 0$. Capacity of each channel = C.
- A group of k multicast sessions S_i for i = 1, 2, 3,..., k. Each session S_i has a source node and a set of destination nodes denoted by $\{s_i, d_{i1}, d_{i2}, ...),$

representing the cardinality of a multicast session i by L_i, which is equal to the number of source and destination nodes in that session.

- By default, every multicast session is assumed to operate at the full capacity of a channel, i.e., at C bps. This will be relaxed later for fractional-capacity sessions.
- Every node is equipped with wavelength converters capable of converting a wavelength to any other wavelength among W channels. These are not separate devices but are a "built-in" property of "opaque" cross-connects.

Variables:

M^i_{mn} = 1 if the link between nodes m and n is engaged for multicast session i;
= 0 otherwise

V^i_p = 1 if node p belongs to multicast session i;
= 0 otherwise

A node belongs to a session if it is either the source or the destination or an intermediate node in the light-tree for the multicast session, e.g., $V^i_{si} = 1$ and $V^i_{di} = 1$.

An integer commodity-flow variable, F^i_{mn}: Each destination node for a session requires one unit of commodity. So, $L_i - 1$ units of commodity run out of the source s_i for session i. F^i_{mn} is the number of units of commodity flowing on the link from node m to node n for session i. F^i_{mn} is also the number of destination nodes in session I downstream of the link between nodes m and n.

Optimize:

The cost of all multicast sessions:

$$\text{Minimize} : \sum_{i=1}^{i=k} \sum_{m,n} w_{mn} M^i_{mn} \tag{7.76}$$

This is a linear objective function used for adding up the cost of individual multicast sessions.

Constraints:

Tree-creation constraints:

$$\forall i, \forall n \neq s_i \quad \sum_m M^i_{ms_i} = V^i_n \tag{7.77}$$

$$\forall i \sum_m M^i_{ms_i} = 0 \tag{7.78}$$

$$\forall i, \forall j \in s_i \quad V^i_j = 1 \tag{7.79}$$

$$\forall i, \forall m \neq d_{ij}, j = 1,2,\dots(L_i - 1) \quad \sum_m M^i_{mn} = V^i_m \tag{7.80}$$

$$\forall i, m \quad \sum_m M^i_{mn} = D_p(m) V^i_m \tag{7.81}$$

$$\forall m, n \quad \sum_m M^i_{mn} \leq P_{mn} W \tag{7.82}$$

Commodity-flow constraints:

$$\forall i \forall m \notin S_i \quad \sum_n F_{nm_i}^i = \sum_n F_{nm}^i \tag{7.83}$$

$$\forall i \forall m = s_i \quad \sum_m F_{s_i n}^i = L_i - 1 \tag{7.84}$$

$$\forall i \forall m = s_i \quad \sum_m F_{n s_i}^i = 0 \tag{7.85}$$

$$\forall i, \forall m \neq d_{ij}, j = 1, 2, \ldots (L_i - 1) \quad \sum_n F_{mn}^i = \sum_n F_{mn}^i + 1 \tag{7.86}$$

$$\forall i, m, n \quad M_{mn}^i \leq F_{mm}^i \tag{7.87}$$

$$\forall i, m, n \quad F_{mn}^i \leq M_{mn}^i N \tag{7.88}$$

Additional constraint:

$$\forall i, m, n \quad F_{mn}^i \leq (L_i - 1) \tag{7.89}$$

The equations are used for making a connected tree of every multicast session. Equation (7.77) indicates that every node belonging to a multicast session (except the source) has one incoming edge. Equation (7.78) represents that the source node has no incoming edge since it is the root of the tree. Equation (7.79) indicates that every source node and destination node of a multicast session belongs to same tree. Equation (7.80) indicates that every node (except destination nodes) belonging to the tree has at least one outgoing edge on the tree. Equation (7.81) indicates that every node with at least one out going edge belongs to the tree. Equation (7.82) limits the number of light-tree segments between nodes m and n by $P_{mn}W$ in either direction [2].

Equations (7.83)–(7.88) represent the flow-conservation equations to create a connected tree with the source having an end-bend connection to every destination in the session. Equation (7.83) indicates that at any intermediate node (which is neither a source nor a destination), the incoming flow is the same as the outgoing flow. Equation (7.84) indicates the outgoing flow at the source node for a session to the number of destinations in the session, whereas equation (7.85) indicates zero incoming flow. Equation (7.86) indicates that the total outgoing flow is one less than the incoming flow for destination nodes. Equations (7.87) and (7.88) indicate that every link occupied by a session contains a positive flow and every link not occupied by the session contains no flow. In equation (7.88), N can be replaced by L_i-1 without altering its meaning. In equation (7.89), it is stated that a flow on any link for a multicast session is limited by the number of destinations in that session [2].

7.12.3 Variation of Problem Formulation with No Converters

In this section, we mention problem formulation of multicast traffic having no converter in which the entire light-tree for a multicast session is on a common wavelength. The problem formulation for a network with the wavelength-continuity constraint (in an all-optical network) is almost similar to the problem with wave-length converters. There are also some additional variables and constraints. In the absence of wavelength converters, proper wavelength assignment of various multicast sessions is required to minimize the overall cost [2,60] . All the parameters in the previous problem are same except that they do not have a relation with the absence of converters [2].

Variables:

$M_{mn}^{ic} = 1$ if the link between nodes m and n is available for multicast session i on wavelength c;

= 0 otherwise

$V_p^i = 1$ if node p belongs to multicast session i; otherwise, Vj

= 0 otherwise

For the source or the destination or an intermediate node in the light-tree for the multicast session, $V_{si}^i = 1$ and $V_{di}^i = 1$.

An integer commodity-flow variable, F_{mn}^i: Each destination node for a session needs one unit of commodity. So, L_i–1 units of commodity run out of the source s_i for session i. F_{mn}^i is the number of units of commodity flowing on the link from node m to node n for session i. F_{mn}^i is the number of destination nodes in session I downstream of the link between nodes m and

$C_c^i = 1$ if multicast session i is on wavelength c

= 0 otherwise

The light-tree for a multicast session inhabits one wavelength as there are no wavelength converters [2].

Optimize:

Cost of multicast session

$$\text{Minimize: } \sum_{i=1}^{i-k} \sum_{c=1}^{c-W} \sum_{m,n} w_{mn} M_{mn}^i \tag{7.90}$$

A linear objective function is used for the cost of individual multicast sessions on dl links and wavelengths.

Constraints:

Tree-creation constraints [2]:

$$\forall i, \forall n \neq s_i \quad \sum_{m,c} M_{mn}^{ic} = V_n^i \tag{7.91}$$

$$\forall i \quad \sum_{m} M_{ms_i}^i = 0 \tag{7.92}$$

$$\forall i, \forall n \in S_i \quad V_j^i = 1 \tag{7.93}$$

$$\forall i, \forall m \neq d_{ij}, j = 1, 2, \dots (L_i - 1) \quad \sum_n M_{mn}^i \geq V_m^i \tag{7.94}$$

$$\forall i, m \quad \sum_{n,c} M_{mn}^i \leq D_p(m) \times V_m^i \tag{7.95}$$

$$\forall m, n \quad \sum_{i,c} M_{mn}^{ic} \leq P_{mn} W \tag{7.96}$$

$$\forall m, n, c \quad \sum_i M_{mn}^{ic} \leq P_{mn} \tag{7.97}$$

Commodity-flow constraints: The first four constraints are the same as equations (7.83), (7.84), (7.85), and (16.86), respectively. Equations (7.87) and (7.88) are, respectively, changed as follows:

$$\forall i, m, n \quad \sum_c M_{mn}^i \leq F_{mn}^i \tag{7.98}$$

$$\forall i, m, n \quad F_{mn}^i \leq N \cdot \sum_c M_{mn}^{ic} \tag{7.99}$$

Wavelength-related constraints:

$$\forall i \quad \sum_c C_c^i = 1 \tag{7.100}$$

$$\forall m, n (m < n) \forall i, c \quad M_{mn}^{ic} + M_{nm}^{ic} = C_c^i \tag{7.101}$$

Equation (7.97) restricts the number of sessions on the same wavelength between a node pair by P_{mn} (effectively ensuring that each fiber link supports no more than W wavelengths). Equation (7.100) indicates a session selecting only one wavelength. Equation (7.101) indicates no link used by a session on the wavelength not selected by it and all links used by a session are on the same wavelength. Other equations provide the same as mentioned earlier. There are $2kN + kN^2 + kWN^2$ number of unknown variables in this system and also an additional $kN + k(W-1)N^2$ number of variables for no full-wavelength conversion capability [2].

7.12.4 VARIATION OF PROBLEM FORMULATION WITH FRACTIONAL-CAPACITY SESSIONS

There are some multicast sessions operating at speeds lower than those of a wavelength channel. Hence, these sessions with different smaller bandwidth signal (such as OC-1, OC-12, or OC-48) are groomed onto a single high-capacity wavelength

(OC-192) on a link. Grooming reduces the overall number of wavelength channels used by the sessions and hence the total cost of operating them. The full channel capacity is indicated by C and BR of low-speed session i is indicated by a fraction fi (of the full capacity C). DXC is needed to groom the signals carrying on bandwidth granularities [61].

Variables:
All variables $M_{mn}^i, F_{mn}^i, V_n^i$ are the same in addition to a new variable f_i, which is the fraction of the channel capacity used by multicast session i. Along with the k multicast sessions, the capacity of each session i is indicated as a fraction of the capacity (C) by f_i.

Optimize:
Cost of all multicast sessions [2],

$$\text{Minimize} \sum_{i=1}^{i=k} f_i \sum_{m,n} w_{mn} M_{mn}^i \qquad (7.102)$$

Constraints:
All constraint equations except equation (7.96) are same. The altered link-capacity constraint is as follows:

$$\forall m,n \quad \sum_{i,c} f_i \cdot M_{mn}^{ic} \leq P_{mn} W \qquad (7.103)$$

Number of Variables and Constraints: There are $k + kN + 2\ k{\cdot}N^2$ number of unknown variables in this system and also additional k variables, which are different from the formulation for a network with full-wavelength conversion capability.

7.12.5 VARIATION OF PROBLEM FORMULATION WITH SPLITTERS CONSTRAINTS

Multicasting in optical form needs optical splitters to have multiple copies of an incoming bit stream transmitted to the destination nodes. Most of the networks have nodes bearing a limited number of splitters with limited splitting capabilities to reduce the cost [62]. Due to restriction in the splitting of a signal at a node, the light-tree is evenly distributed on the network reducing the damage inflicted on the tree because of a node failure [63]. This problem of finding a Steiner tree in a network where the splitter output at a node is limited is known as the degree constraint which is NP-complete [63]. For the all-optical switch architectures (without electronic cross-connects), an array of optical splitters is needed at a node to support several multicast sessions. Each splitter has a finite splitting degree. The number of multicast sessions (requiring splitting) served by a node is restricted by the splitter-bank size [2].

Variables:
   All variables $M_{mn}^i, F_{mn}^i, V_n^i$ used in the previous section remain the same.
   A Boolean variable $A_j^i = \mathbf{1}$ if multicast session i requires a splitter at node j;

= 0 otherwise.

Splitting degree of each splitter is d. Splitter-bank size at each node is B. There is a group of k multicast sessions.

Optimize:

Cost of all multicast sessions [2]:

$$\text{Minimize:} \sum_{i=1}^{i=k} \sum_{c=1}^{c=W} \sum_{m,n} w_{mn} M_{mn}^i \qquad (7.104)$$

Additional Constraints:

Constraint equations used in the previous section remain unchanged and additional constraints are as follows:

$$\forall i, \forall m = d_{ij}, j = 1,2,...(L_i - 1), \quad \sum_c \sum_m M_{nm}^{ic} \le d - 1 \qquad (7.105)$$

$$\forall i, \forall m = d_{ij}, j = 1,2,...(L_i - 1), \quad \sum_c \sum_m M_{nm}^{ic} \le d \qquad (7.106)$$

$$\forall i, \forall m = d_{ij}, j = 1,2,...(L_i - 1), \quad \sum_c \sum_m M_{nm}^{ic} \ge A_n^i \qquad (7.107)$$

$$\forall i, \forall m = d_{ij}, j = 1,2,...(L_i - 1), \quad \sum_c \sum_m M_{nm}^{ic} \le D_p(n) \cdot A_n^i \qquad (7.108)$$

$$\forall i, \forall m \ne d_{ij}, j = 1,2,...(L_i - 1), \quad \sum_c \sum_m M_{nm}^{ic} \ge 2.A_n^i \qquad (7.109)$$

$$\forall i, \forall m \ne d_{ij}, j = 1,2,...(L_i - 1), \sum_c \sum_m M_{nm}^{ic} \le D_p(n) \cdot A_n^i \qquad (7.110)$$

$$\forall n, \quad \sum_i A_n^i \le B \qquad (7.111)$$

Equation (7.105) limits the number of outgoing paths of a destination node n for a session i by $d - 1$, and equation (7.106) finds the same for other nodes in the network. Equations (7.107)–(7.110) represent the following:

- As there is at least one outgoing path of a session at a destination node, a splitter is required at this node for splitting the signal into two or more: one for local drop and the remaining for the downstream destination nodes.
- For all other intermediate nodes (including the source), a splitter is required for the signal having at least two outgoing paths for a session.

Equation (7.111) represents a bound on the number of sessions using splitters available at node n. There are $3kN + kN^2 + kwN^2$ number of unknown variables and also additional number of kN variables compared with the formulation for a network with no wavelength-conversion capability [2]. The number of additional constraint in this MILP formulation is $3kN$, and the bound on the total number of constraints is the same as $O(kN^2W)$.

7.13.6 SIMULATION IN SAMPLE NETWORK FOR MULTICAST TRANSMISSION

For simulation, sample network topology is shown in Figure 7.18, where each link carries two wavelengths. $P_m = 1$ for bidirectional fiber links between adjacent node pairs. A group of five full-capacity multicast sessions ($S_1 = \{F,A,B, C,D\}$, $S_2 = \{C,A,E,F\}$, $S_3 = \{E,A,B,C,F\}$, $S_4 = \{B,D\}$, and $S_5 = \{ A, C\}$) are considered in the network. The first element in the session is the source node, and the remaining are the destination nodes. S_4 and S_5 are, essentially, unicast sessions. The combined optimization problem formulations for RWA of a group of multicast sessions are resolved by standard package MILP solver, CPLEX [57].

Table 7.2 shows the comparison between wavelength and routing results of different multisession, multicast traffics of network, which is illustrated in Figure 7.18. Table 7.2 gives the optimal cost of establishing the above group of multicast sessions in a network with no wavelength converters, and the cost in a network with no wavelength converters is slightly more than the cost in a network having nodes with wavelength converters [2]. Figure 7.18a and b represents the light-trees for the five multicast sessions having nodes with and without wavelength converters, respectively. There is a wavelength conversion from λ_1 (dashed line) to λ_0 (solid line) for the multicast session S_3 at node E in order to reduce the overall cost of establishing all the five sessions. The traffic from session 1 and session **3** uses the same channel (λ_0) on the links A-B and B-C as both the sessions are at half the channel capacity. When the aggregate cost of all the fractional-capacity multicast sessions is minimized, the traffics are groomed

The optimal cost is increased to compensate for limited resource availability, and the total optimal cost of establishing the multicast sessions decreases with an

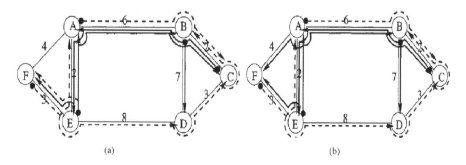

FIGURE 7.18 Sample network having five multicast sessions and each link having two wavelengths λ_0 (solid line) and λ_2 (dashed line) [2] (a) with full-length wavelength converter and (b) with no wavelength converter [2].

TABLE 7.2

Comparison between Results of Different Multisession Multicast Traffic of Network [2] Shown in Figure 7.18 [2]

Full-Capacity Multicast Sessions		Routing and Wavelength Assignment				
Source	Destination Nodes	Full Conversion Route	No Conversion Route	Fractional-Capacity Multicast Session		
				Route	f_i	Wavelength
F	{A, B, C, D}	F-E, E-A, E-D, D-C, C-B Cost = 18	F-E, E-A, E-D, D-C, C-B Cost = 18	F-E, E-A, E-D, D-C, C-B Cost = 7.5	0.5	λ_0
C	{A, E, F}	C-B, E-A, B-A, E-F Cost = 13	C-B, E-A, B-A, E-F Cost = 13	C-B, E-A, B-A, E-F Cost = 13	1.0	λ_1
E	{A, B, C, F}	E-F, E-A, A-B, B-C Cost = 13	E-F, E-A, A-B, B-C Cost = 14	E-F, E-A, A-B, B-C Cost = 6.5	0.5	λ_0
B	{D}	B-D Cost = 7	B-D Cost = 7	B-D Cost = 7	1.0	λ_1
A	{C}	A-B, B-C Cost = 8	A-B, B-C Cost = 8	A-B, B-C Cost = 8	1.0	λ_1

increase in the splitting degree (d) or the size of the splitter bank (B) at a node. It is impossible to establish all multicast sessions with limited network resources ($d = 2$ and $B = 1$), and some sessions will be blocked.

7.13   MULTICAST TREE PROTECTION

Fiber cuts in an optical network provide the disruption of signal transmission, making significant in formation loss in the absence of adequate backup mechanisms. The loss is more when the failed link in a "light-tree" transmits traffic for multiple destinations [2].

7.13.1   PROTECTION SCHEMES

The method of protecting a multicast tree is to determine a link-disjoint backup tree apart from the primary light-tree. This approach provides 1-1dedicated protection where both primary tree and back up tree carry identical bit streams to the destination nodes. After the link fails, the affected destination nodes reconfigure their switches to receive bit streams from the backup tree instead of the primary tree. A more resource-efficient approach is to relax the protection constraint from link-disjointness to directed-link-disjointness, permitting the primary and backup tree to share links, but only in opposite directions. The directed-link-disjointness pre

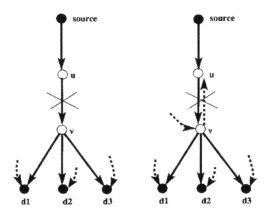

FIGURE 7.19   Fiber cuts along a light-tree from a source node to a set of destination nodes d_1, d_2, and d_3. Solid lines belong to primary tree, and dashed lines form the parts of the backup tree. A fiber cut may disrupt either the primary tree or both primary and backup trees [2].

dedicated protection where the resource for the backup tree is reserved and it is used to carry bit streams when failure occurs [2]. The **directed-link-disjointness** can be successfully used in an optical WDM network (where a failure may disrupt both primary and backup trees) to protect multicast sessions against single fiber cut as follows. Figure 7.19 presents a primary light-tree carrying traffic to three destinations d_1, d_2, and d_3 along a link between nodes u and v [2]. The two cases are as follows:

Case 1: If the backup tree does not inhabit the link $u{\rightarrow}v$ and there is a cut on this link, the affected downstream nodes (d_1, d_2, and d_3 in our case) switch to a different incoming port (shown with dashed lines in Figure 7.19) and continue to receive the bit stream from the backup tree.

Case 2: If the backup tree has the link $v{\rightarrow}u$ (in opposite direction to the primary), a cut in the link u-v shows failure of both the primary tree and the backup tree. Because the backup tree inhibits the link $v{\rightarrow}u$, node v is accessible from the source with the path denoted by the partial dashed lines in Figure 7.19. All disrupted paths to destination nodes downstream of node v are available along the primary tree after the link failure occurs. If the switch at node v is active again for reorganization to provide path for the incoming bit stream from the backup tree (instead of the incoming bit stream from a primary-tree incoming port) to the original (primary) outgoing ports, the affected destinations receive the traffic without having any reconfiguration. Sometimes, a converter is needed at the downstream node (v) to convert the wavelength having a bit stream from a backup-tree wavelength to a primary-tree wavelength if these are different.

7.13.2   General Problem Statement

In the problem of establishing a group of multicast sessions, primary and their backup trees do not share a link in the same direction using a light-tree-based approach on a given physical topology (fiber network) [2]. Here, a general problem statement is written as follows:

1. The network topology is represented by a weighted undirected graph $G_p = (V, E_p)$, where V is a set of network nodes and E_p is the set of links connecting the nodes. Each link in the physical topology is bidirectional and assigned a weight representing the cost (number of hops or equipment operating cost) of moving traffic from one end to the other. Node j is equipped with a $Dp(j) \times Dp(j)$MWRS, where $Dp(j)$, called the physical degree of node j, equals the number of physical fiber links emanating out of (as well as terminating at) node j.
2. Number of wavelength channels on each fiber = W.
3. A group of l_c primary multicast sessions and a binary digit P_i ($i = 1, 2, \dots k$) associated with each of them to indicate whether they require protection or not.

Our goal is to make (if possible) $2k$ primary and backup multicast sessions on the given physical topology by minimizing the total cost, which is the sum of the weights on the physical links occupied by it.

7.13.2.1   Problem Formulation for a Network without λ Continuity

The architecture in Figure 7.20 has full-range wavelength conversion inherent in it. Here, the problem of setting up a group of multicast sessions is formulated for a network equipped with "opaque" switches. The following notations are required in formulation [2]:

- s and d indicate the source node and destination node, respectively, in a multicast session.

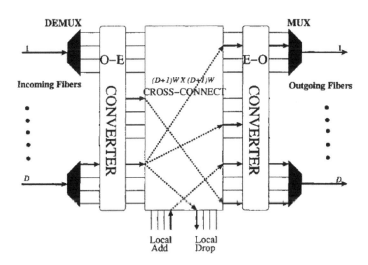

FIGURE 7.20   Opaque switch architecture with $O\text{-}E$ and $E\text{-}O$ converter for multicast session [2].

- m and n indicate the endpoints of a physical link that might occur in a light-tree.
- i represents an index for session number, where $i = 1,2.....k, k+1,.....2k$. Indices 1 through k are considered for the primary trees, and indices $k + 1$ through $2k$ are considered for the backup trees. If a primary session i requires protection, then its corresponding backup tree index is $i + k$. Otherwise, the index is left used and the variables corresponding to that backup tree are ignored.
- N represents the number of nodes in the network.
- W indicates the maximum number of wavelengths per fiber.
- The parameter P_{mn}; $_{Pmn} = P_{nm} = 1$ indicates whether there is a direct physical fiber link between nodes m and n, where $m, n = 1,2,3,...,N$. If there is no fiber link between nodes m and n, then $P_{mn} = P_{nm}, = 0$.
- For physical link between nodes m and n, cost $w_{m,n}$ is associated with a weight.
- C denotes the capacity of each channel $= C$. C represents OC-192=10Gbps.
- A group of k multicast sessions S_i for $i = 1,2,3,..., k$; and $P_i = 1$ or 0 for each session signifying whether session i requires protection or not, respectively. Each session S_i has a source node and a set of destination nodes denoted by $\{s_i, d_{i1}, d_{i2}, ...\}$. The cardinality of a multicast session i is given by L_i, which is equal to the number of source and destination nodes in that session.

The following assumptions are considered:

- Every multicast session is at full capacity of a channel, i.e., at C bps.
- Every node has wavelength converters capable of converting a wavelength to any other wavelength among W channels.

Variables [2]:
Boolean variable, $M_{m,n}^i = 1$, i for the link between nodes m and n occupied by multicast session i.
$= 0$, otherwise
Boolean variable, $V_p^i - 1$, for the node p belonging to multicast session i.
$= 0$, otherwise
A node has a session for either the source or the destination, or an intermediate node in light-tree having multicast session, i.e., $V_{si}^i = 1$ and $V_{si}^i = 1$. $F_{m,n}^i = $ integer commodity-flow variable, for a session needs one unit of commodity. So, there are L_i units of commodity flowing out of source s_i for session i. $F_{m,n}^i = $ the number of units of commodity transmitting on the link from node m to node n for session i. $F_{m,n}^i = $ the number of destination nodes in session I downstream of the link between nodes m and n.

Optimize:
Cost of all multicast sessions:

$$\text{Minimize} \sum_{i=1}^{i=2k} \sum_{m,n} w_{m,n} M_{m,n}^i \qquad (7.112)$$

Constraints:

Tree-creation constraints:

$$\forall i, \forall n \neq s_i, \quad \sum_m M_{m,n}^i = V_n^i \tag{7.113}$$

$$\forall i, \quad \sum_m M_{m,s_i}^i = 0 \tag{7.114}$$

$$\forall i, \forall j \in S_i, j \geq 1, \ V_j^i = 1 \tag{7.115}$$

$$\forall i, \forall m \neq d_{ij}, j \geq 1, \quad \sum_m M_{m,n}^i \geq V_m^i \tag{7.116}$$

$$\forall i, m, \quad \sum_n M_{m,n}^i \leq D_p(m) \cdot V_m^i \tag{7.117}$$

$$\forall m, n \quad \sum_n M_{m,n}^i \leq P_{mn} \cdot W \tag{7.118}$$

Commodity-flow constraints [2]:

$$\forall i, \forall m \notin S_i \quad \sum_n F_{m,n}^i = \sum_n F_{n,m}^i \tag{7.119}$$

$$\forall i, \forall m = s_i \quad \sum_n F_{s_i,n}^i = L_i \tag{7.120}$$

$$\forall i, \forall m = s_i \quad \sum_n F_{n,s_i}^i = 0 \tag{7.121}$$

$$\forall i, \forall m \neq d_{ij}, j \geq 1, \quad \sum_n F_{n,m}^i = \sum_n F_{m,n}^i + 1 \tag{7.122}$$

$$\forall i, m, n \quad M_{m,n}^i \leq F_{m,n}^i \tag{7.123}$$

$$\forall i, m, n \quad F_{m,n}^i \leq N \cdot M_{m,n}^i \tag{7.124}$$

Direct link-disjointness constraints [2]:

$$\forall i = 1, 2 \ldots k \forall m, n \ M_{m,n}^i + M_{m,n}^{i+k} \leq 1 \tag{7.125}$$

The above equations are considered for generating a tree for each multicast session [2,64]. Equation (7.113) indicates every node belonging to a multicast session (except the source) and having at least one in coming edge. Equation (7.114) states that the source node has no incoming edge as a root of the tree. Equation (7.115) indicates every source node and destination node of a multicast session belonging to the tree. Equation (7.116) shows that every node (except the destination nodes) belonging to the tree has at least one outgoing edge. Equation (7.117) states every node with at least one outgoing edge belonging to the tree. Equation (7.118) restricts the number of light paths between nodes m and n by $P_{mn}W$ in every direction. Equations (7.119)–(7.124) represent flow-conservation equations to generate a connected tree with the source having a light path to every destination in the session. Equation (7.119) shows that, at any intermediate node, the incoming flow is the same as the outgoing flow. Equations (7.120) and (7.121) show that outgoing flow at the source node for a session is the number of destinations in the session and the incoming flow is zero. Equation (7.122) indicates total outgoing flow i, which is one less than the incoming flow for destination nodes. Equation (7.123) indicates every link occupied by a session having a positive flow and links not occupied by the session having no flow. In equation (7.124), N can be replaced by L_i without altering its meaning. Equation (7.125) indicates the primary and backup tree sharing a link, if any, only in opposite directions.

7.13.2.2 Problem Formulation for a Network with λ Continuity

In the node architectures having no wavelength converter, the entire light-tree for a multicast session has a common wavelength. The problem formulation for a network needs wavelength-continuity constraint (in an all-optical network), which is almost similar to the problem with wavelength converters. There are some additional variables and constraints needed in the case of no wavelength converters; proper WA of various multicast sessions [64] minimizes the overall cost.

1. Notations s, d, i, m, and n remain the same.
2. Notation c is an index for wavelength of a multicast session.

All other parameters remain the same except for the wavelength-continuity constraint.

Variables:
  Boolean variable, $M_{m,n}^{i,c} - 1$, if the link between nodes m and n is occupied by multicast session i on wavelength c
  = 0, otherwise.
  Other variables V_m^i and $F_{m,n}^i$ remain the same
  A Boolean variable, $C_c^i = 1$, for multicast session i on wavelength c
  = 0, otherwise.
  The light-tree for a multicast session can access only one wavelength as there are no wavelength converters.

Optimize:
  Total cost of all multicast sessions [64]:

$$\text{Minimize:} \sum_{i=1}^{i=2k} \sum_{c=2}^{c=W} \sum_{m,n} w_{m,n} M_{m,n}^{ic} \tag{7.126}$$

Constraints:

Tree-creation constraints [2]:

$$\forall i, \forall n \neq s_i, \quad \sum_{m,c} M_{m,n}^{i,c} = V_n^i \tag{7.127}$$

$$\forall i, \quad \sum_{m,c} M_{m,s_i}^{i,c} = 0 \tag{7.128}$$

$$\forall i, \forall j \in S_i, j \geq 1, \quad V_j^i = 1 \tag{7.129}$$

$$\forall i, \forall m \neq d_{ij}, j \geq 1, \quad \sum_{n,c} M_{m,n}^{i,c} \geq V_m^i \tag{7.130}$$

$$\forall i, m, \quad \sum_{n,c} M_{m,n}^{i,c} \leq D_p(m) \cdot V_m^i \tag{7.131}$$

$$\forall m, n \quad \sum_{i,c} M_{m,n}^{i,c} \leq P_{mn} \cdot W \tag{7.132}$$

$$\forall m, n, c \quad \sum_c M_{m,n}^{i,c} \leq P_{mn} \tag{7.133}$$

Commodity-flow constraints [2]:

$$\forall i, \forall m \notin S_i \quad \sum_n F_{m,n}^i = \sum_n F_{n,m}^i \tag{7.134}$$

$$\forall i, \forall m = s_i \quad \sum_n F_{s_i,n}^i = L_i \tag{7.135}$$

$$\forall i, \forall m = s_i \quad \sum_n F_{n,s_i}^i = 0 \tag{7.136}$$

$$\forall i, \forall m \neq d_{ij}, j \geq 1, \quad \sum_n F_{n,m}^i = \sum_n F_{m,n}^i + 1 \tag{7.137}$$

$$\forall i, m, n \quad \sum_c M_{m,n}^{i,c} \leq F_{m,n}^i \tag{7.138}$$

$$\forall i, m, n \quad F_{m,n}^i \leq N \cdot \sum_c M_{m,n}^{i,c} \tag{7.139}$$

Direct link-disjointness constraints:

$$\forall i = 1,2\ldots k \forall m,n \quad M_{m,n}^{i,c} + M_{m,n}^{i+k,c} \leq 1 \tag{7.140}$$

Wavelength-related constraints:

$$\forall i \quad \sum_c C_c^i = 1 \tag{7.141}$$

$$\forall m,n(n > m)\forall i,c \quad M_{m,n}^{i,c} + M_{n,m}^{i+k,c} \leq C_c^i \tag{7.142}$$

Equations (7.127)–(7.132) are already explained earlier. Equation (7.133) restricts the number of sessions on the same wavelength between a node pair by P_{mn} (effectively ensuring that each fiber link supports no more than W wavelengths) [2]. Equations (7.134)–(7.140) are already explained as same as earlier. Equation (7.141) shows a session choosing only one wavelength. Equation (7.142) indicates no link I occupied by a session on the wavelength not chosen by it and all links occupied by a session are on the same wavelength [2]

7.13.3 Network Having Protection Based on Light-Trees

Here, a 15-node network is shown in Figure 7.21 [2], where each link of the network carries $W = 4$ wavelengths in both directions. $P_{mn} = 1$ indicates bidirectional fiber links between adjacent node pairs. We consider a group of five multicast sessions: $S_1 = \{0,1,2,4,5,6,10,11,12,13,14\}$, $S_2 = \{9, 1, 2, 3, 4,5, 6, 10,11\}$, $S_3 = \{12,0,5,8,9,10,1\}$, $S_4 = \{14,1,2,3,4\}$, and $S_5 = \{7,0,6\}$ to be set upon the network. Sessions S_1, S_4, and S_5 need single-link failure protection; i.e., each of them requires a backup light-tree having session identifiers S_6, S_9, and S_{10}, respectively. The formulations for RWA for a group of primary and their backup sessions are made by standard ILP [2]. Figure 7.21 shows an optimal RWA of the above eight sessions in the absence of the wavelength-continuity constraint. Primary trees are represented by the solid lines, and backup trees are represented by the dotted lines [2].

7.13.4 Other Protection Schemes [2]

The directed-link-disjoint backup tree is difficult to estimate after determining a primary tree. It is not impossible to protect each segment in the primary tree by determining a segment-disjoint path. A segment is basically a sequence of edges from the source or any splitting point (on a tree) to a downstream splitting point. A destination node acts as a segment end node, which is a splitting point where a portion of a signal is dropped locally, and the remainder continues [65]. Figure 7.22 shows a primary tree represented by the dotted lines, which is found along edges $s \rightarrow u$, $s \rightarrow v$, $v \rightarrow d_1$, and $v \rightarrow d_2$. The node v is a splitting point having four segments, viz., $s \rightarrow u$, $s \rightarrow v$, $v \rightarrow d_1$, and $v \rightarrow d_2$. There is no other path, besides the paths along the primary tree, from source node s to either destination node d_1 or d_2; hence, it is impossible to estimate a directed-link-disjoint backup tree. However, each of the three segments

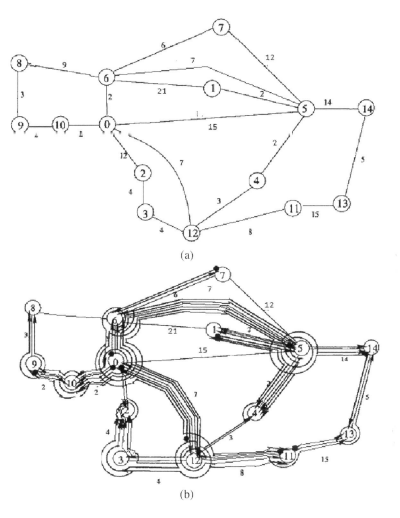

FIGURE 7.21 (a) Network having 15 nodes and with bidirectional physical links having cost (b) Optimal RWA of the group of seven multicast sessions with full-wavelength conversion. The dark circles represent the source nodes, and the nodes where arrows terminate are the destination nodes [2].

$s - v$, $v - d_1$, and $v -,d_2$ of the tree is protected by segment-disjoint paths $<s\text{-}w\text{-},v>$, $<v\text{-}d_2\text{-}d_1>$, and $<v\text{-},d_1\text{-}d_2>$, respectively. The segment-disjoint paths share their paths with existing primary tree or another backup segment (Figure 7.22a) in which an arc-disjoint backup tree cannot be obtained. Transmitting signal having information from a source node to all destination nodes needs a path available from the source to each destination node. In order to protect each path, a backup path from the source node to the destination node is needed. A link-disjoint path pair from the source node to every destination node handles any link failure in a directed graph. Figure 7.22b shows link-disjoint paths from s to d_1 (primary path $<s\text{-}u\text{-}d_1>$ and

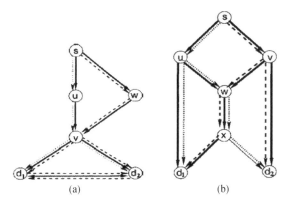

FIGURE 7.22 (a) Using shared segment-disjoint approach to protect a multicast session where directed-link-disjoint tree approach fails. (b) Using optimal path-pair approach to protect a multicast session where directed-link-disjoint tree approach fails [2].

backup path $<s-v-w-z-d_1>$ represented by the dotted and dashed lines, respectively) and link-disjoint paths from s to d_2 (primary path $<s-v-d_2>$ and backup path $<s-u-w-z-d_2>$). Both backup paths for d_1 and d_2 share a link $(w-z)$ [66,67].

In the WA problem, a multicast tree is given, and the problem is to select the wavelengths on the links in the tree and the intermediate nodes for wavelength conversion. Another issue of light-tree routing is power budget constraint. The light-tree routing normally uses a number of optical power splitters for high-power budget constraints. So, it is required to find out the constraints on the end-to-end paths in order to guarantee an adequate signal quality and to ensure a measure off air ness among the destination nodes [68,69].

7.14   PROTECTION OF TRAFFIC GROOMING-BASED OPTICAL NETWORK

A wavelength channel capacity is STS-192, and it is now STS-768. The bandwidth requirement of a typical connection request is STS-1 or lower. For an efficient use of the network resources, sub-wavelength-granularity connections are groomed into direct optical transmission channels, or wavelength channels. The failure of a network element may damage the failure of several connections, which results in large data and revenue loss. Fault-management protection schemes repair the services against such failures. Different low-speed connections want different bandwidth granularities along with protection (dedicated or shared). In survivable traffic-grooming problem, sub-wavelength-granularity connections need to be protected. For a static traffic matrix and the protection requirement of each connection request, the work in Ref. [70] provides an integer linear program and a heuristic for satisfying the bandwidth and protection requirements of all the connection requests while minimizing the network cost in terms of transmission cost and switching cost. There are three schemes: protection-at-light path (PAL) level, mixed protection-at-connection (MPAC) level, and separate protection-at-connection (SPAC) level for dynamically provisioning of shared-protected sub-wavelength-granularity connection requests against fiber-cut failures. The characteristics of these schemes are realized under a generic grooming-node architecture and design-efficient heuristics. This section discusses the partial-grooming switch architectures, which are already discussed in Chapter 2. There are two types of resource constraints: wavelengths and grooming ports [25]. More the number of wavelengths the network has, the less the number of grooming ports a node needs, and vice versa. For the studies of these approaches, we consider every edge corresponds to a bidirectional fiber; each fiber has two wavelengths considering the wavelength capacity of STS-192 [71]; every node has three grooming ports.

7.14.1 PROTECTION-AT-LIGHTPATH (PAL) LEVEL

PAL exhibits protection with respect to light path after failure occurrence [2]. Under PAL, a connection is set up along with a sequence of protected light paths, or p light paths. A p-light path is taken as a working path, whereas a link-disjoint path as backup path [2]. The working path approaches a grooming-add port at the source node and a grooming-drop port at the destination node of a p-light path, and the working path of a p-light path by passes any intermediate nodes along its path. In case the working path fails, protection occurs at light path level and the backup path is set up as a light path by using the grooming ports previously used by the working path. Figure 7.23a shows a method of provisioning c_1 under PAL with the arrival of the first connection request c_1, (0, 2, STS-12c, t_1). Connection c_1 is set up with p-light path l_1, where (O-1-2) is used as working path and path (0–5-4-2) is used as backup path. The p-light path l_1 consists of a grooming-add port at node 0 and a grooming-drop port at node 2. The free capacity of p-light path l_1 is STS-192. Considering c_1 in the network, the second connection request c_2, (0, 3, STS-3c, t_2), comes to the network. Figure 7.23b shows another method of provisioning c_2 under the current network state [2]. Connection c_2 is set up via p-lightpath l_2, which has light path (O-1-3) as working path and path (0–5-4-3) as backup path. For the third connection request c_3, (4, 3, STS-48c, t_3), Figure 7.23c shows another provisioning c_3 under the current network.

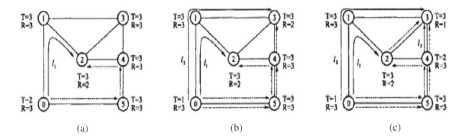

(a)   (b)   (c)

FIGURE 7.23   PAL: provisioning connections c_1 ((*O*, 2, STS-12c, t_1)), c_2 ((0, 3, STS-3c, t_2)), and c_3 ((4, 3, STS-48c, t_3)): (a) after provisioning c_1, (b) after provisioning c_2, and (c) after provisioning c_3 [2].

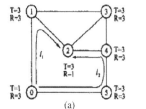 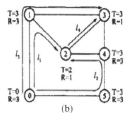 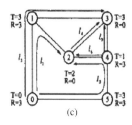

(a)   (b)   (c)

FIGURE 7.24   MPAC: provisioning connections c_1 ((O,2, STS-12c, t_1)), c_2((0,3,STS-3c, t_2)), and c_3((4,3,STS-48c, t_3)): (a) after provisioning c_1, (b) after provisioning c_2, and (c) after provisioning c_3.

7.14.2   MIXED PROTECTION-AT-CONNECTION (MPAC) LEVEL

MPAC provides the protection for the connections. A connection is set up through link-disjoint working and backup paths with sequence of light paths. The backup light path is shared among multiple backup paths where their corresponding working paths are link disjoint. The capacity of one wavelength can be used by both working paths and backup paths. MPAC accommodates individual connections and therefore can pack connections efficiently. Considering the first connection request c_1, (0, 2, STS-12c, t_1), Figure 7.24a presents one way of provisioning c_1 under MPAC. The working and backup paths of connection c_1 have light paths l_1 and l_2, respectively. The free capacity of both light paths l_1 and l_2 is STS-180 [2]. The backup capacity in light path l_2 is STS-12c. Both light paths l_1 and l_2 have a grooming-add port at node 0 and a grooming-drop port at destination node 2. Having connection c_1 in the network, another connection request c_2, (0, 3, STS-3c, t_2), reaches. The setup of c_2 under MPAC is shown in Figure 7.24b. Connection c_2 is set up with the two-light path sequence (l_2, l_4) as working path and backup path, respectively. The working path transmits with either of the two paths, light path l_3. The free capacity of both light paths l_3 and l_4 is STS-189. The free capacity of light path l_2 is STS-177 since light paths l_1 and l_3 transmit common link (0, 1). The backup capacity on light path l_2 is STS-15 (STS-12c capacity is used to protect the working path of connection c_1, and STS-3c capacity is used to protect the working path of c_2). The backup capacity on light path $l4$ is STS-3c, and it is used to protect the working path of connection c_2. Having both connections c_1 and c_2 in the network, connection request c_3, (4, 3, STS-48c, t_3), comes to the network and Figure 7.24c shows a way of provisioning c_3 under MPAC in this case.

7.14.3   SEPARATE PROTECTION-AT-CONNECTION (SPAC) LEVEL

SPAC gives protection of each connection established in the network. Under SPAC, an established connection is set up via link-disjoint working and backup paths. In this approach, the capacity of a wavelength can be used by either working paths or backup paths, but not both. A grooming-add port at the source end of the link and a grooming-drop port at the destination end of the link are recorded for each reserved wavelength because multiple backup paths groomed on to the same wavelength on a link go to different next hops [2].

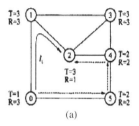 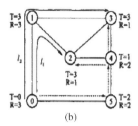 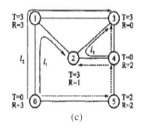

(a)							(b)							(c)

FIGURE 7.25 SPAC: provisioning connections $c_1((0, 2, STS-12c, t_1))$, $c_2((0, 3, STS-3c, t_2))$, and $c_3((4, 3, STS-48c, t_3))$: (a) after provisioning c_1, (b) after provisioning c_2, and (c) after provisioning c_3.

We consider the same network shown in Figure 7.24 for the analysis of SPAC approach. Considering connection request c_1, $(0, 2, STS-12c, t_1)$, Figure 24.25a presents the setup c_1 under SPAC. The working path of connection c_1 is light path l_1, and the backup path is path (0–5–4–2). The free capacity of light path l_1 is STS-180. Having a connection c_1 in the network, the connection request c_2, $(0, 3, STS-3c, t_2)$, reaches. Figure 7.25b presents the working path of connection c_2 to be light path l_2, and the backup path to be path (0–5–4–3). Considering link (0, 1) failure shows connection c_1 to be rerouted along (0–5–4–2), and connection c_2 needs to be rerouted along (0–5–4–3). Considering arrival of connections c_1 and c_2 in the network, Figure 7.25c shows provisioning of a new arrival c_3 under SPAC. The working path of connection c_3 is light path l_3 and the backup path is path (4, 3). The free capacity of light path l_3 is STS-144. STS-48c of the entire backup capacity a long link (4, 3) is used to protect the working path of c_3; out of this STS-48c capacity, STS-3c is also used to protect the working path of c_2 [2].

SUMMARY

This chapter mentions the fault-management mechanisms used in deploying a survivable optical network in both SONET/SDH rings and mesh-based architectures using OXCs. Various protection and restoration schemes are mentioned considering the concepts of stacked rings and protection cycles, routing algorithms for computing a pair of link-disjoint paths, ILP formulations of path-protection schemes to optimize the total number of bandwidth used for a given set of static traffic demands, the techniques to maximize share ability for the shared-protection schemes, and dynamic restoration.

We then show some advanced network survivability in terms of service availability, service reliability, RT, and service restorability. A framework for cost-effective availability- aware connection provisioning was mentioned to provide differentiated services in WDM mesh networks.

Other than protecting provisioned bandwidth, there is a parallel body of research on "Internet Protocol (IP) resilience". The objective of IP resilience is to pre compute or dynamically discover alternate routes for data packet traffic, to effectively deal with network failures.

This chapter also presents two switch architectures (opaque and transparent) for supporting multicasting in WDM networks. The concept of "light-tree" was mentioned, and we also show how a light-tree can reduce network-wide average packet hop distance and the total number of transceivers for unicast and broadcast traffic. Efficient algorithms are mentioned in this chapter to reduce network resources for establishing multicast connections using light-trees. The *directed-link-disjointness, segment protection,* and *path protection* were shown for protecting multicast trees.

EXERCISES

7.1. There is a network topology having only dedicated protection shown in Figure Exercise 7.1.

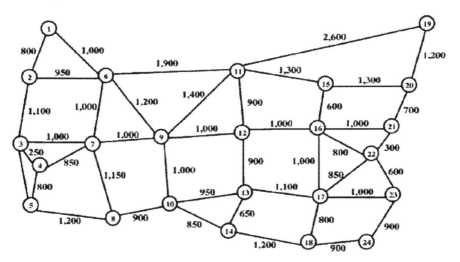

EXERCISE 7.1   Network having 24 nodes.

a. Find the shortest primary path from source node 3 to destination node 23.

b. Find the shortest backup path from node 3 to node 23 so that this path is link disjoint to the path found in part (a). Calculate the total cost of this two-step approach (sum of primary and backup paths) for dedicated path protection between nodes 3 and 23.

c. We now describe a one-step algorithm for finding the primary and backup paths concurrently for dedicated path protection:
   • Reverse the links of the primary path found in part (a). We will refer to this path as reverse path.
   • Find another shortest path from the source node to the destination node on the updated network with the above reverse path.
   • Remove the links on this newly found shortest route which uses links on the *reverse* path.
   • Reverse the links of the *reverse* path.

Use the above one-step algorithm to find two link-disjoint paths between nodes **3** and **23**. Compare the cost of the one-step algorithm with the two-step approach. Is the new backup path found by your node disjoint to its primary?

7.2. Consider the same network topology having shared protection in the previous problem.

  a. Consider another connection from node 6 to node 16. Use the one-step algorithm to find the primary and backup paths between nodes 6 and 16. What is the total cost of establishing the two connections (3–23 and 6–16) with dedicated path protection?

  b. Find the primary path for the connection from node 6 to node 16. Also find its backup while allowing it to share links with the backup path of the connection between nodes 3 and 23. What is the cost of establishing the two connections (3–23 and 6–16) with shared-path protection?

7.3. Consider the same network topology with shared-link protection shown in Figure Exercise 7.1. Show the backup routes for protecting the two primary paths (between nodes 3 and 23 and between nodes 6 and 16 using shared-link protection). Compare the cost of shared-link protection with that of dedicated-link protection.

7.4. Consider N:1 protection (N primary paths protected by 1 shared backup) in Figure Exercise 7.2. Calculate the possibility that N:1 backup is insufficient (number of failed path 2).

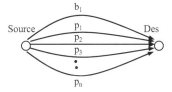

EXERCISE 7.2 N:1 backup.

7.5. Consider the network topology in Figure Exercise 7.3. A light path is setup between nodes 1and 8 along the link $1\rightarrow3\rightarrow6\rightarrow8$. If the link $3\rightarrow6$ goes down, which of the following restoration strategies will lead to a path with minimum number of hops? Find the number of via in each case 1: N backup path restoration.

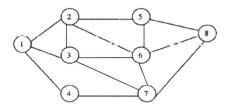

EXERCISE 7.3   Network.

7.6. Consider the network topology in Figure Exercise 7.3, where all link shaves the same cost. Connection requests arrive between nodes 1-5 and nodes 5-4. Design a protection scheme for these two connection requests while minimizing the total number of wavelengths used (for primary as well as backup).

7.7. A transparent switch, with two optical switches, has 12 incoming/outgoing fibers, having 8 wavelengths in each fiber. The sizes of the larger and the smaller optical switches are 98×102 and 20×20, respectively. Find the splitter-bank size. Assume that the number of local add/drop ports in the larger and smaller switches is 2 and 4, respectively. What is the splitting degree of each splitter?

Consider the modified switch having one OXC, and find the size of the switch.

7.8. An MILP formulation of multicast traffic with wavelength converters has 160 variables. Find the number of nodes in the network if there are 4 multicast sessions. Assume that each fiber link has 6 wavelengths. What is the upper bound of the number of constraints in the above MILP? How does the number of variables change, assuming no wavelength converter? Also, calculate the number of constraints.

7.9. Consider the MILP formulation of multicast traffic with grooming. Find the number of variables and the number of constraints. How does the result change if you have only **5** splitters at each node, with each having a splitting degree of 4? Consider the data found in Problem 7.8.

7.10. Consider the network topology in Figure 3.9. There is only one direct physical bidirectional fiber link between any two nodes. Each link carries one wavelength. We need to establish **5** full-capacity multicast sessions $S_1 = \{E, A, B, C\}$, $S_2 = \{D, E, F, C\}$, $S_3 = \{C, F, A, D\}$, $S_4 = \{A, E, F\}$, and $S_5 = \{F, E, D\}$. Solve the RWA problem using an MILP. What is the total cost?

Consider the network topology in Figure 3.9. There is one direct physical bidirectional fiber link between any two nodes. Each link carries **2** wavelengths. We need to establish **5** full-capacity multicast sessions $S_1 = \{E, A, B, C\}$, $S_2 = \{D, E, F, C\}$, $S_3 = \{C, F, A, D\}$, $S_4 = \{A, E, F\}$, and $S_5 = \{F, E, D\}$. Solve the RWA problem using an MILP. What is the total cost? Assume

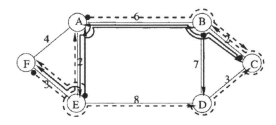

EXERCISE 7.4 Optimal routing of the group of five multicast sessions with full-range wavelength converters. The dark circle represents the source node of a session, and the nodes where arrows terminate are the destination nodes. Sessions on wavelengths λ_0 and λ_1 are represented by the solid lines and the dashed lines, respectively.

an o-X conversion network. Solve this problem for a full A-conversion network.

7.11.   Consider the network topology in Figure Exercise 7.1. There is only one direct physical bidirectional fiber link between any two nodes. Each link carries only **2** wavelengths. We need to establish 6 multicast sessions $S_1 = \{1.0,6,1,5,4\}$, $S_2 = \{0.25,0,6,10,2\}$, $S_3 = \{1.0,8,6,7,5,14\}$, $S_4 = \{1.0,1,6,8,5\}$, $S_5 = \{0.5,12,4,5,8,9,10,0,2\}$, and $S_6 = \{0.5,7,6,5,1, 14,13\}$. Find the RWA solution by using a MILP. Considering a full λ conversion network, how many wavelength converters are needed? At which nodes, these are placed? Find the total cost for the network having no converter and with converters.

7.12.   Make a MILP formulation of a dual problem for multicast tree protection without λ continuity.

REFERENCES

1. R. Ramaswami and K. Sivarajan, *Optical Networks: A Practical Perspective*, 2nd ed., San Francisco: Morgan Kaufmann Publishers, 2001.
2. B. Mukherjee, *Optical WDM Networks*, Berlin: Springer-Verlag, 2006.
3. M. To and P. Neusy, "Unavailability analysis of long-haul networks," *IEEE Journal on Selected Areas in Communications*, vol. 12, pp. 100–109, January 1994.
4. S. Rarnamurthy, *Optical Design of WDM Network Architectures*, Ph.D. Dissertation, University of California, Davis, 1998.
5. S. Ramamurthy and B. Mukherjee, "Survivable WDM mesh networks, part I - protection," *Proceedings, IEEE INFOCOM '99*, New York, pp. 744–751, March 1999.
6. K. Zhu and B. Mukherjee, "On-line approaches for provisioning connections of different bandwidth granularities in WDM mesh networks," *Proceedings, OFC '02*, Anaheim, CA, pp. 549–551, March 2002.
7. J. Zhang and B. Mukheriee, "A review of fault management in WDM mesh networks: Basic concepts and research challenges," *IEEE Network*, vol. 18, no. 2, pp. 41–48, March/April 2004.
8. W. Goralski, *SONET/SDH*, 3rd ed., New York: McGraw-Hill, October 2002.
9. G. Ellinas, A. Hailemariam, and T. E. Stern, "Protection cycles in mesh WDM networks," *IEEE Journal on Selected Areas in Communications*, vol. 18, pp. 1924–1937, October 2000.
10. O. Gerstel and R. Ramaswami, "Optical layer survivability: A services perspective," *IEEE Communications Magazine*, vol. 38, pp. 104–113, March 2000.
11. O. Gerstel, R. Ramaswami, and S. Foster, "Merits of hybrid optical networking," *Proceedings, OFC '02*, Anaheim, CA, pp. 33–34, March 2002.
12. O. Gerstel, "Opportunities for optical protection and restoration," *Proceedings, OFC'98*, San Jose, CA, pp. 269–270, February 1998.
13. O. Gerstel, "Optical networking: A practical perspective (tutorial)," *IEEE Hot Interconnects*, August 2000.
14. G. Li, D. Wang, C. Kalmanek, and R. Doverspike, "Efficient distributed path selection for shared restoration connections," *Proceedings, IEEE INFOCOM '02*, New York, pp. 140–149, June 2002.
15. R. Ramamurthy, et al., "Capacity performance of dynamic provisioning in optical networks," *IEEE/OSA Journal of Light wave Technology*, vol. 19, pp. 40–48, January 2001.
16. S. Ramamurthy, L. Sahnsrabuddhe, and B. Mukherjee, "Survivable WDM mesh networks," *IEEE/OSA Journal of Lightwave Technology*, vol. 21, pp. 870–883, April 2003.

17. H. Zang and B. Mukherjee, "Connection management for survivable wavelength-routed WDM mesh networks," *SPIE Optical Networks Magazine*, vol. 2, no. 4, pp. 17–28, July 2001.

18. S. Gowda, M. Sivakumar, and Krishna M. Sivalingam, "Protection mechanism for optical WDM networks based on wavelength converter multiplexing and backup path relocation technique," *IEEE INFOCOM*, vol. 1, pp. 12–21, 2003.

19. C. Ou, et al., "Sub-path protection for scalability and fast recovery in optical WDM mesh networks," *IEEE Journal on Selected Areas in Communications*, vol. 22, no. 11, pp. 1859–1875, November 2004

20. H. Mouftah, et al., "A framework for service-guaranteed shared protection in WDM mesh networks," *IEEE Communications Magazine*, vol. 40, pp. 97–103, February 2002.

21. V. Anand, S. Chauhan, and C. Qiao, "Sub-path protection: A new framework for optical layer survivability and its quantitative evaluation," Technical Report, Department of Computer Science and Engineering, State University of New York at Buffalo, January 2002.

22. C. Ou, K. Zhu, H. Zang, L. H. Sahasrabuddhe, and B. Mukherjee, "Traffic grooming for survivable WDM networks – Dedicated protection," *OSA Journal of Optical Networking*, vol. 3, pp. 50–74, January 2004.

23. A. Fumagalli, I. Cerutti, and M. Tacca, "Optimal design of survivable mesh networks based on line switched WDM self-healing rings," *IEEE/ACM Transactions on Networking*, vol. 11, pp. 501–512, June 2003.

24. W. D. Grover, J. Doucette, M. Clouqueur, D. Leung, and D. Stamatelakis, "New options and insights for survivable transport networks," *IEEE Communications Magazine*, vol. 40, pp. 34–41, January 2002.

25. H. Zang, C. Ou, and B. Mukherjee, "Path-protection routing and wavelength-assignment (RWA) in WDM mesh networks under duct-layer constraints," *IEEE/ACM Transactions on Networking*, vol. 11, no. 2, pp. 248–258, April 2003.

26. J. Zhang, K. Zhu, S. J. B. Yoo, L. Sahasrabuddhe, and B. Mukherjee, "On the study of routing and wavelength assignment approaches for survivable wavelength-routed WDM mesh networks," *SPIE Optical Networks Magazine*, vol. 4, no. 6, pp. 16–28, November/December 2003.

27. J. W. Suurballe and R. E. Tarjan, "A quick method for finding shortest pairs of disjoint paths," *Networks*, vol. 14, pp. 325–336, 1984.

28. J. Wang, W. Cho, V. R. Vemuri, and B. Mukherjee, "Improved approaches for cost-effective traffic grooming in WDM ring networks: ILP formulations and single-hop and multi hop connections," *IEEE/OSA Journal of Light wave Technology*, vol. 19, no. 11, pp. 1645–1653, November 2001.

29. G. Mohan, C. S. R. Mmthy, and A. K. Somani, "Efficient algorithms for routing dependable connections in WDM optical networks," *IEEE/ACM Transactions on Networking*, vol. 9, no. 5, pp. 553–566, October 2001.

30. J. Zhang, K. Zhu, and B. Mukherjee, "Backup re provisioning to remedy the effect of multiple link failures in WDM mesh networks," *IEEE Journal on Selected Areas in Communications*, vol. 24, no. 8, pp. 57–67, 2005.

31. C. Qiao, Y. Xiong, and D. Xu, "Novel models for efficient shared-path protection," *Proceedings, OFC '02*, Anaheim, CA, pp. 546–547, March 2002.

32. A. Fumagalli, A. Paradisi, S. M. Rossi, and M. Tacca, "Differentiated reliability (DiR) in mesh networks with shared path protection: Theoretical and experimental results," *Proceedings, OFC '02*, Anaheim, CA, pp. 490–492, March 2002.

33. A. Fumagalli, M. Tacca, F. Unghvary, and A. Farago, "Shared path protection with differentiated reliability," *Proceedings, IEEE ICC '02*, New York, pp. 2157–2161, April 2002.

34. A. Hac, "Improving reliability through architecture partitioning in telecommunication networks," *IEEE Journalon Selected Areas in Communications*, vol. 12, pp. 193–204, January 1994.

35. W. Wen, B. Mukherjee, and S. J. B. Yoo, "QoS-based protection in MPLS-controlled WDM mesh networks," *Photonic Network Communications*, vol. 4, pp. 297–320, July 2002.

36. A. Banerjee, G. Kramer, and B. Mukherjee, "Fair queuing using service level agreements (SLAs) for open access in an Ethernet passive optical network (EPON)," *Proceedings, IEEE International Conference on Communications (ICC) '05*, Seoul, Korea, May 2005.

37. K. S. Trivedi, *Probability and Statistics with Reliability, Queuing, and Computer Science Applications*, Englewood Cliffs, NJ: Prentice-Hall, 1982.

38. G. Ellinas, S. Rong, A. Hailemariam, and T. E. Stern, "Protection cycle covers in optical networks with arbitrary mesh topologies," *Proceedings, OFC '00*, Baltimore, MD, pp. 213–215, March 2000.

39. C. Ou, et al., "Sub-path Protection for Scalability and Fast Recovery in Optical WDM Mesh Networks," *IEEE Journal on Selected Areas in Communications*, vol. 22, no. 11, pp. 1859–1875, November 2004.

40. M. Kodialam and T. V. Lakshman, "Dynamic routing of bandwidth guaranteed tunnels with restoration," *Proceedings, IEEE INFOCOM '00*, Tel Aviv, Israel, pp. 902–911, March 2000.

41. P. P. Sahu and R. Pradhan, "Reduction of blocking probability in protected optical network using Alternate routing and wavelength converter," *Journal of Optical Communication*, vol. 29, pp. 20–25, 2008.

42. P. P. Sahu, "New traffic grooming approaches in optical networks under restricted shared protection," *Photonics Communication Networks*, vol. 16, pp. 223–238, 2008.

43. P. P. Sahu, "A new shared protection scheme for optical networks," *Current Science Journal*, vol.91, no. 9, pp. 1176–1184, 2006.

44. P. P. Sahu and R. Pradhan, "Reduction of blocking probability in restricted shared protected optical network," *Proceedings of XXXIII OSI Symposium on Optics and Optoelectronics*, Tezpur, India, pp. 11–14, 2008.

45. M. Clouqueur and W. D. Grover, "Availability analysis of span restorable mesh networks," *IEEE Journal on Selected Areas in Communications*, vol. 20, pp. 810–821, May 2002.

46. S. Paul, *Multicasting on the Internet and Its Applications*, Boston, MA: Kluwer Academic Publishers, 1998.

47. C. K. Miller, *Multicast Networking and Applications*, Reading, MA: Addison-Wesley, 1999.

48. R. Malli, X. Zhang, and C. Qiao, "Benefit of multicasting in all-optical networks," *Proceedings, SPIE Conference on All Optical Networking*, vol. 2531, pp. 209–220, November 1998.

49. S. Sankaranarayanan and S. Subramaniam, "Comprehensive performance modeling and analysis of multicasting in optical networks," *IEEE Journal of Selected Areas in Communications*, vol. 21, no. 9, pp. 1399–1413, November 2003.

50. K.-C. Lee and V. O. K. Li, "A wavelength-convertible optical network," *IEEE/OSA Journal of Light wave Technology*, vol. 11, no. 516, pp. 962 970, May/June 1993.

51. C. Chen and S. Banerjee, "Optical switch configuration and light path assignment in wavelength routing multi hop light wave networks," *Proceedings, IEEE INFOCOM '95*, Boston, MA, pp.1300–1307, June 1995.

52. V. Kompella, J. Pasquale, and G. Polyzos, "Multicast routing for multimedia communications," *IEEE/ACM Transactions on Networking*, vol. 1, no. 3, pp. 286–292, 1993.

53. J. He, S. H. G. Chan, and D. H. K. Tsang, "Routing and wavelength assignment for WDM multicast networks," *Proceedings, IEEE Globe com '01*, San Antonio, TX, pp. 1536–1540, 2001.

54. F. K. Hwang and D. S. Richards, "Steiner tree problems," *Networks*, vol. 22, no. 1, pp. 55–89, 1992.

55. M. R. Garey and D. S. Johnson, *Computers and Intractability: A Guide to the Theory of NP-Completeness*, New York: W. H. Freeman and Company, 1979.

56. L. H. Sahasrabuddhe and B. Mukherjee, "Light-trees: Optical multicasting for improved performance in wavelength-routed networks," *IEEE Communications Magazine*, vol. 37, no. 2, pp. 67–73, February 1999.

57. CPLEX, http://www.ilog.com.

58. S. Ramesh, G. N. Rouskas, and H. G. Perros, "Computing blocking probabilities in multiclass wavelength-routing networks with multicast calls," *IEEE Journal on Selected Areas in Communications*, vol. 20, no. 1, pp. 89–96, January 2002.

59. S. Sankaranarayanan and S. Subramaniam, "Comprehensive performance modeling and analysis of multicasting in optical networks," *IEEE Journal of Selected Areas in Communications*, vol. 21, no. 9, pp. 1399–1413, November 2003.

60. Y. Sun, J. Gu, and D. H. K. Tsang, "Multicast routing in all optical wavelength routed networks," *Optical Networks Magazine*, vol. 2, pp. 101–109, July/August 2001.

61. K. Zhu and B. Mukherjee, "Traffic grooming in an optical WDM mesh network," *IEEE Journal on Selected Areas in Communications*, vol. 20, no. 1, pp. 122–133, January 2002.

62. R. Libeskind-Hadas, "Efficient collective communication in WDM networks with a power budget," *Proceedings, Ninth, IEEE International Conference on Computer Communications and Networks (ICCCN)*, Las Vegas, NV, pp. 612–616, October 2000.

63. F. Bauer and A. Varma, "Degree-constrained multicasting in point-to-point networks," *Proceedings, IEEE INFOCOM '95*, Boston, MA, vol. 4, pp. 369–376, April 1995.

64. X. Jia, D. Du, X. Hu, M. Lee, and J. Gu, "Optimization of wavelength assignment for QoS multicast in WDM networks," *IEEE Transactions on Communications*, vol. 49, no. 2, pp. 341–350, February 2001.

65. X. Zhang, J. Wei, and C. Qiao, "Constrained multicast routing in WDM networks with sparse light splitting," *Proceedings, IEEE INFOCOM '00*, Tel Aviv, Israel, pp. 1781–1790, 2000.

66. N. K. Singhal, C. Ou, and B. Mukherjee, "Cross-sharing vs. self-sharing trees for protecting multicast sessions in mesh networks," *Computer Networks Journal*, vol. 50, no. 02, pp. 200–206, 2005.

67. N. K. Singhal, L. H. Sahasrabuddhe, and B. Mukherjee, "Provisioning of survivable multicast sessions against single link failures in optical WDM mesh networks," *IEEE/OSA Journal of Light wave Technology*, vol. 21, no. 11, pp. 2587–2594, November 2003.

68. Y. Xin and G. N. Rouskas, "Light-tree routing under optical power budget constraints [Invited]," *Journal of Optical Networks*, vol. 3, no. 5, pp. 282–302, May 2004.

69. M. Ali and J. Deogun, "Power-efficient design of multicast wavelength routed networks," *IEEE Journal Selected Areas in Communications*, vol. 18, no. 10, pp. 1852–1862, October 2000.

70. A. Lardies, R. Gupta, and R. A. Patterson, "Traffic grooming in a multi-layer network," *SPIE Optical Networks Magazine*, vol. 2, no. 3, pp. 91–99, May/June 2001.

71. G. Keiser, *Optical Fiber Communication*, New York: McGraw-Hill, 1999.

8 Restoration Schemes in the Survivability of Optical Networks

Due to failures occurring in a network, the transmission of signals is disrupted [1–3] as mentioned in Chapter 7. During construction or destructive natural calamities, such as earthquakes, flood, etc., a cut in a fiber cable (having a large number of optical fibers) can lead to a huge loss in data and revenue. There are mainly two failures which may disrupt transportation – traffic-node failure and link failure [1] – as shown in the ring network in Figure 8.1. A node can also fail because of disastrous events such as fire or flooding; this is called as node failure. Apart from node and link failures, there is another type of failure called channel failure in optical WDM networks. A channel failure is basically because of the failure of the transmitting or receiving equipment operating on that channel. For survivability against these failures, there are two types of approaches – protection and restoration. In protection, backup paths are formed during the establishment of a connection initially. Different protection schemes were already discussed in Chapter 7. Although resource utilization is improved by considering different sharing schemes (as discussed in Chapter 7), still a solution to the problem of poor utilization of resources [2] (especially wavelength channel and hardware (add/drop mux, WDM, optical switch, OXC, etc.)) needs to be found. Restoration may be a better solution to the resource utilization problem. In this chapter, we shall discuss different restoration schemes [4].

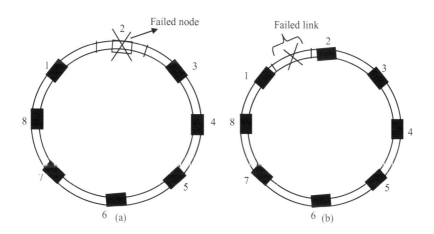

FIGURE 8.1 Restoration of ring topology against (a) node failure and (b) link failure.

8.1   RESTORATION NETWORKS

Restoration of every service and function is made in the IP layer. There are techniques for rerouting demand flows around failures by considering dynamic changes of routing table in the IP layer or by dynamic routing and robust trunk group network design in the switched traffic condition for transient service-layer disruptions of few minutes. The difficulty of IP restoration is that one link cut gives a large number of simultaneous logical link failures. In this direction, priority-based restoration and transmission-level restoration (e.g., SONET, WDM) provide the solution to these difficulties. The restoration in ring network is slightly less complicated than that in mesh network [2].

8.1.1   RING TOPOLOGY

Survivability in ring topology is achieved by using simple shared protection schemes which can restore the network if or when a failure occurs. Rings' restoration is the general class of survivable network designs. There are two types of ring restoration – SDH multiplex-section shared-protection ring (MS-SP Ring) and SONET bidirectional line-switched ring (BLSR) (which has already been discussed in Chapter 7) – requiring the protection bandwidth share of the backup ring to restore all span cuts in the working path of the ring through very fast (50–150 ms) line-level loop back switching [2]. Traffic is restored very easily by considering double-ring topology under node failure and link failure as shown in Figure 8.1. In Figure 8.1a, the node-2 fails, and due to this, the ring is configured by connecting links of the ring near to neighboring node-1 and node-3. In Figure 8.1b, the link fails, and due to this, the ring is configured by connecting links of the ring nearnode-1 and node-2.

8.1.2   MESH TOPOLOGY RESTORATION

Mesh topology restoration is more complex than any other restoration. The mesh topology is an irregular structure where nodes are distributed arbitrarily to cover a nationwide area or metropolitan area [2]. In mesh restoration, rerouting can use many diverse paths throughout the network, using relatively small amounts of spare capacity on each span. The spare capacity in a mesh network is used again in more ways than that in ring-based networks. So mesh restorable networks are more efficient in comparison with centralized control [5]. There are mesh restorations considered on the basis of restoration speed.

1. **Restoration in optical network:** This concept uses a new generation of optical cross-connect systems (OXCS's) with fast restoration switching.
2. **Distributed preplanning:** It is a slow distributed restoration algorithm (DRA), but it performs very fast restoration through "distributed preplanning" (DPP) [6]. The DPP works in the real-time phase of restoration. The real-time phase provides very fast restoration and activation.
3. **Capacity design for path restoration:** In span-restorable networks, path-restorable in networks can perform well in terms of the low total redundancies (as low as 21%) requiring 100% restorability. A distributed path restoration protocol makes the restoration more efficient.

8.2 PARAMETERS CONSIDERED IN RESTORATION

Service restorability in a network can provide survivability against a particular failure. The restorability $R_f(i)$ of f-order ($f > 1$) failure in ith scenario is basically a fraction of failed working capacity that is restored using a specified approach with the spare capacity in a network [7]. The overall restorability R_f of a network is the average value of $R_f(i)$ over f-order failure scenarios. The ratio between restorable capacity and failed capacity overall single-failure (dual-failure) scenarios is called the single-failure (dual-failure) network restorability. Different parameters such as disruption rate, restoration time, restorability, capacity efficiency, and algorithm scalability are discussed below.

8.2.1 DISRUPTION RATE

Service disruption rate is a function of the failure rate and operation policies. It is basically the number of traffic disturbed per unit time due to failure occurrences in the network. Both restoration and protection are functions of disruption rate. The successful backup rearrangement is not sure especially when the network load is high, so non-reverting may cause unpreferred service degradation. Hence, operation policies are needed for restoration according to customers' demand and network resource utilization [2].

8.2.2 RESTORATION TIME

Service restoration time is also called as restoration-switching time which is arisen according to different fault-management schemes [1,2]. It is the time required to restore traffic through predesigned protection assets. In dedicated (link, path, or sub-path) protection, OXCs on the backup paths can be pre-designed when the connection is set up. Then, no OXC configuration is needed when the failure occurs. This type of recovery can be very fast. If traffic is not transmitted on both paths, the destination node needs to wait until the source node is notified, and it switches traffic to the backup path. So, the restoration-switching time comprises the time for failure detection, failure announcement, and the propagation delay. The OXCs on the backup paths cannot be designed until the failure occurs. For dynamic restoration [2], the service restoration time comprises the time for route computation and resource discovery, failure detection time, notification time, OXC reconfiguration time, and propagation delay.

There are two steps – to divide a large network into smaller domains and then to restore each connection when a failure occurs. Only the affected domain will be operated in its protection *sub-path* so that the restoration time is reduced due to the reduced path length.

8.2.3 RESTORATION SPEED

Restoration speed depends on elapsed time span between a failure occurrence moment and the traffic restoration moment [1]. *Restorability* depends on failure

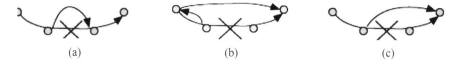

(a)   (b)   (c)

FIGURE 8.2   (a) Link restoration, (b) path restoration, and (c) sub-path restoration [8].

types, i.e., link, node, or multiple failures [1], whereas *scalability* criterion is a function of the time to obtain the routes as well as the restoration time [1].

The restoration speed is one of the issues in optical networks in which it is required how speedy restoration is possible while traffic flow is disrupted due to failures. Many restoration schemes have been reported to restore disrupted traffic due to failures in generic mesh networks. There are two steps – route computation and execution – for restoration in networks. In link-based restoration methods, disrupted traffic is rerouted around the failed link (Figure 8.2a), whereas path-based rerouting reinstates the whole path between the two endpoints of a demand as shown in Figure 8.2b. In case of sub-path restoration, part of the path is replaced by rerouting the affected area of the path (as shown in Figure 8.2c) while failure occurs. The link-based method needs the capability to identify a failed link. This approach has difficulty in restoration against a node failure. It has also limitation in the choice of restoration paths. The pre-estimated approach computes restoration paths before a failure happens, and prior availability of rerouting information is required at the affected nodes where actions have been taken after the detection of the failure and hence allows fast restoration. The latter needs time to compute the alternate route after detection of the failure. Centralized restoration estimates restoration (and primary) paths for all demands at a central controller in which updated information related to the network is available [8].

8.2.4   CAPACITY EFFICIENCY

Capacity efficiency depends on the following:

- the fraction of demands is restored for a network capacities, point-to-point and primary route.
- the network capacities for 100% restoration.
- the total network capacities for primary and restoration routes providing point-to-point 100% restorability.

Span-restorable networks are more efficient than rings in terms of the use of resources but not efficient when each failed demand is replaced end to end across the network from source to destination. The capacity placement in a path-restorable network may provide 100% restorability by using multi-commodity maximum flow (MCMF) solution [9].

8.2.5   RESOURCE SUCCESS TIME

Restoration success rate (RSR) is the ratio between the number of successfully restored connections and total number of disrupted connections after a network failure occurs [9].

8.2.6   AVAILABILITY

Availability is the probability that a connection is found to be in the operating state at a random time after restoration in the future. Connection availability depends on the failure rate and failure repair rate. An availability analysis for a connection is made according to the type of the schemes – dedicated path protected or shared path restoration in a WDM mesh network. Analysis parameters are the uptime known as mean time to failure (MTTF) and the repair time known as mean time to repair (MTTR) and are not dependent on memory processes with known mean values. The procedure of memory process is known as an *alternating renewal process.* A network component's availability is determined on the basis of its failure characteristics. This procedure is known as an alternating renewal process. Consequently, the availability of a network component j (denoted as a_j) can be estimated as follows [10]:

$$a_j = \frac{\text{MTTF}}{\text{MTTF} + \text{MTTR}} \tag{8.1}$$

where MTTF of a fiber link is a function of the distance and is determined as per measured fiber-cut statistics. MTTR is normally considered to be 2 hours for equipment and 12 hours for cable cut.

8.2.7   END-TO-END PATH AVAILABILITY

The availability of ith path (A_i) can be estimated according to the availabilities of the network components along the route. The ith path is obtainable when all the network components along its route are accessible. We consider a_j to be the availability of network component j and G_i to be the set of network components used by path i. Then, A_i is written as [2,11]

$$A_i = \prod_{j \in G_i} a_j \tag{8.2}$$

Particularly, the distance-related fiber-cut rates are determined by using fiber-cut statistics. If connection t uses one single path for a given route of the path, the availability of $t(\Lambda_p)$ is estimated as per known availabilities of the network components along the route. Connection t *is* obtainable when all the network components along its route are accessible. We consider a_j to be the availability of network component j and G_t to be the set of network components used by path t. Then A_t is determined as given below [2]:

$$A_t = 1 + \left(1 - A_p\right)\left(1 - A_b\right) \tag{8.3}$$

where A_p and A_b are the availabilities of primary path p and backup path b, respectively. As mentioned, the availability of connection $t(A_t)$ depends on the size of S_t and the availabilities of the connections in S_t if shared path restoration is used. Here we discuss a preliminary connection availability analysis for a shared path connection for restoration. A shared path restored connection t is obtainable if p is accessible or p is inaccessible, b is accessible, and p can access the backup resources when the other paths in the sharing group S_t are unsuccessful. Therefore, A_t can be obtained as [2]

$$A_t = A_p + \left(1 - A_p\right) A_b \sum_{i=0}^{N} \delta_t^i p_i \qquad (8.4)$$

where N = the size of S_t, δ_t^i = the probability that t uses the backup resources when both p and the other i primary paths in S_t are unsuccessful, and p_i = the probability that exactly i primary paths in S_t are engaged. δ_t^i A continuous time Markov chain is required to determine δ_t^i [2].

8.2.8 RELIABILITY

Reliability is defined as the probability of a system being active without any disruption for a period of time. Service reliability indicates the number of hits or disruptions in a period of time. Availability and reliability are used as measures for service quality analysis. The reliability is estimated by using a reliability model in which it depends on all the paths used for restoration, primary path, and the number of paths required to be restored. The estimation of reliability for protection schemes has already been discussed in Chapter 7.

8.3 RESTORATION SCHEMES FOR MESH TOPOLOGY

The restoration approach is normally distributed, pre-computed, path based, and failure independent. The restoration paths are basically failure dependent having more capacity utilization. However, a large number of restoration paths for each failure scenario are pre-computed. The restoration problem is a scalability problem [2]. Many restoration schemes are reported for generic mesh networks [2,4]. There are many restoration schemes such as link-based restoration and path-based restoration. The link-based restoration performs rerouting of disrupted traffic around the failed link, whereas path-based rerouting rejects the whole path between the two endpoints of a demand.

The link-based approach needs the ability to recognize a failed link at two ends of the link. The restoration is difficult to implement in the event of a node failure. The pre-computed approach estimates restoration paths before a failure occurs, and the real-time approach [9] performs the same. The former needs prior availability of rerouting information to the nodes where actions are taken after the failure is detected, and the latter needs time to compute the alternate route after failure is detected, and hence it is slower.

Centralized restoration methods [9] determine the restoration (and primary) paths for all demands reaching a central controller and also provide up-to-date information concerning the network to be available at the controller; the routes determined by this central controller may then be sent to the nodal databases for primary and restoration route tables. These algorithms are generally path based. Due to the difficulty in fast failure isolation in optical networks, this approach is not good. These algorithms usually provide higher capacity utilization. But it requires updated network and capacity information at a central controller through frequent communication between the nodal databases and the central controller. Moreover, centralized computation is difficult to scale for large networks with large numbers of demands. The distributed approach is an alternative for computing restoration routes for optical mesh networks. Distributed methods use pre-computed reroute tables or real-time discovery of capacity and routes. Real-time capacity discovery-based methods [12,13] distribute flooding messages to search for available link capacities after a link failure occurs and then find paths for the affected traffic around the failed link. The advantages of these link-based systems are simplicity and distributed nature, and the disadvantages are slowness, inefficient capacity utilization, and restorability limited to single-link failures [12,13]. Several distributed pre-computed approaches have been reported [14–16]. In most of the cases, restoration paths are failure dependent providing high capacity utilization. A large number of restoration paths (one set for each failure traffic) are pre-calculated and stored in memory. Failure-independent restoration uses the disjoint path idea– the service route and its restoration route are diverse (node- and link-wise). Thus, this scheme restores any single link failure or node failure. Since every service route uses a restoration route, no dedicated capacity needs to be retained for the restoration route, resulting in capacity savings [17]. In their approaches, restoration routes are either provisioned or computed earlier with a simple centralized shortest-path algorithm [9], and capacity sharing is determined for each possible failure in a centralized way.

8.3.1 PATH RESTORATION ROUTING PROBLEM

Span restoration is one of the path restoration procedures in which the replacement signal paths are set up between the end nodes of the failed path, yielding the maximum total amount of replacement capacity and saving the finite number of spare links on each span [9]. There are $O\left(\overline{d}-1\right)^{h-1}$ loop-free distinct routes of up to hops h between any two nodes in a network of average nodal degree $\overline{d}$. For path restoration, the routing problem is a multi-commodity flow problem in a discretely capacitated multi-graph reflecting the logical nature of a SONET STS or WDM wavelength-managed transport network, and the path restoration *routing* problem is formulated as follows [9,18].

The failure network is denoted as graph $G(N,E,s)$, where N = the set of nodes, E = the set of spans, and s = the vector of spare capacities on each span. For a failure such as a single span cut, or a node loss, D_i is denoted as the number of O–D pairs misplacing one or more units of demand. These O–D pairs are indexed to be restored by $r \in (1,\dots,D_i) X_i^r$ = the number of demand units lost by an O–D pair for a failure where X_i^r keeps out of non-restorable demands whose end nodes are failure nodes. The objective functions are written as [9]

$$\text{Min} \sum_{r=1}^{D_i} \sum_{p=1}^{P_i^l} f_i^{r,P} \qquad \forall (i) \tag{8.5a}$$

$$\text{Subject to} \sum_{p=1}^{P_i^l} f_i^{r,P} = X_i^r \qquad \forall (r \in 1,\ldots, D_i) \tag{8.5b}$$

$$\sum_{r=1}^{D_i} \sum_{p=1}^{P_i^l} \delta_{i,j}^{r,P} \cdot f_i^{r,P} \le s_j \qquad \forall (j \in E) \tag{8.5c}$$

$$f_i^{r,P} \ge 0, \text{integer} \qquad \forall (r,p) \tag{8.5d}$$

where

$f_i^{r,P}$ = quantity used for traffic flow to the pth route accessible for restoration of O–D pair, r for failure scenario i

$P_i^{r,P}$ = number of eligible restorable routes accessible for the restoration of O–D pair, r for failure scenario i.

$\delta_{i,j}^{r,P}$ = input variable that is 1 if span j is in the pth eligible route for restoration of O–D pair r in the event of failure scenario i and 0 otherwise.

Every path for restoration is restricted by the maximum hop and distance which represent $\sum_{j \in E} \delta_{i,j}^{r,P} < h$, $\forall (i,r,p)$ while pre-processing the network for distinct eligible routes. The objective function maximizes all restoration flow assignments for distinct eligible routes between each pair of affected end nodes in the reserve network the under condition that no O–D pair is restored to an extent greater than what is required [9].

A span-DRA produces end-to-end paths operating for a given Sender–Chooser node rather than those nodes directly adjacent to the physical span cut in which any span-DRA can be curved into a rudimentary path-DRA by operating the span-DRA for all affected source–destination demand pairs [9]. The problem with using a span-DRA is how pre-path restoration takes place. The total recovery depends on the sequence in which the span-DRA is carried out for each failed O–D pair. The MCMF problem provides a mutual capacity issue whose solution is basically obtained from path restoration. The solution of the MCMF problem can theoretically be obtained by using centralized computation. There are two parameters – path number efficiency (PNE) and restorability. The difference between PNE and restorability for a path-DRA is as follows [9]:

$$\text{PNE} = \frac{1}{F} \sum_i \sum_r \frac{\text{Number of paths found for } r\text{th O–D pair by DRA in failure scenario } i}{\text{Number of paths found for } r\text{th O–D pair by MCMF in failure scenario } i}$$

$$\text{Restorablity} = \frac{1}{F}\sum_{i}\sum_{r}\frac{\text{Number of paths found for } r\text{th O–D pair by DRA in failure scenario } i}{\text{Number working demands unit for } r\text{th O–D pair in test network}}$$

Where F is the number of failure scenarios. In this restoration, a k-shortest paths (kSP) routing process generates PNE in the presence of the trap than the corresponding max-flow path set. Here, a shortest path is greedily detected in building up a path set, making one or more subsequent paths, which are present in the max-flow solution, infeasible under the simpler kSP procedure. The heuristic principle based on these arguments is to maximize restoration when the collective path set has a minimum interference number over the constituent isolated path sets. Mathematically, the interference number of a path set is written as [9]:

$$I_i^{r,P} = \sum_{\substack{x=1 \\ x \neq r}}^{D_i} \sum_{y=1}^{P_i^x} \sum_{\substack{r,j \in E \\ r,j \notin i}} \left(\partial_{i,j}^{r,P} \cdot \partial_{i,j}^{x,y} \right) \qquad \forall (i,p,r) \qquad (8.5e)$$

where

$I_i^{r,P}$ = interference of pth prospective restoration path for relation with respect to all other path prospectives pth sub-sets for other relation affected with scenario i.

$\partial_{i,j}^{r,P} = 1$, if interference of pth prospective restoration path for demand pair after failure i uses span j

$= 0$, otherwise.

D_i, P_i^r, E have already been explained.

The development of a path-DRA depends on a distributed version of the interference heuristic which is used as a heuristic for synthesizing composite path sets; this is like the MCMF problem in a centralized polynomial-time algorithm. The procedure is as follows [9].

```
For every failure scenario do
        {
                (Optionally) Release the surviving portions of
all failed paths into the reserve network.
                        // (i.e., Initiate "stub release")
                While restoration paths are feasible for any
unrestored O-D pair r affected by failure
                scenario i
                        do {
                        For every unrestored O-D pair r
affected by failure scenario i do {
                                Find the set of successively
shortest routes within the reserve network for O-D
                                pair in  isolation}
                //end "for" loop (on line 4)
```

```
              For every unrestored O-D pair r affected by failure
scenario i
                    do {
                         For every path in the isolated pathset
from O-D pair r do {
                            Compute I_i^{r,P} in the presence of all
other prospective paths from all other O-D
pairs }
        // end "for" loop  (on line 8)
                    }
        //end "for" loop  (on line 7)
           Implement that path with the lowest I_i^{r,P}
              Remove the spare capacity for that path
implemented in line 12
                    }
        // end "while" loop (on line 3)
         Record total number of restoration paths formed
              Reset the reserve network to its no failure
state
                    }
              //end "for" loop (on line 1)
```

The computation time is mostly determined in steps 7–11 of this algorithm. Time complexity for estimating the shortest paths within a reserve network of nodes for O–D pairs is $O(n \log n)$ by using a standard shortest-path algorithm [9]. The number of O–D pairs in a network is $O(n^2)$. The complexity to obtain $I_i^{r,P}$ values for a single failure scenario i is $O(n^3 \log n)$ with this centralized procedure [9].

8.3.2 OPERATION FLOW

For other restoration schemes, there are generic tasks for establishing restorable connections in the optical network under a distributed signaling environment. The normal operation flow used in optical network with clients attached through a user network interface (UNI) is as follows [9]:

1. Periodically distribute link-state information with channel availability over the network. There are three fundamental channel states – in service, unassigned, and reserved for restoration.
2. Carry out connection admission control (CAC) and establish accepted connections.
3. Identify network failures and restore the failed connections.
4. Periodically carry out a capacity planning process to find where additional new channels are set up to get projected service and network restoration objectives, for WDM/channel installation (often many months). In this direction, the network service objective is to obtain a condition on the proportion of connection requests rejected due to insufficient capacity.
5. Periodically re-optimize restoration paths with a centralized algorithm.

Step 2 (CAC):

i. An OXC gets a connection request to set up connections from customer equipment over the UNI. The connection request has attributes having bandwidth and restoration requirements.
ii. Estimate a working path for the connection (in OXCs) where sufficient channels are there. Connection requests are in excess if no such path exists.
iii. If a restorable connection is required, then it is set up and set aside. For that, sufficient channels are required for (later) restoration so that the network restoration objectives are fulfilled. For insufficient capacity case, the connection request should be rejected.

Step 3 (detection of network failures and restoration of the failed connections):

i. Identify the failure, send a failure indication to the related nodes, and carry out the restoration on the failed connections.
ii. After identification, repair the failure.
iii. Normalize the network after repairing the failure, i.e., return the connections to the original or re-optimized paths while minimizing service disruption.

There are four categories of restorations: Category-1 includes the use of dedicated (1 + 1) restoration connections in large-scale applications 1 + 1 protection with restoration in higher layers (e.g., IP). The shortest-path algorithm is taken to estimate both the service and restoration paths in which the pre-estimated restoration path are available from the service path. The pre-planned cross-connect maps are employed in category 2, whereas for category 3, the channels are chosen after the failures occur. Category 4 is generally referred to as fully dynamic restoration. Its advantages are it requires less pre-planning and is dynamic to different failure events, and its disadvantage is that the restoration time is necessarily slower. Category4 includes fully distributed restoration protocols that find paths in the network by flooding bandwidth path requests having no current counterparts and protocol reuse or adaptation. Among all categories, category 3 restoration is a practical method for optical network services as category 3 uses capacity efficiently. Category 3 restoration method is a Robust Optical Layer End-to-End X-Connection for minimizing restoration time by incorporating these aspects. The following procedures are used for the computation of restoration paths in execution step-5 of operation flow.

Each node stores the execution of operations in step 5 to reoptimize all network restoration paths upon receiving every connection request in a centralized system. Updated restoration data must be sent to all relevant nodes after receiving every connection request [18].

Computation of a simplified restoration path: A restoration path is identified with step 5 being run frequently to optimize the restoration resource allocations.

Computation of a hybrid (approximate) restoration path [9]: A hybrid approach between the above two approaches uses the parameter S_{kf} which is the number of channels needed for restoration over link k for a given failure event f. To optimize restoration paths for the chosen service and restoration path of a new connection, accurate estimation is needed during connection provisioning. Finally, the S_{kf}s are sent for updates by path setup messages sent along the restoration path during connection provisioning.

8.3.3 RESTORATION PROBLEM

There are three main issues for restoration and provisioning.

- **Problem 1: maximum restoration:** In a network having spare capacity and used capacity, find restoration paths for as many demands as possible under the link capacity constraints.
- **Problem 2: minimum capacity restoration:** With the input given in problem 1, find restoration paths for demands (having 100% restoration capability) with the minimization of the capacity requirement.
- **Problem 3: joint optimization:** In problems 1 and 2 for a set of point-to-point demands, find both the primary route and the restoration route for each demand with the minimization of network capacity requirement.

8.3.3.1 Maximum Restoration Problem

The problem 1 is NP-complete and a disjoint path problem [2,9]. Here, the main aim is to estimate mutually node-disjoint paths between multiple $s-d$ pairs in a network. When an artificial node is added to the network, then this node is connected to all the nodes in the $s-d$ pairs and denotes the two-hop paths through this node between these $s-d$ pairs as primary paths. As all primary paths share the common fault, no capacity is shared between alternate paths. The centralized approach uses a heuristic approach based on distributed computation for solving this problem. A distributed heuristic approach is discussed here. As mentioned earlier, the problems 1 and 2 use the same basic algorithms. Problem 2 includes an additional step for finding the best locations to use the new capacity to estimate restoration routes in case of unrouted demands with minimum additional capacity. The approach involves distributed pre-estimation of restoration routes and permits progressively better capacity utilization. The source node of each demand searches for its restoration route independently. Following are the four basic issues:

- How does the origin node of a demand search its route?
- How does the source estimate whether the link has sufficient spare capacity, if the source node of a demand uses a link on the restoration route of that demand?
- How can deadlocks be disallowed when multiple demands are simultaneously contending for link capacities during their searches (for restoration routes)?
- How can the capacity utilization be optimized?

Here the algorithm uses four basic steps:

- route (path) search
- link capacity control
- concurrent resource contention locking
- capacity optimization.

By using these basic procedures as sub-routines, the outline of the algorithm is stated below [9]:

1. The source node of each demand performs contention-locking procedure to lock out its contenders.
2. The source waits for a random amount of time, when the lock is unsuccessful; otherwise, it starts searching for the restoration route.
3. When searching, the source node uses a distributed breadth-first search (BFS) algorithm to obtain a route from the source to the destination using only links with residual spare capacity.
4. If a route is found, the source records it as the restoration route for the demand. Otherwise, it activates the optimization procedure for releasing some link capacities critical to the route construction of the current demand by changing the restoration routes of previously routed demands.

8.3.3.2 Restoration Route (Alternate Path) Search Procedure

There is a primary path for the demand of each connection. The search procedure is to obtain a valid restoration path of an $s-d$ pair [9]. Two criteria for the validity of a path are as follows:

- It should only use nodes and links that are not for the primary path of the demand.
- It should use the links having sufficient spare capacity and contending demands already routed on the same links.

The procedure is realized by a distributed BFS over a residual network. The residual network is obtained from the original network by not considering hop nodes and links of the primary path and also links with insufficient capacities after enquiring using the link control procedure.

8.3.3.3 Link Capacity Control Procedure

This approach finds the path where the link should have sufficient capacity to contain a demand of restoration route. The following considerations [9] are required: total spare capacity of the link, all the demands of restoration routes, and contention relationship among those demands and new demand.

The connections are *contended* if their service routes use a common node or a common link. The service routes having a common node or link are under failure at the same time with a single fault. The restoration routes are to be active simultaneously contending for the same capacities. A simple way to estimate the restoration

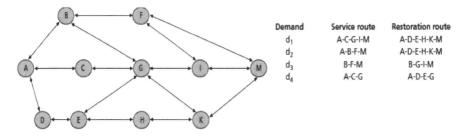

FIGURE 8.3 Restoration based on capacity sharing [9].

capacity requirement on a given link is to find all possible failures one by one, considering the maximum across all failures (Figure 8.3).

8.3.3.4 Concurrent Contention-Locking Procedure

Although the algorithm provides services to multiple demands simultaneously, no two demands are in contention, and there may be deadlocks during path searches. This is just like a collision problem in Ethernet. In this problem, resource contention occurs in both time dimension and the space (network) dimension. The deadlock contention problem must be solved before the path search begins. Thus, one demand at the most out of all contending demands will be sought for a restoration route at a given time. The non-contending demands conduct their path searches at the same time. This non-concurrent-contention condition is imposed by the contention-locking procedure given below.

The following steps are taken in the contention-locking procedure [9].

- Before a demand searches for a restoration route, the source node of the demand transmits a "locking" message along its primary route to the destination.
- This message collects tokens from every node and link along the path. It turns round the trip by encountering the first node or link whose token is not available, and then it releases all the tokens collected on its way back to the source node.
- Upon message arrival, the source node is delayed for a random amount of time and again transmits the locking message. When the message successfully reaches the destination node, the destination node understands that there is currently no other contending demand searching for a restoration route, so that all the nodes and links on the demand's primary route have been locked.
- Either source or the destination finishes its route search by sending an "unlocking" message to the other end along the primary route. This message releases a token back to every node and link it is transmitted through along the service route.

8.3.3.5 Optimization Algorithm

Due to distributed characteristics and lack of coordination of search procedure [18,19], the resulting routes require more capacity than that of a centralized algorithm. Here, more demands may get their restoration routes with the same link capacities that the above route search procedure accommodates. The optimization procedure is used for the improvement of capacity utilization by obtaining restoration routes for more demands. When a demand fails to find a restoration route in the route search procedure, the optimization procedure is activated for a restoration route by changing the restoration routes of previously routed demands. Both centralized and distributed computation are used here. The centralized case is used first. The mathematical description of the problem is made by an integer programming formulation. We then consider two heuristic algorithms for problem 2–the first algorithm uses a Lagrangian relaxation of the integer programming formulation, and the second algorithm uses local rerouting.

8.3.3.5.1 Problem Formulation

The following notation is considered for the formulation [9].

- N is number of nodes, $L \subset N \times N$ is number of links, and $D \subset N \times N$ is number of demands.
- For each link $(ij) \in L$, C_{ij} (a positive real number) is the capacity weight for a link considered to be a measure of capacity consumption per wavelength on the link.
- For a given design solution, W_{ij} is the resulting capacity requirement on a link in terms of the number of wavelengths.
- The objective function $\sum\limits_{(ij) \in L} C_{ij} W_{ij}$ is minimized.

  If $C_{ij} = 1$ for all the links, then the objective is to minimize the total number of wavelengths, which is equal to the mileage of the link.

In capacity optimization, the total network infrastructure cost including the link cost and node cost is to be minimized. Link cost [9] whereas Node cost involving the cost of nodal equipment such as multiplexers, switches, and ports.

For integer programming formulation, the following parameters are considered [9]:

- F = the set of all possible (single) faults to be either a failed node or a failed link, $F \cup L$
- L_f = the set of links affected by fault for each $f \in F$. If f is a link fault, then L_f contains only the failed link. If f is a node fault, L_f contains the set of links subtending to the failed node, i.e., $L_f = \{(ij) \in L : j = f\}$. For each demand $d \in D$, s_d and t_d be the source node and destination nodes of demand d, respectively.

Binary variables are as follows [9]:

$X_{ij}^d = 1$, if demand d's primary path traverses link ij; $X_{ij}^d = 0$, otherwise.
$Y_{ij}^d = 1$, if demand d's alternate path traverses link ij; $Y_{ij}^d = 0$, otherwise.

$Z^d_{iif} = 1$, if demand d is rerouted through link ij under fault f; $Z^d_{iif} = 0$, otherwise. W_{ij} is a non-negative integer representing the number of wavelengths traversing link ij.

Problem-2 (minimum capacity restoration) can be formulated as follows [9]:

$$\text{Minimize} \sum_{(ij) \in L} C_{ij} W_{ij} \tag{8.6}$$

where

$$\sum_{J:(ij) \in L} X^d_{ij} - \sum_{J:(ij) \in L} X^d_{ji} = \begin{cases} 1 & i = s_d \\ -1 & i = t_d \ \ d \in D \\ 0 & \text{otherwise} \end{cases} \tag{8.7}$$

$$\sum_{J:(ij) \in L} Y^d_{ij} - \sum_{J:(ij) \in L} Y^d_{ji} = \begin{cases} 1 & i = s_d \\ -1 & i = t_d \ \ d \in D \\ 0 & \text{otherwise} \end{cases} \tag{8.8}$$

$$\sum_{J:(ij) \in L,(ij) \in L_f} Z^d_{ijf} - \sum_{J:(ij) \in L,(ij) \in L_f} Z^d_{jif} = \begin{cases} \displaystyle\sum_{(kl) \in L_f} X^d_{kl} & i = s_d \\ \displaystyle\sum_{(kl) \in L_f} X^d_{kl} & i = t_d, d \in D, f \in F \\ 0 & \text{otherwise} \end{cases} \tag{8.9}$$

$$Z^d_{ijf} \leq Y^d_{ij} \qquad d \in D,(ij) \in L, f \in F,(ij) \notin L_f \tag{8.10}$$

$$\sum_{d \in D} X^d_{ij} + \sum_{d \in D} Z^d_{ijf} \leq W_{ij}, \ \ (ij) \in L, f \in F,(ij) \notin L_f \tag{8.11}$$

The objective function in the equation (8.6) contributes total weighted capacity requirement minimized during computation. Equations (8.7) and (8.8) indicate the flow conservation constraints for demand d's service route and restoration route, respectively. Equation (8.9) represents the constraint of logical relationship where the restoration routes get through the link capacity if and only if the primary/service route faces the fault. Equation (8.10) indicates that the restoration route of a demand is independent of the failure. Equation (8.11) estimates the link capacity requirement [9]. The constraint (8.11) is the bundling constraint that bundles different demands together, and the constraint (8.11) indicates that the problem disintegrates into a number of independent sub-problems. This disintegration would significantly reduce the computational complexity of the original problem. A common technique for relaxing such constraints is Lagrangian relaxation method [9]. Although this relaxation simplifies the problem, the relaxed problem provides a lower bound to the original problem, and a solution to the relaxed problem is not necessarily a feasible solution to the original problem. The Lagrangian relaxation approach is used for the relaxing constraint (8.11). A solution for the relaxed problem is made into a feasible solution for the original problem by including the Lagrangian multiplier $\lambda_{ijf} \geq 0$ for the link (ij) with fault f and adding the penalty term $\sum_{(ij) \in L, f \in F, (ij) \in L_f} \lambda_{ijf} \sum_{d \in D} \left(X_{ij}^d + Z_{ijf}^d - W_{ij} \right)$ to the objective function (8.6) [9]. For an algebraic manipulation, the relaxed problem is made meaningful only by adding the constraint $\sum_{f \in F, (ij) \in L_f} \lambda_{ijf} = C_{ij}$ to the Lagrangian

multipliers in addition to their non-negativity constraints. The original problem is divided into a series of sub-problems, one for each demand. For each demand d, we get the objective of the sub-problem in the following form:

$$\sum_{(ij) \in A_d} \sum_{f \in F_P} \lambda_{ijf} + \sum_{(ij) \in P_d} C_{ij}$$

where P_d indicates a primary path (route) connecting s_d and t_d, A_d indicates also an alternate path connecting s_d and t_d, and P_d and A_d are node disjoint except at s_d and t_d where F_P represents all possible faults that bring down route P. Even for only one demand involved in problem (IP$_d$), the capacity sharing between the restoration routes of different demands is represented by the Lagrangian multipliers [9]. Since $\sum_{f \in F, (ij) \in L_f} \lambda_{ijf} = C_{ij}$, the link cost C_{ij} for the restoration routes is represented as being split across different faults in which the fraction of C_{ij} for a restoration route using the link ij has faults affecting the corresponding service route. More specifically, λ_{ijf} is represented as the cost of a restoration route if this route has link (ij) and the service route of the demand has fault f. To establish the relationship between (IP$_d$) and (IP), the following theorem is used [9].

Theorem: For each set of feasible values of λ_{ijf}, let v_d be the optimum value of (IPd), the objective function value of its optimal solution for each $d \in D$, and v be the optimum value of (IP). Then, $\sum_{d \in D} v_d \leq v$.

Proof: $\overline{X_{ij}^d}, \overline{Y_{ij}^d}, \overline{Z_{ijf}^d}$, and $\overline{W_{ij}}$ are possible solutions of problem (IP). For each demand, d, P_d and A_d are the primary path and alternate path determined by $\overline{X_{ij}^d}$ and $\overline{Y_{ij}^d}$, respectively, considering P_d and A_d being disjoint due to sharing fault f. Then under fault f, either constraint (8.10) or constraint (8.11) is not satisfied. Thus, P_d and A_d are disjoint. Therefore, they are feasible solutions of problem (IPd). Now,

$$v = \sum_{(ij)\in L,} C_{ij} \overline{W_{ij}} \geq \sum_{d\in D} C_{ij} \max_{f\in F} \left(\sum_{d\in D} \overline{X_{ij}^d} + \sum_{d\in D} \overline{Z_{ijf}^d} \right)$$

$$= \sum_{(ij)\in L} C_{ij} \sum \overline{X_{ij}^d} + \sum C_{ij} \max_{f\in F} \sum_{d\in D} \overline{Z_{ijf}^d}$$

$$= \sum_{d\in D} \sum_{(ij)\in L} C_{ij} \overline{X_{ij}^d} + \sum_{(ij)\in L} C_{ij} \max_{f\in F} \sum_{d\in D} \overline{Z_{ijf}^d}$$

where $\displaystyle\sum_{d\in D} \overline{Z_{ijf}^d} = \sum_{d\in D, f\in P_D, (ij)\in A_d} 1$ and $\displaystyle\sum_{f\in F, (ij)\in L_f} \lambda_{ijf} = C_{ij}$ Substituting these in the above equation, we can write as

$$v \geq \sum_{(ij)\in L} \sum_{(ij)\in L} C_{ij} \overline{X_{ij}^d} + \sum_{//////(ij)\in L} \sum_{f\in F, (ij)\in L_f} \lambda_{ijf} \max_{f\in F} \sum_{d\in D, f\in P_d, (ij)\in A_d} 1$$

$$= \sum_{d\in D} \sum_{(ij)\in L} C_{ij} \overline{X_{ij}^d} + \sum_{d\in D} \sum_{(ij)\in A_d} \sum_{f\in P_d} \lambda_{ijf} = \sum_{d\in D} \sum_{(ij)\in P_d} C_{ij} + \sum_{d\in D} \sum_{(ij)\in A_d} \sum_{f\in P_d} \lambda_{ijf}$$

Since P_d and A_d are a feasible solution to (IPd), $\displaystyle\sum_{(ij)\in P_d} C_{ij} + \sum_{(ij)\in A_d} \sum_{f\in F_{Pd}} \lambda_{ijf} \geq v_d$. Therefore, $\displaystyle\sum_{d\in D} v_d \leq v$.

For each fixed set of λ values, $v(\lambda) = \displaystyle\sum_{d\in D} v_d$, the more appropriate value of the relaxed problem. $v(\lambda)$ provides a lower bound to the original problem. We need to adjust λ in such a way that $v(\lambda)$ increases for an increase in lower bound. $v(\lambda)$ is a linear precise function of λ_{ijf}. The parameter λ is used in the adjustment procedure which is called a *sub-gradient optimization*. The sub-gradient of $v(\lambda)$ is used as an array which is a function of λ:

$$\nabla v(\lambda) = a_{ijf}, \quad (ij) \in L, (ij) \notin L_f, f \in F$$

where $a_{ijf} = \displaystyle\sum_{d\in D} \left(X_{ij}^d + Z_{ijf}^d \right) = \sum_{d\in D, (ij)\in P_d} 1 + \sum_{d\in D, (ij)\in A_d, f\in F_{Pd}} 1$

Considering constraint $\displaystyle\sum_{f\in F, (ij)\notin L_f} \lambda_{ijf} = C_{ij}$, the projected sub-gradient can be written as [9]

$$\nabla v(\lambda) = a_{ijf} - \frac{\displaystyle\sum_{f\in F, (kl)\notin L_f} a_{klf}}{m - 3}$$

where m = the total number of nodes and links and $m - 3$ is the total number of faults that do not affect a given link. The performance of the algorithm depends on the iteration on λ values by sub-gradient optimization to increase the lower bound so that after each iteration, solutions of all the sub-problems are feasible to the original problem (IP). The ascending sequence of lower bounds indicates the merits of those feasible solutions. The best feasible solution is the final solution. The solution of sub-problem (IP$_d$) is easily obtained, but (IP$_d$) is an NP-complete problem where the following algorithm solves sub-problem (IP$_d$) [9].

Step-1: Get the shortest node-disjoint pair of paths between s_d and t_d as the initial solution. Take the shorter path, Pd, as the service route and the longer one, A_d, as the restoration route. Estimate the objective function $v(P_d, A_d)$ of (IP$_d$), where (P_d, A_d) is the best feasible solution.

Step-2: Compute a set of K shortest paths between s_d and t_d in ascending order of path cost where K is a parameter.

Step-3: For each path P in the path table, choose it as a service route. With P fixed, find the shortest restoration route A as the shortest disjoint path using the new link cost metric: for link (ij), $R'_{ij} = \sum_{f \in F_{Pd}} \lambda_{ijf}$. A may not exist.

Determine the objective value $v(P, A)$. If it is less than the best value, substitute the best value and the best feasible solution by (P, A) and $v(P, A)$, respectively.

Step-4: If the path cost is not less than the best value for some path in the path, end the algorithm. The current best feasible solution is optimal. If this never happens, use the best feasible solution as the final solution; use the cost of the last path in the path set as a lower bound; take the shorter path, P_d, as the service route and the longer one, A_d, as the restoration route; break the tie arbitrarily; and estimate the objective function $v(P_d, A_d)$ of (IP$_d$) where (P_d, A_d) is called the best feasible solution, and $v(P_d, A_d)$ is the best value.

The solution found by the above algorithm is always the best solution if the value of K becomes large. Still it is needed to obtain the feasible solution as a fallback alternative as K shortest-path computation is too time consuming for large networks with large K. After obtaining the solution of all the sub-problems, a new lower bound for (IP) is determined as either the optimal value of (IPd) or its lower bound. S is a feasible solution of (IP), and the $C(S)$ is the objective function value (total capacity requirement) of S. $C(S)$ is the efficiency obtained by using the following procedure.

The outline of algorithm G is as follows [9],

Step-1: Initialization: Consider all λ values equally and scale them so that $\sum_{f \in F, (ij) \notin L_f} \lambda_{ijf} = C_{ij}$. For each demand $d \in D$, resolve the sub-problem (IF$_d$) by using the above algorithm. All the sub-problem solutions provide an initial feasible solution S_0 to the original problem (IP). Set the best feasible solution (thus far) $S^* = S_0$.

Step-2: At the kth iteration, renew λ values as

$$\left| \lambda_{ijf}^{k+1} \right| = \left| \lambda_{ijf}^{k} \right| + \alpha \nabla_p v \left| \lambda_{ijf}^{k} \right|$$

where α is the step size. α is chosen in such a way that λ^k is still non-negative and adaptive to the improvement of the last iteration. Solve all sub-problems (IPd), and obtain a feasible solution S_{k+1}. If $C(S^*) > C(S_{k+1})$, set $S^* = S_{k+1}$.

Step-3: Repeat step-2 until either $v(\lambda)$ cannot be improved or the value of $C(S^*)$ does not decrease for a predefined number of iterations.

Step-4: S^* is the best feasible solution obtained thus far and can be further improved by a local rerouting procedure as discussed below.

Then the solution is improved by revising the service and restoration routes of demands one at a time with the following algorithm L [9].

Step-1: For each demand, choose the shortest node-disjoint pair of paths between s_d and t_d as the initial solution. Determine the capacity requirement on each link.

Step-2: For each demand $d \in D$, with its fixed primary path P_d, revise its restoration route, A_d, as follows:
- First, remove A_d and renew link capacities.
- Second, have a link cost metric: For each link (ij), run the link capacity control procedure to check whether its capacity increases if it is used by d's restoration route. If so, $R_{ij} = C_{ij}$ If not, $R_{ij} = 0$. Consider $R_{ij} = \alpha$ for all the links for assuring P_d disjointness.
- Third, take the shortest path between s_d and t_d under the new cost metric as a new restoration route for d.

Step-3: For each demand, $d \in D$, with its restoration route, A_d, fixed, reroute its primary route, P_d, in the following way:
- First, take P_d and renew link capacities, although A_d is unchanged, in the absence of P_d.
- Second, take the link cost metric: For each link (ij), run the link capacity control procedure to check each link on A_d, and determine whether its capacity increases if the three faults i, j, k are introduced to d's service route. Let L_{ij} be the set of links on A_d whose capacity increases. Describe $R_{ij} = C_{ij} + \sum_{(kl) \in L_{ij}} C_{kl}$. Set $R_{ij} = \alpha$ for all the links for A_d to assure disjointness.
- Third, find the shortest path between s_d and t_d under the new cost metric. If the total cost is improved, choose this new path as the new primary route for d.

Step-4: Repeat step-2 and step-3 till no further improvement is obtained.

The advantages of this algorithm are speed, simplicity, and easy accommodation of additional constraints. The use of a link by a restoration path does not increase capacity requirement. Each demand chooses its route very fast. Two demands rerouted at the same time may conflict their link capacity queries of their restoration routes due to changes in the fault of their service routes. In order to keep away from the conflict, contention-locking procedure of problem 1 is adjusted where a demand selects its (fixed) restoration path before it starts to search for its (new) service path. In this locking mechanism, all demands doing service path computation are guaranteed disjoint restoration paths. Step-2 and step-3 cannot be carried out simultaneously for two different demands. Therefore, some global synchronization is necessary. To avoid this, the following three stages of execution [9] are required:

- **Stage-1:** Demands perform step 1 in parallel.
- **Stage-2:** Demands perform step 2 asynchronously.
- **Stage-3:** All demands perform step 3 asynchronously.

Synchronization between stages is obtained by broadcasting or centralized locking.

8.4 RESTORATION ACTIVATION ARCHITECTURES

In this section, restoration activation architectures are discussed [9]. As the restoration route of each demand is pre-estimated, the route details are recorded at the destination node. As no reserved capacity is assigned for the restoration route, the route is not established beforehand. The actual route connection is carried out in real time after the occurrence of a failure.

Once the destination node finds the failure of the service route, it begins cross-connect operations at every node on the restoration route to connect the path and then signals of the source node to shift the traffic to this route and switches the signal selector for the demand to find the restoration route. The activation step needs the following functions [9]:

- failure detection at the destination
- inter-node message delivery and processing
- cross-connect operation, traffic shifting
- signal choice.

Before the introduction of the activation architectures, these functions were carried out by the network hardware and software architectures. There are two activation architectures-sequential activation architecture [9] and parallel activation architecture [9].

8.4.1 Sequential Activation Architecture

Figure 8.4 shows the schematic of sequential activation architecture. The upper source–target (S–T) path is the primary route, whereas the lower S–T path is the restoration route. The dashed lines indicate flows of restoration activation messages. The destination node begins the restoration route activation, and the route connection is established at the destination node in the following steps [9]:

- After finding the failure of primary route of a demand, the destination or target node T first finds an incoming port of the restoration route and controls its signal selector to get a signal on this port.

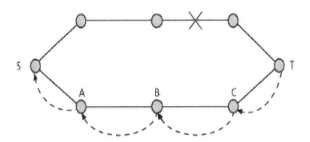

FIGURE 8.4   Sequential architecture [9].

- It detects the corresponding outgoing port connected to its preceding neighbor node C on the restoration route.
- Finally, it transmits a cross-connect message to node C. Two information items are enclosed in the message – the outgoing port number for the cross-connection to be made at node C and the restoration route topology together with demand ID.
- After getting the message, node C finds the incoming port number and carries out the cross-connection. It then transmits a message having the same type of information to its preceding neighbor node B.
- After getting the message, node B also takes the same action and transmits the same type of message to the next node. This step is repeated till the source node S gets the same message.
- After node S receives this message, it finds the demand by the demand ID with source and destination. It then passes the signal of the given demand to the outgoing port whose number is found from the message.

8.4.2   PARALLEL ACTIVATION ARCHITECTURE

To increase the speed of the setting up of the restoration route, the parallel activation approach is used, allowing parallel operation of cross-connects at different nodes. The main problem of obtaining parallelism is the excess messaging for port assignment. There are two types of messages – type 1 messages from the destination node to all other nodes on the restoration route and type 2 messages between neighboring nodes on the restoration route. Figure 8.5 presents message flows where the dotted links indicate type 1 message flows whereas the dashed lines indicate type 2 message flows. The upper path of S–T is the primary route, and the lower path of S–T is the restoration route. The following procedure is used for restoration activation [9]:

- After finding the failure of the primary route of a demand, the destination node T first regains the restoration route for affected demands and transmits a type 1 message to each node on the restoration route. Each type 1 message has three pieces of information – the message type, the demand ID, and the node ID of the node immediately preceding the receiving node on the restoration route.

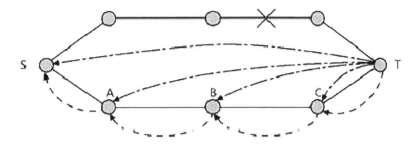

FIGURE 8.5   Example of parallel activation architecture [9].

- Node T finds an incoming port number for the alternate path and controls its signal selector to get a signal on this port. Finally, it finds the corresponding outgoing port number at its preceding neighbor node C on the restoration route and transmits a type 2 message to node C. Each type 2 message has also three pieces of information – the message type, the demand ID, and the outgoing port number.
- After getting a type 1 message, each intermediate node does the following actions.
    Action-1: It finds an incoming port number for its cross-connect operation and finds the corresponding outgoing port number at its preceding neighboring node.
    Action-2: It then transmits a type 2 message to its preceding neighboring node.
    Action-3: After getting both a type 1 and a type 2 message, each intermediate node makes the cross-connection.
    Action-4: Once the source node S gets a type 2 message from node A, it recognizes the demand by seeing demand ID and sends the signal of the demand to the outgoing port whose number is found from the message.

There are two ways to speed up failure detection and activation of restoration routes. It provides reduction of message processing and exchanges, reduction of cross-connect activations, and dedication of signaling channels.

8.4.2.1   Message Processing and Exchange Reduction

This method provides the reduction of nodal message processing time and inter-node message exchange by pre-assigning channel numbers to the restoration routes of the demands on a link-by-link basis. The pre-assignment is same as backup path assignment, but the difference is that the path is restored after getting a failure. Channel pre-assignment on a link is achieved if each restoration route has a dedicated capacity reservation on the link.

There are three demands d_1, d_2, and d_3 using their restoration routes on a given link l. These demands are in contention, but no single fault can fetch all three demands. Each link has only two channels. If two channels are pre-assigned to three demands, two of the demands d_1 and d_2 are assigned to the same channel.

For channel pre-assignment on link l, we need three units of capacity. Although three units also happens to be the capacity required by the restoration routes without capacity sharing, the channel pre-assignment capacity is usually much less. In fact, based on past experience with similar problems in *SONET/SDH* rings, a fast heuristic algorithm is developed to assess capacity penalty on channel pre-assignment. Once channels are pre-assigned to demands, port numbers need not be communicated between nodes at the time of restoration activation.

8.4.2.2 Cross-Connect Reduction

The restoration speed depends on the number of cross-connect requests [20] at a small number of nodes. The speed-up of restoration requires the release of some load from the maximally loaded (critical) nodes. The approach is to make cross-connection pre-wiring at these critical nodes. The pre-wiring made at each node is independent of individual demands. The following steps should be taken:

- The demands of restoration routes should be found for every node.
- All possible pairwise link combinations should be examined.
- For each such pair of links I_1 and I_2, the minimum restoration traffic flow passing through both I_1 and I_2 under all failures should be determined.
- If the minimum flow is k units, k channels transmit signals from link I_1 to I_2. At the time of activation, k demands are effectively disconnected from this node's cross-connect load.

8.4.2.3 Dedicated Signaling Channels

A dedicated signaling channel is another approach to speed up restoration [9] for each demand for faster messaging. There are two ways in this approach – a channel provisioned at a lower layer to bypass the protocol stack and to avoid significant software processing or the messages travel faster using the dedicated channel. Figure 8.6 shows the operation of the approach. Here the upper path is its service route represented as "P", and the lower path is its restoration route, indicated as "R". The two dashed lines "A" and "B" are the two dedicated signaling channels for this demand along its restoration route. Once the node suffers failure of the primary route, it sends restoration messages through signaling channel B to all the intermediate nodes on the restoration route and to the source node.

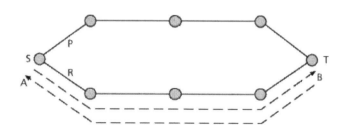

FIGURE 8.6   Dedicated signaling channel per demand.

8.4.3   OPTIMIZATION PERFORMANCE OF RESTORATION APPROACHES

In this section, we discuss the evaluation of the performance of the restoration algorithms in terms of capacity utilization optimization [9]. The performances of distributed algorithm and centralized algorithms are different because of their stochastic elements. The distributed algorithm should be evaluated by a complete simulation model which computes the all the steps of the algorithm. A simpler simulation model is used for analyze the dynamics of demand contention locking. After giving the demand ordering, a deterministic procedure is used for the computation. As non contending demands make their path searches in parallel, this ordering is not necessarily performed in a single sequence. However, this parallelism does not add any complexity to the deterministic procedure.

We define the following terms and notation for the simulation:

d denotes a demand that does not identify a restoration route in the search.

s_d and t_d are d's source node and destination node, respectively.

V_d and W_d are accessible sub-set and its complement sub-set which are partitioned sets.

C_d is a set of links connected between V_d and W_d in which any path from s_d to t_d must transmit at least one of the links.

Q_l is the set of demands that are d's critical contenders over l, as determined by the link capacity control procedure. The path removal of any one of the demands in Q_l from l would release capacity for d.

$B_d = \bigcup_{l \in C_d} Q_l$ is known as the set of cut contenders of d.

The basic idea of the optimization algorithm is to change the restoration route of one of d's cut contenders to release some critical links for d. The following are the steps of this algorithm [9].

Step-1: For each $k \in B_d$, find the restoration route A_k of k and release capacity for all the links on the route.

Step-2: Execute the route search procedure for d. If no route is found, restore route A_k for k, go to step 1, and continue the loop with a different k. Otherwise, the route A_d be employed. Update capacity tables for all the links on A_d and go to the next step.

Step-3: Execute the search procedure for finding a restoration route for k. If one is found, the end the procedure. We have found restoration routes for both d and k. Break the loop and stop. Otherwise, calculate cut C_k as a by-product of the route, search for k, and go to the next step.

Step-4: Find A_d and release capacity for all the links on the route. For each link in the network that has capacity for d except links in C_k whose cost is high, assign a unit cost. Run a shortest-path algorithm to find a minimum-cost route A_d. Update the capacity for all the links on A_d.

Step-5: Execute the route search procedure again for k. If a route is found, end the procedure. Break the loop and stop. Otherwise, restore route A_k for k, go to step 1, and continue the loop with a different k.

When demand *d* is negotiating with demand *k*, other demands contending either of them have to be locked out. This can easily be made by adjusting the basic contention-locking procedure to lock both service routes. The optimization performance of the distributed algorithm is analyzed by the number of restoration routes to be found in comparison with those of the optimum solution for fixed link capacities.

8.4.3.1 Centralized Algorithms

Centralized algorithms [21,22] are used to determine the primary capacity and the restoration capacity. The primary capacity is the capacity used by the primary routes, whereas the restoration capacity is used by the restoration routes. Two ways are used to determine restoration capacity – sharing capacity among restoration routes and assuming that each demand has dedicated capacity reserved on its restoration route. Restoration without restoration capacity sharing is considered to be 1 + 1 restoration. If capacity overbuilding is of no concern, 1 + 1 restoration offers the fastest possible restoration and the simplest possible implementation architecture. Even when the network capacity is not abundant for 1 + 1 protection [9] everywhere, it may be desirable to partition demands into different priority traffic classes. The capacity-dedicated restoration routes may be used for traffic with higher priority and capacity-shared restoration routes [9] to protect traffic with lower priority.

8.4.4 SCALABILITY AND APPLICATION TO SERVICE LAYER RESTORATION

A simple approach to scalability is to obtain the groups among the demands between a pair of nodes into groups of size $\leq K$ and consider each group as a single demand for routing purposes. The algorithms scale by taking K proportional to the total number of demands in the whole network. The total number of demands is a function of n, where n is the number of nodes and its value lies between cn and n^2, where c is a constant. For the distributed algorithms, in a conceptual sense, the computational complexity at each node is proportional to the demand density – that is, the average number of demand terminations at a single node or the average number of demands carried by a single link (by shortest-path routing). While the number of demands increases linearly with n, the demand density remains a constant or a linear function of n. The scalability of our distributed algorithms can be estimated by increasing the number of nodes.

8.4.4.1 Call Admission Control for Restorable Connections

For restoration, an optimal path is found on the basis of the network restoration objective, and for that, an algorithm is used to access to a database of connections, their service, and restoration paths for call admission control. The restoration route selection algorithm uses iterations through each failure event and estimates failure to determine if sufficient channels have been retained to get the restoration objective. Three general alternatives are given below [9].

- *Compute an Optimized Restoration Path*
  Each node gets all of the information mentioned above and operates to re-estimate all restoration paths against every connection request.

Updated restoration information must be distributed to all relevant nodes after every connection request.

- *Compute a Simplified Restoration Path*

  A simple restoration path is taken (e.g., the shortest-hop path physically diverse from the service path), and the operation steps are run frequently to optimize the restoration resource allocations. This approach is considered as "provision and pray".

- *Compute Hybrid (Approximate) Restoration Path*

  A hybrid approach is a combination of the above two approaches. The parameters provide the number of channels required for restoration over link k for a given failure event. The estimation of restoration paths for a new connection requires accurate calculation. The parameters used here are updated by path setup messages sent along the restoration path during connection provisioning [9].

  Summary

  In this chapter, design and restoration algorithms for optical ring and mesh networks are discussed. Initially we mentioned the need for restoration in which the restoration uses less resources in comparison to protection schemes. We have started with restoration in optical ring network. We have defined different parameters used in restoration and estimated those for performance in different restoration schemes.

  Restoration algorithms are of two types – distributed and centralized. The restoration architecture, contention-locking mechanism, and restoration capacity sharing scheme make an algorithm robust, fast, and capacity efficient.

EXERCISES

8.1. Consider the network topology in Figure Exercise 8.1. A lightpath is set up between nodes 1 and 8 along the link $1 \rightarrow 3 \rightarrow 6 \rightarrow 8$. If link $3 \rightarrow 6$ goes down, which of the following restoration strategies will lead to a path with minimum number of hops? Find the number of hops in each case.
   a. Path restoration
   b. Sub-path restoration
   c. Link restoration

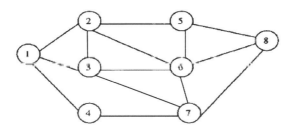

FIGURE EXERCISE 8.1   Network.

8.2. Consider the network topology in Figure Exercise 8.1. A lightpath is set up between nodes 1 and 8 along the link $1 \rightarrow 3 \rightarrow 6 \rightarrow 8$. If link $3 \rightarrow 6$ goes down, which of the following restoration strategies will lead to a path with minimum number of hops? Find the number of paths between nodes 3 and 8 and between nodes 3 and 7 by using shared-link restoration if link $3 \rightarrow 6$ fails. Compare the cost of shared-restoration with that of dedicated-restoration.

8.3. Formulate shared-mesh restoration using the above example considering failures of three links: $3 \rightarrow 6$, $2 \rightarrow 5$, and $4 \rightarrow 7$.

8.4. Formulate shared-mesh restoration using the above example considering three node failures – node 6, node 2, and node 4. Formulate dedicated restoration considering the same failures. Compare the performances in both cases in terms of resource capacity.

REFERENCES

1. J. Zhang and B. Mukheriee, "A review of fault management in WDM mesh networks: Basic concepts and research challenges," *IEEE Network*, vol. 18, no. 2, pp. 41–48, 2004.

2. B. Mukherjee, *Optical WDM Networks*, Springer-Verlag, Berlin, 2006.

3. R. Ramaswami and K. Sivarajan, *Optical Networks: A Practical Perspective*, 2nd ed., Morgan Kaufmann Publishers, Burlington, MA, 2001.

4. S. Gowda and K. M. Sivalingam, "Protection mechanism for optical WDM networks based on wavelength converter multiplexing and backup path relocation technique," *IEEE INFOCOM*, pp. 12–21, 2003.

5. W. Stallings, *Data and Computer Communication*, Prentice Hall Inc, Upper Saddle River NJ, 1997.

6. R. R. Iraschko and W. D. Grover, "A highly efficient path-restoration protocol for management of optical network transport integrity," *IEEE Journal on Selected Areas in Communications*, vol. 18, pp. 779–794, 2000.

7. M. Clouqueur and W. D. Grover, "Availability analysis of span restorable mesh networks," *IEEE Journal on Selected Areas in Communications*, vol. 20, pp. 810–821, 2002.

8. J. Wang, L. Sahasrabuddhe, B. Mukherjee, "Path vs. subpath vs. link restoration for fault management in IP-over-WDM networks: Performance comparisons using GMPLS control signaling," *IEEE Communications Magazine*, vol. 40, pp. 80–87, 2002.

9. B. T. Doshi, S. Dravida, P. Harshavardhana, O. Hauser, and Y. Wang, "Optical network design and restoration," *Bell Labs Technical Journal*, vol. 4, pp. 58–84, 1999.

10. K. S. Trivedi, *Probability and Statistics with Reliability, Queuing, and Computer Science Applications*, Prentice-Hall, Englewood Cliffs, NJ, 1982.

11. Y. Xin and G. N. Rouskas, "Light-tree routing under optical power budget constraints [Invited]," *Journal of Optical Networks*, vol. 3, no. 5, pp. 282–302, 2004.

12. W. D. Grover, "The self-healing network: A fast distributed restoration technique for networks using digital cross-connect machines," *Proceedings of IEEE GLOBECOM '87*, Tokyo, pp. 1090–1095, 1987.

13. C. H. Yang and S. Hasegawa, "Fitness: Failure immunization technology for network services survivability," *Proceedings of the IEEE GLOBECOM '88*, Hollywood, FL, pp. 1549–1554, November 1988.

14. K. Struyve, P. Demeester, L. Nederlof, and L. Van Hauwermeiren, "Design of distributed restoration algorithms for ATM meshed networks," *Proceedings of IEEE* 3rd *Symposium on* Communications and Vehicular Technology, Eindhoven, Netherlands, pp. 128–135, October 1995.

15. K. Murakami and H. S. Kim, "Optimal capacity and flow assignment for self-healing ATM networks based on line and end-to-end restoration," *IEEE/ACM Transaction on Networking*, vol. 6, no. 2, pp. 207–221, 1998.

16. J. Anderson, B. Doshi, S. Dravida, and P. Harshavaradhana, "Fast restoration of ATM networks," *IEEE Journal of Selected Areas in Communications*, vol. 12, pp. 128–138, 1994.

17. R. Kawamura, K. Sato, and I. Tokizawa, "Self-healing ATM networks based on virtual path concept," *IEEE Journal of Selected Areas in Communications*, vol. 12, no. 1, pp. 120–127, 1994.

18. R. R. Iraschko, W. D. Grover, and M. H. MacGregor, "A distributed real-time path restoration protocol with performance close to centralized multi-commodity maxflow," *Proceedings of* 1st *International Workshop on Design of Reliable Communications Networks*, Brugge, Belgium, paper O-9, May 1998.

19. B. A. Coan, W. E. Leland, M. P. Vecchi, A. Weinrib, and L. T. Wu, "Using distributed topology update and preplanned configurations to achieve trunk network survivability," *IEEE Transactions on Reliability*, vol. 40, pp. 404–416, 1991.

20. C. Palmer and F. Hummel, "Restoration in a partitioned multibandwidth cross-connect network," *Proceedings of IEEE GLOBECOM '90*, San Diego, CA, pp. 81–85, December 1990.

21. M. Stoer and G. Dahl, "A polyhedral approach to multicommodity survivable network design," *Numerische Mathematik*, vol. 68, pp. 149–167, 1994.

22. C.-W. Chao, P. M. Dollard, J. E. Weythman, L. T. Nguyen, and H. Eslambolchi, "FASTAR: A robust system for fast DS3 restoration," *Proceedings of IEEE GLOBECOM '91*, Phoenix, AZ, pp. 1396–1400, December 1991.

9 Network Reliability and Security

Network reliability represents a probability that a network node or network element can be functioned satisfactorily for a given time interval under particular operations. The four key factors here are probability, satisfactory performance, time and specified operating conditions [1,2]. The need for network reliability is given below [2]:

1. Probability is used for the measure of the performance. When the devices function under similar conditions, failures may take place at different points in time; the occurrence of the failures can be expressed in terms of probability theory.
2. The network performance shows accomplishment of the system having a combination of qualitative and quantitative factors. These are performance parameters such as bit error rate (BER), throughput, and delay.
3. Time is an important factor for the evaluation of reliability measurement showing the probability of a system being operational.
4. The operating conditions are geographical location of the system, weather extremes in which the system is exposed (e.g., temperature, wind, humidity, vibration, shock and potential damage).

The topological connectivity measures the network reliability. The connectivity is written in terms of path or node failure respectively [2]. From a detailed point of view, reliability is represented by the rate of breakdown of lines and nodal equipment along with the mean time that is required to repair these faults. Several techniques are used for achieving fault-tolerant networks [2].

9.1 CONNECTIVITY USING REDUNDANCY

To obtain high reliability under line or nodal equipment failure, the network requires line and equipment redundancy where network can continue to function after losing a few lines or nodes, possibly at a lower performance level. To analyze the redundancy in a network, graphical representation is required in which the network nodes and their connecting lines are collected in order to plot a graph. We consider the details of using graph theory for network optimization where a network will connect splitting of two or more parts upon the removal of certain nodes of lines. The network in Figure 9.1 is disconnected if we remove nodes 3 and 6. The network has a connectivity of node 2 in which nodes 1, 5 and 4 are connected with node 2.

If the links are removed instead of nodes, the network will disconnect upon the removal of links. The measure of the number of links that need to be removed to split the network is called the arc connectivity.

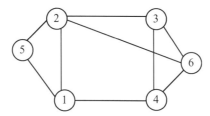

FIGURE 9.1   Representation of a network [2].

The arc connectivity of a network is an important parameter in the network in which the minimum of lines are cut from one node to other nodes. If the two most weakly connected nodes have only K-arc-disjoint paths (i.e., paths having no arcs in common between them) then the network has a connectivity of k or alternatively the network is k-arc-connected.

9.1.1   MIN-CUT MAX-FLOW THEOREM

The concept of a cut can be used for finding the mean information flow in a network. In the network shown in Figure 9.2 the number next to each line indicates its capacity. We consider nodes 1 and 10. However not all possible cuts are drawn where cut 3 is the minimum cut. Since all the information that flows between nodes 1 and 10 must pass through the three links indicated by cut 3, the maximum possible flow between nodes 1 and 10 is 9 units of data (the capacity of the minimum cut). This Min-Cut Max-Flow Theorem indicates the maximum flow between two nodes that cannot exceed the capacity of the minimum cut separating these two nodes.

A variety of algorithms are used for the estimation of the maximum flow and the minimum cut has been developed and reported for the same [2].

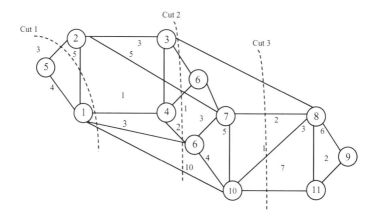

FIGURE 9.2   Possible cuts in a network [2].

9.1.2 THE CUT-SATURATION ALGORITHM

The cut-saturation algorithm [3] is a heuristic method that can be used in the net-work-connectivity problem in which a large number of cuts are present in the net-work. As the traffic increases in a network the lines in one of these cuts will move toward saturation. There are individual lines in which the traffic flow cannot be increased without creating excessive delay. To deal with this problem, one can make the capacity of network larger by increasing the capacity of one or more links within this cut, or adding more lines to the cut. The cut saturation algorithm [3] is used to obtain the network throughput within specified bounds while iteratively dropping the overall line costs, reliability constraints etc. which is as follows:

1. Order all the links according to the volume of traffic they carry where some lines will be highly utilized (80–90% is a suitable figure) whereas others have little traffic.
2. Conceptually remove links in sequence from the network beginning with the most utilized ones, until the network has been separated into two disjoint groups of nodes.

The step of the cut-saturation algorithm has been adding and deleting links between the two disjoint parts of the network that are separated by the saturated cuts. If we choose to replace the lowest-cost link this will usually rejoin primary nodes only. In practice, this is found to shift the cuts slightly without greatly improving net-work performance. We can use adding lines by the cut-saturation method until the throughput exceeds the network requirements. At this point, we can use link dele-tion. If the network is fully connected but expensive, links can be deleted from the cut. The two main criteria used for line removal are utilization and cost. So we can take away the most costly, least-used link. This link is the one that maximizes the quantity [3]

$$E_i = \frac{D_i(C_i - f_i)}{C_i} \tag{9.1}$$

where D_i, C_i and f_i are respectively the cost, capacity and flow in link i. The quantity $(C_i - f_i)/C_i$ is the relative excess capacity of link i. In removing links, we place a two-connectivity constraint on the network to ensure reliability.

9.2 PROBABILITY OF CONNECTIVITY

In reliable networks, the objective is to maximize the probability to get the network active with specific probability of failure for each of its nodes and lines [2]. We require the techniques for assessing these probabilities. A network having L links and N nodes is considered where there are at least two-connected. Here, for a given network with known failure probabilities of its nodes and lines, it is essential to find the probability that at least one complete path exists between a source–destina-tion (SD) pair. The computation of the reliability measure requires the details of all simple paths between all SD pairs. The complexity of this problem augments with network size and topological connectivity [2]. We consider the following [2]:

- All communication link failures and node breakdown are statistically independent.
- The links in the network fail with a probability p, and nodes break down with a probability q.

9.2.1 Node Pair Failure Probability

First, we attempt to estimate the probability of one node in the network becoming isolated [2]. The node pair failure probability is the probability of the network becoming disconnected. Here, we consider line and node failures separately. We also assume that [2]

- The probability of node failure q is zero.
- The probability of line failure p is zero.

We derive the probability of successful communication between node pairs. The node failures are negligible compared to line failures ($p \gg q$). $A_L(i)$ can be designated in the number of ways that a network can be connected when i links are operational and the remaining $L - i$ links have failed. The probability of any particular set of i links being operational is then $(1 - p)^i P_{L-i}$. For all values of i, the probability $P_c(p)$ of successful communication between any pair of operating nodes I is then approximated by [4]

$$P_c(p) = \sum_{i=0}^{L} A_L(i)(1 - p)^i P^{L-i} \qquad (9.2)$$

When failures occur in any of the $N - 2$ nodes, they can be traversed by paths linking two arbitrary nodes i and j. Since line failures p are negligible compared to node failures $q(p \ll q)$, the probability that k of the $N - 2$ nodes are operational while the remaining $N - 2 - k$ nodes have failed is $(1 - q)^k q^{N-2-k}$. Then, the probability of successful communication is

$$P_c(q) = \sum_{k=0}^{N-2} A_N(k)(1 - q)^k q^{N-2-k} \qquad (9.3)$$

where the coefficients $A_N(k)$ denote the number of combinations of k nodes such that if they are operational (and the others have failed), at least one communication path exists between nodes i and j.

Alternately, we can estimate the probability of a communication and failure between any pair of operative nodes. The probabilities corresponding to line and node failures, $P_f(p)$ and $P_f(q)$, respectively, are given by [5]

$$P_f(p) = \sum_{i=0}^{L} C_L(i)p^i(1-p)^{L-i} \tag{9.4}$$

$$P_f(q) = \sum_{k=0}^{N=2} C_N(k)q^k(1-q)^{N-2-k} \tag{9.5}$$

Here, the coefficients $C_L(j)$ and $C_N(j)$ represent the number of ways that a network can become disconnected for j links and nodes, respectively. Thus, $C_L(j)$ is equal to the total number of cuts of size j with respect to nodes i and j. For any network having a line connectivity equal to d, $C_L(i) = 0$ for all $i \leq d$. Therefore, the most prominent term for $P_f(p)$ in equation (9.4) is given by

$$P_f^d(p) = C_L(d)p^D(1-p)^{L-d}$$

Since for small values of p (much less than 0.5), this term is reduced for increasing values of d, the ideal network should be maximally connected. While analyzing the reliability of a network, it is desirable to calculate the minimum value of the connectivity probability or the maximum value of the failure probability. For the simplicity of discussion, let us concentrate here on the line failure probability given in equation (9.4). To find this, we need to compute the parameter $C_L(i)$. The exact calculation of the maximum value of $P_f(p)$ requires the examination of all $i-j$ cuts for all $i-j$ node pairs. This can only be done with algorithms run on a computer [5]. However, by using some simplifying arguments, upper and lower bounds on $P_f(p)$ can be established [5]. For given N nodes, $N-1$, links are required to connect them. If there are fewer links, we have that the number of ways the network can be disconnected of i links 3 are not operational is given by [5]

$$C_L(i) = \binom{L}{i} = \frac{l!}{i!(L-i)!} \quad \text{for} \quad i = 0,1,2,\ldots,N-2 \tag{9.6}$$

The network can also obviously be disconnected for $i = N-1$, but we will ignore this here since we are looking for a lower bound. Using equation (9.6) in equation (9.4) then gives a lower bound of

$$P_f(p) \geq \sum_{i=0}^{N-2} \binom{L}{i} p^{L-i}(1-p)^i \tag{9.7a}$$

We can also write

$$p_f(p) \geq \sum_{k=L-N+2}^{L} \binom{L}{k} p^k(1-p)^{L-k} \tag{9.7b}$$

The upper bound is considered where M is the minimum number of links that need to be removed for the network to be disconnected. We consider

$$C_L(i) = 0 \quad \text{for} \quad i = 0,1,2,\ldots,M-1 \tag{9.8}$$

We note here that these M links are generally not an arbitrary selection of lines and, in addition, more than M links may be needed to disconnect the network. The network may not be disconnected for a random value of $i = M, M+1,\ldots, L$ links. We discount these possibilities here in seeking an upper bound on $P_f(p)$. Using equations (9.4) and (9.8), we obtain

$$P_f(p) = \sum_{k=M}^{L} \binom{L}{k} p^k (1-p)^{L-k} \tag{9.9}$$

For $p \sim 1$, $P_f(p)$ is approximated by equation (9.7), and for $p \sim 0$, $P_f(p)$ is approximated as in equation (9.9).

A variety of other reliability analyses that go beyond the simple examples given here are carried out. These analyses include the consideration of simultaneous link and node failures [6,7], network performance bounds based on link reliability [2], evaluation of networks having dependent failures [8] and reliability measures based on system availability [9].

9.3   RELIABILITY MODEL

For reliability and the probability of network connectivity, the reliability and availability of a network should be analyzed with various failure and repair states. The simple time-dependent evaluation should be made for a network. For this purpose, we assume:

- The networks have two state elements (node connections, transmission lines, terminals and other devices), where an element is made to have the states – if it either operates or fails.
- The system can be repaired if elements in it fail. One considers network repair whenever the average repair cost of a piece of equipment is a fraction of its initial cost.
- If such a system can be quickly returned to service, the effect of a failure is minimized.

The functions are the reliability function, the mean time to failure (MTTF), the mean time to repair (MTTR), the mean time between failures (MTBF) and the mean time to function.

9.3.1   RELIABILITY FUNCTION

In network active at $t = 0$, the reliability is a probability that is active over the time interval [0, t]. A number of component failures take place after $t = 0$, but the network remains ready through the interval $[0, t]$. The reliability function $R(t)$ is written as [10,11]

$$R(t) = 1 - F(t) \tag{9.10}$$

Here, $F(t)$ is the probability of the system having unsuccessful by time t and is thus called the failure distribution function or the unreliability function. The temporal distribution of failures can be written in terms of a failure density function $f(t)$ as

$$f(t) = \int_0^t f(t)\,dt \tag{9.11}$$

It follows from equations (9.10) and (9.11) that

$$f(t) = \frac{dF(t)}{dt} = \frac{dR(t)}{dt} \tag{9.12}$$

To obtain the reliability function, a Markov state-transition model is used when estimating message traffic flow. Figure 9.3 represents the state-transition diagram for the simple reliability model, where the transition is between the initial operating state 0 and the final failed state 1. The failure rate (the transition rate from state 0 to state 1) is given by $\lambda(t)$.

The differential equations describing the transition between states are [2]

$$\frac{dP_o(t)}{dt} + \lambda(t)P_0(t) = 0 \tag{9.13a}$$

$$\frac{dP_1(t)}{dt} - \lambda(t)P_0(t) = 0 \tag{9.13b}$$

Using the boundary conditions $P_0(0) = 1$ and $P_t(0) = 0$, the reliability function is written as

$$R(t) = P_0(t) = 1 - P_1(t) = \exp\left[-\int_0^t \lambda(\tau)\,dr\right] \tag{9.14}$$

Considering the failure rate $\lambda(t)$ at a constant rate λ, we can write

$$R(t) = e^{\lambda t} \tag{9.15}$$

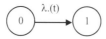

FIGURE 9.3   State transition diagram of reliability model [2].

Here, the arrivals of failures are Poisson. We will consider that failures occur at a constant rate λ.

9.3.2 RELIABILITY MEASURES

Three common measures of reliability are MTTF, MTTR and MTTB. MTTF is the average time in which the element attains he failed state, provided that it was active at time $t = 0$. So MTTF is the mean life of the element. MTTF depends on the reliability function, which is written as [2]

$$\text{MTTF} = \int_0^\infty R(t)\,dt = \frac{1}{\lambda} \tag{9.16}$$

where we have assumed that the failure rate is constant.

Analogous to the failure rate derivation, we consider that repair times are exponentially distributed and repairs are made at a constant rate μ. MTTR for a single element is written as [2]

$$\text{MTTR} = \frac{1}{\mu} \tag{9.17}$$

The average time between successive failures of that element is basically MTBF, which is written as

$$\text{MTBF} = \text{MTTF} + \text{MTTR} \tag{9.18}$$

9.3.3 AVAILABILITY FUNCTION

The availability function shows performance benefits of repair in a system having tolerance of shutdown times. The availability function $A(t)$ represents the probability that an element (or system) is active at time t. In contrast, the reliability function $R(t)$ represents the probability that the element is active over the interval $[0,t]$. Normally, $R(t) \le A(t)$. To derive the availability function, we consider the state-transition diagram shown in Figure 9.4. The element passes from an operating state to a failed state at a constant rate λ, and through repair, it transitions back to the operating state at a rate μ.

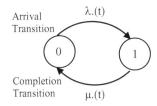

FIGURE 9.4   State transition diagram of availability function model [2].

From this Markov model, the state transitions are represented using constants λ and μ as [2]

$$\frac{dP_0(t)}{dt} + \lambda P_0(t) = \mu P_1(t)$$

$$\frac{dP_1(t)}{dt} + \mu P_1(t) = \lambda P_o(t) \quad (9.19)$$

Using the boundary conditions $P_0(0) = 1$ and $P_1(0) = 0$, the availability which is the probability $P_0(t)$ that the element is active at time t is written as [2]

$$A(t) = P_o(t) = \frac{\mu}{\lambda + \mu} + \frac{\lambda}{\lambda + \mu}\exp[-(\lambda + \mu)t] \quad (9.20)$$

The variation between $A(t)$ and $R(t)$ is represented by their steady-state characteristics. For large value of t, $R(t)$ approaches zero and $A(t)$ reaches some nonzero steady-state value, which is written as [2]

$$A_{ss}(t) = \lim_{t \to \infty} A(t) = \frac{\mu}{\lambda + \mu} \quad (9.21)$$

The same can be written as [2]

$$A_{ss}(t) = \frac{\text{MTTF}}{\text{MTTF} + \text{MTTR}} \quad (9.22)$$

9.3.4 SERIES NETWORK

The overall reliability is occurred in ring or bus network where a number of elements are connected in series. The series configuration is one of the simplest methods used for analysis. For the series-connected units shown in Figure 9.5, all the components in the network must work if the system is to operate properly. If these components are not interactive, the failures are not dependent and the system reliability $R(s(t))$ is a product of the reliabilities of the individual constituent units [2].

$$R_s(t) = \prod_{i=1}^{k} R_i(t)$$

$$= \prod_{i=1}^{k} [1 - F_i(t)] \quad (9.23)$$

FIGURE 9.5 Network elements in series configuration [2].

where $R_i(t)$ is the reliability function, and $F_i(t)$ is the failure distribution function of the ith system unit. If the ith component has a constant failure rate λ_i, then we can write

$$R_s(t) = \prod_{i=1}^{k} e^{-\lambda_i t} = \exp\left[-\left(\lambda_1 + \lambda_2 + \cdots + \lambda_k\right)t\right] \tag{9.24}$$

This provides the overall reliability of a k-element series-connected network for a specified time period t. For the series configuration, MTTFs can be written as [2]

$$\text{MTTF}_s = \int_0^{\infty} R_s(t)\,dt = \left(\sum_{i=1}^{k} \lambda_i\right)^{-1} \tag{9.25}$$

9.3.5 PARALLEL NETWORK

In a parallel network, alike individual components are connected in parallel. Although this may give redundancy, it increases network reliability in which the system will only fail if all its components break down. To obtain the reliability function for the parallel network shown in Figure 9.6, the following two assumptions are considered:

- Units are operational sharing the network load.
- Components are statistically not dependent.

For non-identical components, the failure distribution function at time t is [2]

$$F_p(t) = \prod_{i=1}^{k} F_i(t) \tag{9.26}$$

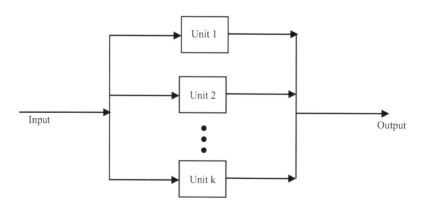

FIGURE 9.6   Network elements in parallel configuration [2].

where $F_i(t) = 1 - R_i(t)$ is the failure distribution of the ith component. In the same way, the parallel-configuration reliability is [2]

$$R_p(t) = 1 - F_p(t)$$

$$= 1 - \prod_{i=1}^{k} [1 - R_i(t)] \qquad (9.27)$$

MTTF$_p$ for the parallel network is formulated from the relationship:

$$\text{MTTF}_p = \int_0^\infty R_p(t)\, dt \qquad (9.28)$$

We consider a parallel system consisting of two nonidentical units having constant failure rates λ_1 and λ_2. Then, we have

$$R_p(t) = 1 - \left(1 - e^{-\lambda_1 t}\right)\left(1 - e^{-\lambda_2 t}\right)$$

$$= e^{-\lambda_1 t} + e^{\lambda_2 t} - e^{-(\lambda_1 + \lambda_2)_i t} \qquad (9.29)$$

So we can write

$$\text{MTTF}_p = \int_0^\infty \left[e^{-\lambda_1 t} + e^{-\lambda_2 t} - e^{-(\lambda_1 + \lambda_2)_i t} \right] dt$$

$$= \frac{\lambda_1 + \lambda_2}{\lambda_1 \lambda_2} - \frac{1}{\lambda_1 + \lambda_2} \qquad (9.30)$$

9.3.6 RELIABILITY IMPROVEMENT TECHNIQUES

The reliability and availability of a network is increased by making elements active in the transmission path as minimum as possible, reducing the number of repeaters or line amplifiers between stations and having distributed the control unit throughout the network. Reliability can also be enhanced by increasing idleness of transmission paths and nodal equipment. Specially in the ring network, two fundamental techniques are used for improving reliability by a redundant standby loop in parallel [12,13]. The implementations of these schemes have been carried out [33]. The first method is known as the bypass technique in which standby rings send in the same direction. When a failure takes place, a bypass element is used to route the signal stream onto the standby ring. The second method is a self-heal technique in which the main and standby loops send the signals in opposite directions [2]. For a failure occurrence, a reconfiguration switch reroutes the signal stream so that it bends back on itself via the standby loop at either side of the failure. A hybrid network using a combination of either the bypass or the self-heal method can also be implemented. In this case, the standby loop must have bidirectional capability.

9.3.7 AVAILABILITY PERFORMANCE

The availability A of a particular element in a network is the probability that the element is active. The probability D that a unit is not active is then given by [2]

$$D = 1 - A \qquad (9.31)$$

where D is the part of time that an element has failed and is also represented as the downtime ratio or DTR (the ratio of the length of time the element has failed to a specific time duration the network is in use) [2]. It is used as a parameter for comparing the relative performance of different reliability enhancement techniques. In Figure 9.7 a variety of impairments are obtained in the network – a terminal could fail within the dashed box denoted by E at node j in which since this breakdown does not affect the performance of the other terminals on the link, the DTR parameter D_E associated with it is small compared to failures that disrupt the entire ring and a failure could also occur at the terminal-to-ring interface [2] where loss of this interface is disastrous since it stops all communications on the ring. A failure in the transmission line between two stations has exactly the same effect on network performance as a breakdown at a node.

Failures in the network controller are analyzed through the parameter D_{NC}, which represents the fraction of time that is taken that node component is failed. This DTR encompasses both faults related to control of the ring and all impairments in the network controller that cut the entire ring connection from the outside world. We include the concept of a terminal-attachment point and associate a downtime ratio $D_p(m)$ with it. A terminal-attachment point is active and available for use if

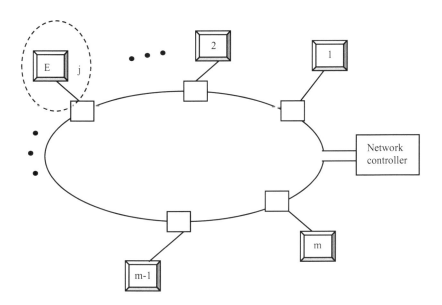

FIGURE 9.7   General ring topology [2].

communications can be obtained from that point to any other point of the ring or the outside world. The factor $D_p(m)$ is then given by [2]

$$D_p(m) = 1 - (1 - D_{NC})(1 - D_E)(1 - D_r)_m \qquad (9.32)$$

where $(1 - D_{NC})$ is the availability of the network controller, $(1 - D_E)(1 - D_r)$ is the probability that the terminal connection point becomes active and $(1 - D_r)m^{-1}$ is the probability that there are no disastrous failures for the remaining m^{-1} terminals of the ring [2].

The expression $D_p(1)/D_p(m)$ compares the failure-induced degradation of an m-terminal network with respect to a one-terminal system. This degradation is shown in Figure 9.7 for the case where $D_E \ll D_T$, $D_{NC} \ll D_r$ and $D_T = 10^{-4}$. The parameters D_E and D_{NC} were chosen small relative to D_T to avert their values from the terminal degradation effect for small m.

The failure-bypass technique is shown in Figure 9.8a. The special bypass devices called reconfiguration units are included into the loop in the figure. Here, L units are added to the ring so $L - 1$ groups of X terminals are formed, where [2]

$$X = \frac{m}{L - 1} \qquad (9.33)$$

Two different failure possibilities are taken into account. There are failures of nodes within a terminal group and failure of the reconfiguration unit. If a failure occurs in a terminal group, each group is disabled and must be bypassed. If failures occur in several different groups, then each of these groups must be bypassed. For the reconfiguration unit, for the simplicity of analysis, any failure is considered to be catastrophic. That is, loss of this unit disrupts entire ring activities. Thus, we considered D_s as the downtime ratio associated with a failure in the reconfiguration unit

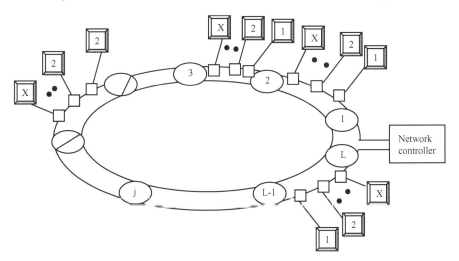

FIGURE 9.8a Byepass technique in ring topology [2].

(Continued)

or switch. We can now define the DTR of a terminal-attachment point of a ring with failure-bypass capability. A terminal-attachment point is considered to be operational if a communication link can be established between it and any other functioning node. To achieve this, the following conditions must be considered [2]:

- The terminal-attachment point is surely active with the probability $(1 - D_E)(1 - D_T)$.
- The probability that there are disastrous failures among the other x^{-1} terminals in the same group is $(1 - D_T)x^{-1}$.
- All reconfiguration units are operational with a probability $(1 - D_s)^L$.
- The network controller is operated with an availability $(1 - D_{NC})$.

These factors provide the availability A_B for a terminal-attachment point [2]:

$$A_B = (1 - D_E)(1 - D_T)(1 - D_T)^{X-1}(1 - D_s)^L(1 - D_{NC}) \qquad (9.34)$$

where the subscript B refers to a ring with bypass capability. The probability that a terminal-attachment point is not active is

$$D_B(L,m) = 1 - A_B$$

$$= 1 - (1 - D_E)(1 - D_T)^X(1 - D_s)^L(1 - D_{NC}) \qquad (9.35)$$

For the improvement offered by the bypass technique in the simple ring network, we analyze the ratio $D_p(m)/D_B(L,m)$. We consider $D_T = D_S = 10^{-4}$, $D_E \ll D_T$ and $D_{NC} \ll D_T$. For each m, the optimum values L_{opt} are obtained at which a minimum D_B occurs. The value of L_{opt} is given by

$$L_{opt} = \left[m\frac{\ln(1 - D_T)}{\ln(1 - D_s)} \right]^{(1/2)} + 1 \qquad (9.36)$$

Using the above equations, we obtain

$$D_{B.opt}(m) = 1 - \exp\left\{ -2\left[(mD_TD_S)^{1/2} + D_s \right] \right\} \exp\left[-(D_E + D_{NC}) \right] \qquad (9.37)$$

There are two factors: first, for few reconfiguration units, the bypass features are insufficiently applied since a failure in a terminal-attachment point isolates a large portion of the network and if there are many reconfiguration units relative to the number of terminals, then the probability that one of these units fails begins to degrade the advantage of the bypass mechanism. This course depends on the value of D_s. Second, as the failure probability of the reconfiguration unit decreases, L_{opt} increases for a fixed number of terminals. In fact, for a highly reliable bypass unit, L_{opt} is equal to m; i.e., the minimum value of $D_{B.opt}$ is reached when each terminal has a bypass unit [2].

9.3.8 THE SELF-HEAL TECHNIQUE

Figure 9.8c shows the self-heal technique handling different failure cases. Reconfiguration switching units are employed analogous to the bypass technique. In this scheme, however, a reconfiguration unit loops the signal path back on itself via the auxiliary channel on each side of a failure.

As shown in Figure 9.8b, if a terminal-attachment point fails, the entire group to which the terminal belongs is disabled. Second, if terminals in two different groups are impaired, then not only are the two affected groups disabled, but also the entire ring section between the two groups is avoided, as shown in Figure 9.8b. Figure 9.8d shows the breakdown of a reconfiguration unit also influencing the two adjacent terminal groups. Figure 9.8d shows the effect of losing two reconfiguration units. Here, all terminal groups between the affected units plus the two adjacent terminal groups are inactivated.

The worst-case downtime ratio in the self-heal technique is formulated for a terminal that has the largest number of reconfiguration units between it and the network controller showing this worst-case DTR, which is given by [2]

$$D_{SH}(L,m) = 1 - (1 - D_E)(1 - D_{NC})$$

$$\times \left[2(1 - D_T)^{(m/2)L/(L-1)}(1 - D_S)^{l/2} - (1 - D_T^{m})(1 - D_S)^L \right] \tag{9.38}$$

For L even, and

$$D_{SH}(L,m) = 1 - (1 - D_E)(1 - D_{NE})$$

$$\times \left[(1 - D_T)^{(m/2)(L+1)(L-1)}(1 - D_S)^{(L+5)/2} + (1 - D_T)^{m/2}(1 - D_S)^{(L+3)/2}(1 - D_T)^m (1 - D_S)^L \right]$$

$$\tag{9.39}$$

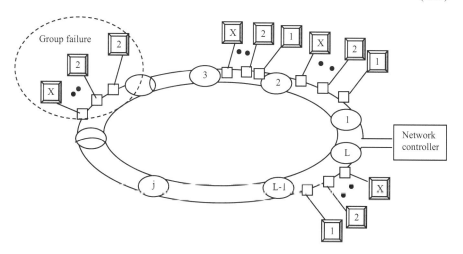

FIGURE 9.8b (CONTINUED)   Self-healing technique for group failure in ring topology [2].

(Continued)

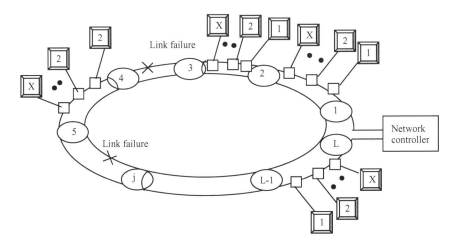

FIGURE 9.8c (CONTINUED)   Self-healing technique for link failure in ring topology [2].

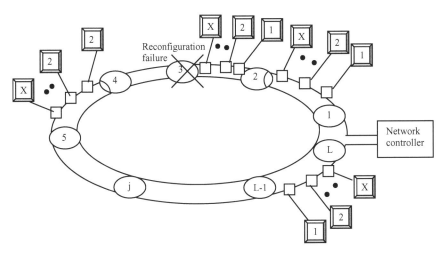

FIGURE 9.8d (CONTINUED)   Self-healing technique for reconfigurable unit failure in ring topology [2].

For odd L, odd, where the subscript SH refers to a ring with self-heal capability.

The improvement of the self-heal technique over that of a primitive loop is given by $D_p(m)/D_{SH}(L,m)$, where $D_p(m)$ is represented by equation (9.32). Again, there is a unique value of L_{opt}, which minimizes D_{SH}.

9.3.9   FAIL-SAFE FIBER-OPTIC NODES

The fail-safe nodes have been used in fiber-optic local area networks for reliable transmission [34–42]. The characteristics of these nodes have a string of active repeaters in the network that keeps functioning when power is lost at one or more nodes. Figure 9.9 shows a ring network consisting of a transmitter and receiver pair,

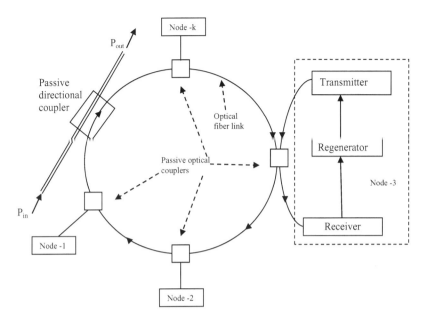

FIGURE 9.9   Optical fiber ring network having connection with another bus network [2].

a regenerator connecting this pair and a passive lightwave coupler. The nodes are connected by optical fibers. In normal operation, each node actively monitors the transmission line and regenerates information flowing in the network. While a node transmits, its regenerator remains switch off and inserts its information into the network. Thus for the node failures or removal of components for maintenance, the lightwave coupler shows the optical-line continuity for uninterrupted network operation. Here, for the operation, the fail-safe network needs the following sensitivity and interference constraints [2]:

- The sensitivity constraint states that any receiver must be sensitive enough to detect the signal from a preceding transmitter when there are several nodes between the transmitter and the receiver. The exact allowed number of adjacent failed nodes is derived from the system requirements.
- The interference constraint states that when two nodes are transmitting simultaneously, a downstream node should receive the signal from the closer node. This means that the less intense signal from the more distant transmitter must be below the threshold level of the receiver.

Each fail-safe node has one or two optical couplers, depending on the particular operation desired. Figure 9.9 shows the concept of coupler node. Here, P_{in} is the input optical signal incident in the coupler from the ring, whereas P_t is the transmitted optical signal of the coupler. The receiver must be off when the transmitter is on to avoid saturation of the receiver. This is accomplished with a signal format having a duty cycle less than 50%, and having the transmitter operate out of phase from the receiver [2].

The coupler is characterized by a four-port device as discussed in Chapter 3. Its basic parameters are C_T, which represents the fraction of optical signal power removed from the ring; F_i, which is the fraction of optical signal power lost in the coupler and E, which is the coupling efficiency of optical power onto the ring. For the simplicity of analysis, here we will ignore the factor F_C, fraction of power loss at each port of the coupler (which is mainly connector loss).

Under the off condition of the transmitter, the optical power P_r entering the receiver is given by [2]

$$P_r = C_T P_{in} \tag{9.40}$$

The output power of the coupler entering the ring is

$$P_{out} = (1 - F_i)(1 - C_T)P_{in} + EP_t \tag{9.41}$$

where P_{in} and P_t fall in different time slots. There are Q adjacent upstream nodes having failed or off. The fibers between nodes have a length L and an attenuation α_f; the power received after these Q nodes is $P_S + P_t$. P_S is written as [2]

$$P_S = C_T \alpha_f^{Q+1} (1 - F_i)^Q (1 - C_T)^Q EP_t \tag{9.42}$$

The power received from the closest active transmitter is written as

$$P_t = \sum_{j=Q}^{\infty} C_T \alpha_f^{j+2} (1 - F_i)^{j+1} (1 - C_T)^{j+1} EP_t$$

$$= \frac{\alpha_f^{Q+2} \left[(1 - F_i)(1 - C_T)^{Q+1} \right] C_T EP_t}{1 - (1 - F_i)(1 - C_T)\alpha_f} \tag{9.43}$$

The powers from all the other upstream active transmitters can cause interference. Equation (9.43) estimates the worst case of interference. The sensitivity and interference constraints are fulfilled when the effective received power is equal to the difference between P_S and P_t, which is greater than the sensitivity S of the receiver [2]:

$$P_S - P_t \geq S \tag{9.44}$$

9.4 NETWORK SECURITY

Due to having the rapid growth of inter-networking, the security and integrity of the information transmitted through these networks are important issues to draw the attention among the researchers [14–17]. The networks are particularly susceptible to security compromise since these are designed to provide many user access points for the connection of remote networks. To reduce the number of components and information in the network especially from unauthorized users, network security is required for the administrative, physical and technical issues. It needs to estimate threat environment and assess the security risks in the network. The main issue is how to physically protect the hardware from accidental or intentional damage and access and to ensure only trusted users accessing the network; the appropriate security techniques can be applied against those.

9.4.1 Network Security Problems

The concepts of security have a variety of interfaces to the network, including individually attached users, multiple users connected through a host computer; a bridge connecting a similar network; a gateway connecting a dissimilar network or public network and other equipment such as printers, memory banks and shared computers [2]. The transmission media connecting the various elements and networks are twisted-pair wires, coaxial cables, optical fibers, leased phone lines, microwave or satellite links [2].

9.4.1.1 Threats

Considering the volume of the information in the network, other untrusty party intercepts the information transmitted in the network. This interception is basically through wiretapping. In network, an intruder interrupts at a point of the network through which all information of interest must pass. There are two types of wiretapping: passive and active types.

In passive wiretapping, an intruder simply monitors the messages at some point without interfering with the information flow. The illegal supervision of information is basically leakage of message content and basically passive wiretapping. The objective is to determine the contents of the messages. Furthermore, the wiretapper determines the quantities, lengths and frequencies of the message transmissions. These types of passive attacks are considered for traffic analysis.

In active wiretapping, message processing is performed on the information stream, but at access tapping point, the messages are selectively revised, removed, deferred, sorted out and duplicated either immediately or passed through unaffected portion of the network. In addition, fake messages can be generated with original message. Here, the adversary can discard all messages that pass through the connection point or can delay all the messages going in either one or both directions. Attacks of this nature are classified as unauthorized denial of message service.

Both active wiretapping and passive wiretapping require different protection mechanisms. The passive attacks of message contents and traffic usually cannot be detected but those are required to be prevented. The attacks of message stream modification, denial of message service and message playbacks cannot be debarred. Secure communications are dependent on the cryptography technique to protect the confidentiality and integrity of messages transmitted between machines. Cryptography is derived from the Greek word "kryptos", which means "hidden" and from graphy, which means "to write", and it is the science of secret communication. The original message is cleartext and the transformed message is known as ciphertext, which is also sometimes called a cryptogram. Initially, military and intelligence organizations have used some cryptographic techniques to keep the message sensitive and secret over the years. The message-encoding methodology was first reported by Julius Caesar [2] when sending messages. The process of changing plaintext to ciphertext is called enciphering or encryption. The inverse operation converting ciphertext to plaintext is known as deciphering or decryption.

The encryption is made in TCP/IP [2]. There are two fundamental approaches to network security: link-oriented and end-to-end encryption measures. The general scheme is shown in Figure 9.10a where encryption is performed independently on each communication link between successive modems. The encryption is carried out by means of a function called a key. Each link corresponds to a data-link layer association in the TCP/IP model [2], which can cover origin-to-destination information flow patterns.

The end-to-end security provides security of each message along its entire route from source to destination, as shown in Figure 9.10b. Thus, messages are sent through transmission links of the network in which local computers, intermediate node and switches in an encrypted form are provided by the encryption device at the message originator [2].

Encryption is end-to-end security usually made in host-to-host layer [2]. Since the encryption is performed in higher level, the greater security is required for the user. Some caution is needed for end-to-end security because of the flexibility in identifying exactly where the security measures are required to maintain.

9.4.2 Data Encryption

One should know first the basic concepts of data encryption and then describe two fundamental data encryption techniques [17–21]: transposition ciphers and

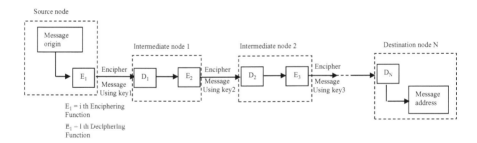

FIGURE 9.10A (CONTINUED)   Link-oriented security measures [2].

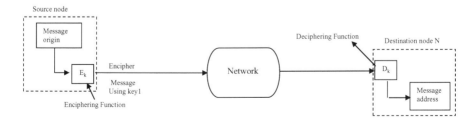

FIGURE 9.10B (CONTINUED)   End-to-end security scheme using enciphering with key k [2].

simple-substitution ciphers. These are product ciphers as exemplified by the Data Encryption Standard (DES), and exponential ciphers by the Rivest–Shamir–Adleman (RSA) scheme.

9.4.2.1 Basic Concepts

The fundamental encryption and decryption processes are pictorially represented in Figure 9.11. In encryption function, there are two inputs: plaintext and a key. The key consists of a finite number of bits represented as decimal, hexadecimal, or alphanumeric character strings. Although the length of the key may be of the same order of magnitude as the plaintext, it is commonly eight characters long for practical reasons.

As shown in Figure 9.11 the function of the encryption key K_E is to carry out mathematical operation (denoted by E) on the plaintext to transform it into ciphertext. There is a decryption algorithm D, which is a reverse process of the encryption transformation used to recover data with the appropriate decryption key KD. We consider the followings for the implementation of encryption and decryption processes [2]:

$$M = \text{plaintext message}$$
$$C = \text{ciphertext}$$
$$K = \text{key}$$
$$E = \text{encryption function}$$
$$D = \text{decryption function}$$

Then after encryption operation on plaintext, we can write

$$C = E_k(M) \tag{9.45}$$

After using decryption operation, we can get

$$= D_K\left[E_k(m)\right] \tag{9.46}$$

The efficiency of the encryption technique does not depend on the encryption algorithm steps, which makes the information to be kept secret. The information security depending on degree of complexity needs deciphering it without having knowledge

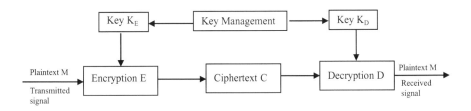

FIGURE 9.11   Scheme having encryption and decryption processes.

of the decryption key. A property of the decryption key is not known by anyone having a knowledge of the encryption key, the ciphertext [2].

The analysis for finding a secret from either ciphertext or encryption key without the authorization of the sender is called as cryptanalysis [22–23]. A cryptanalyst has one of the three following kinds of attacks:

1. **Ciphertext:** There are one or more cryptograms that are known to have been encrypted with the same key by cryptanalyst.
2. **Known Plaintext:** Cryptanalyst should have one or more plaintexts and the resulting cryptograms are known to have been created by the same key.
3. **Chosen Plaintext:** For any desired plaintext, cryptograms are obtained by cryptanalyst using the same key from which the undeciphered cryptogram of interest was created.

Any encryption system should secure the plaintext against a chosen-plaintext attack. If it is possible, then it must secure against a known-plaintext attack. Failing this, it should secure also at least to immune to a ciphertext-only attack. The ciphers become either secret or public [24]. In a conventional (secret) cipher, the same key is used to encrypt and decrypt a message. Such a key must obviously remain secret. For a public key cipher (which is also known as an asymmetric or a two-key cryptosystem), different keys are used to encrypt and decrypt a message [25]. These keys have a pair of transformations, each of which is the inverse of the other and neither of which is derivable from the other. The encryption key is made publicly known, whereas the corresponding decryption key remains secret.

There are different keys: master, primary, or key encryption keys. Keys used during the course of a single association between communication parties are known as working, secondary, or data encryption keys. Master keys are used to verify the authenticity of the communicating parties and to safely transmit working keys. Working keys are used exclusively to encrypt information on a single association. An important aspect is the management of these keys [26,27]. Key management includes generating, distributing, storing, entering, using and destroying or archiving cryptographic keys.

9.4.2.2   Transposition Ciphers

Transposition ciphers have the rescheduling of each character in the plaintext message to produce a ciphertext where the encryption involves reversing the entire message, restructuring the message into a geometrical shape, reorganizing the plaintext by scrambling of columns and periodically permuting the characters of the plaintext [2].

We consider that the following encryption using message reversal shows that the plaintext is represented as backward to produce a ciphertext. If the plaintext message [2] is "NETWORK SECURITY", then the encrypted message reads "YTIRUCES KROWTEN".

The security may not be strong to decipher it one recover the ciphertext in reverse [2]. First, the plaintext message is written into the figure according to a particular pattern. The ciphertext is then created by taking the letters off the figure according to a different path. We consider that the plaintext word RGEKHTENSTGINER is written into a 3×5 matrix by rows as follows [2]:

Column number	1	2	3	4	5
Ciphertext	R	G	E	K	H
	T	E	N	S	T
	G	I	N	E	R

If the letters are arranged by columns in the order 24135, the resulting ciphertext is GKREHESTNTIEGNR [2]. In columnar transposition, the plaintext message is transposed into a rectangular structure by columns. The columns are next reorganized and the letters are then taken off in a horizontal fashion.

A negative aspect of columnar transposition ciphers is that entire matrices of characters must be made encrypted and decrypted. The characters of the plaintext are permuted for a predetermined period d. The function f is a permutation of a block of d characters, whereas the encryption key is $K(d,f)$. Thus, a plaintext message [2]

$$M = m_1 m_2 \ldots m_d m_{d+1} \ldots m_{2d} \ldots \tag{9.47}$$

where m_i are the individual characters encrypted as [2]

$$E_k(M) = m_{f(1)} m_{f(2)} \ldots m_{f(d)} m_{d+f(1)} \ldots m_{d+f(d)} \ldots \tag{9.48}$$

where $m_{f(1)} m_{f(2)} \ldots m_{f(d)}$ is a permutation of $m_1 m_2 \ldots m_d \cdot d = 5$ and f permutes the sequence i: 1 2 3 4 5 into $f(i) = 3\ 5\ 1\ 4\ 2$. In this case, plaintext word "GROUP" becomes "OPGUR".

9.4.2.3 Substitution Ciphers

In substitution enciphering, four basic classes of substitution ciphers are obtained [2,19–20]:

- Simple substitution: Every character is substituted by a corresponding character of ciphertext having one-to-one mapping from plaintext to ciphertext.
- Homophonic substitution: Each character is encrypted with a variety of ciphertexts having the mapping from plaintext character to ciphertext with one-to-many.
- Polyalphabetic substitution: Here, multiple alphabets are used to change plaintext to ciphertext having mappings of one-to-one with simple substitution, which is used in change within a single message.
- Polygram substitution: General ciphers permit arbitrary substitutions for groups of characters in plaintext.

Here, simple substitution ciphers are taken in which the following notations are considered [2]:

- Plaintext of n-character alphabet A is $\{a_0, a_1, a_2, \ldots, a_{n-1}\}$.

- In a simple substitution cipher, each character of A is replaced by a corresponding character from an ordered cipher alphabet C denoted by $\{f(a_0), f(a_1), ..., f(a_{n-1})\}$. So the function f is a one-to-one mapping of each character of A to the corresponding character of C.
- A plaintext message is taken as

$$M = m_1 m_2 ... \tag{9.49}$$

So it is then written in ciphertext as [2]

$$E_k(m) = m_{f(1)} m_{f(2)} ... \tag{9.50}$$

where m_i is a character of A. C is simply a reorganization of the characters in A. One of the simple-substitution ciphers is Morse code having a substitution of letters of the alphabet by a series of dots and dashes. The ASCII code is an example of simple-substitution cipher. In this code, the letter A is represented by the binary number 1000001 or 65 in decimal notation, B is given by 1000010 or 66 and so forth up to 1011010 or 90 for Z.

The simple-substitution cipher is also made a shifted alphabet in which the letters of the alphabet are shifted to the right by k positions, modulo the size of the alphabet. If a represents both a letter of A and its position in the alphabet [19,20], then

$$f_{(a)} = (a + k) \bmod n \tag{9.51}$$

where $n = 26$ considering English alphabet. This type of cipher is also known as a Caesar cipher in which Julius Caesar used it with $k = 3$ to send messages to his general.

If $k = 3$, then we have [19,20]
Plaintext:

Alphabet A B C D E F G H I J K L M N O P Q R S T U V W X Y Z

Ciphertext:

Alphabet D E F G H I J K L M N O P Q R S T U V W X Y Z A B C

Using Caesar cipher, the plaintext word "LIECHTENSTEINER" becomes OLHFKWHQVWHLQHU in ciphertext.

9.5   DATA ENCRYPTION STANDARDS (DES)

DES is used first in unclassified United States Government applications [28]. The encryption algorithm made at IBM was represented as a consequence of the LUCIFER cipher [29]. The DES algorithm swaps over blocks of 64 bits of plaintext to 64 bits of ciphertext. The key contains 64 binary digits having 56 bits for the encryption algorithm and the other 8 bits for error detection [2]. A block of data has an initial permutation having a complex key-dependent computation involving 16 iterations of a function f that combines substitution and transposition. In this direction, the product ciphers and block ciphers are mentioned below [2].

9.5.1 Product Cipher

A product cipher is to engross a combination of transposition (permutation) and substitution to produce a ciphertext [2] having cryptography made with information theory. The products of the form $B_1 M B_2 M \ldots B_n$ are used where M is a mixing permutation and B_i's are used as simple cryptographic transformation. A product cipher is the application of a sequence of n enciphering functions $f_1, f_2, \ldots, f_n$, where f_i is the permutation cipher P on the substitution cipher S [2].

Figure 9.12 presents an example of a 12-bit message $M = (m_1 m_2 \ldots m_{12})$. But in practice, longer block is taken [2]. The enciphering scheme alternately applies k substitutions S_i and $k - 1$ permutations P_1, yielding

$$C = E_k(M) = S_k P_k S_{k-1} \ldots S_2 P_1 S_1(M) \tag{9.52}$$

where each S_i is a function of the key K.

The 12-bit plaintext has four 3-bit sub-blocks (each acted as a different invertible 3-bit to 3-bit mapping or substitution cipher S_{ij}). The resulting 12 bits are scrambled within the permutation box P_i and input to the next round of enciphering [2]. The permutation mixes bits from different S_{ij} boxes to avoid the overall transformation from degenerating into a substitution on 3-bit blocks.

9.5.2 Block Ciphers

Block ciphers make encrypting and decrypting in blocks of information bits [20]. The M is taken as a plaintext message having a block cipher breaking into successive blocks $M_1, M_2, \ldots$ and enciphers of each M_i with the same key K, i.e., [2]

$$E_k(M) = E_k(M_1) E_k(M_2) \ldots \tag{9.53}$$

Here, simple substitution is considered where the block is one character long. Block ciphering has the same basic structure as block coding for error control. Both transform a block of uncoded or plaintext message into a block of coded or encrypted

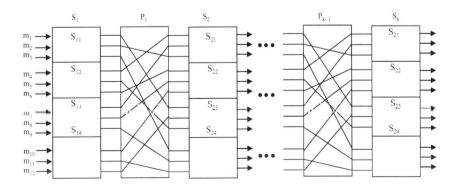

FIGURE 9.12 Product enciphering applied to a 12-bit message.

message. The difference between the two methods is that block ciphering is done with encryption keys, whereas most block coding schemes rely on parity checking [2].

The level of secrecy in message transmission is made with block ciphering having increased by [2]

- Dividing the plaintext into blocks, and then encrypting and decrypting are made separately.
- Repeating or iterating the block encryption a number of times using a combination of both of the above.

The basic concept of block ciphering having partitioning and iteration is illustrated in Figure 9.13 [2]. A block of message is distorted iteratively $i = 1, 2,..., r$ times and fragmented equally into left and right halves with L_i and R_i. The block is n bits long, with L_i and R_i of $n/2$ bits. Encryption and decryption are made by means of the set of iteration-dependent keys K_{i+1} and a transformation function f. The transformation function depends on R_i and K_{i+1} for encryption operation and on L_{i+1} and K_{i+1} for decryption operation. As shown in Figure 9.13 for the $(i + 1)$th iteration, the encryption yields [2]

$$L_{i+1} = R_i$$

$$R_{i+1} = L_i \oplus f(K_{i+1}, R_i) \tag{9.54}$$

where $\oplus$ denotes the modulo-2 addition. For decryption, the order of K_{i+1} is reversed, i.e.,

$$L_i = R_{i+1} \oplus f(K_{i+1}, L_{i+1})$$

$$R_i = L_{i+1} \tag{9.55}$$

9.5.3 THE DES ALGORITHM

In case of encryption scheme called DES algorithm, there are six operations: partitioning, iteration, permutation, shifting, selection and modulo-2 addition [2].

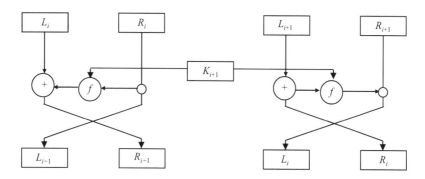

FIGURE 9.13   Basic concept of block enciphering with portioning and iteration [2].

Figure 9.14 shows the basic steps of the enciphering computation algorithm used for both enciphering and deciphering [2]. The algorithm processes input blocks of 64 bits. An input message $T = t_1, t_2, \ldots, t_{64}$ is initially transposed under an initial permutation IP providing $T_0 = IP(T)$. This permuted input is then going through 16 iterations of a cipher computation function having substitution and permutation, where a different key K_i is used for each iteration i, where $i = 0, 1, \ldots, 16$. The matrix is transposed under the inverse initial permutation IP^{-1} to give the output ciphertext [2].

Initial permutation: Figure 9.15 shows DES algorithm where the 64 bits of the input block T are first transposed as per initial permutation IP with particular pattern shown in Table 9.1. It begins from top to bottom and from left to right. The permuted block T_0 has 58 bits of the input T as its first bit, bit 50 as its second bit and so on with bit 7 as its last bit, i.e., [2]

$$T_0 = t_{58}t_{50}t_{42}\ldots t_{23}t_{15}t_7 \tag{9.56}$$

This permuted block is the input to the iterative enciphering process.

The iteration function: $f(R_{i-1}, K_i)$. The 64-bit block T_0 is broken into two parts: a right block R_0 and a left block L_0 of 32 bits each [2], i.e.,

$$T_0 = L_0 R_0 \tag{9.57}$$

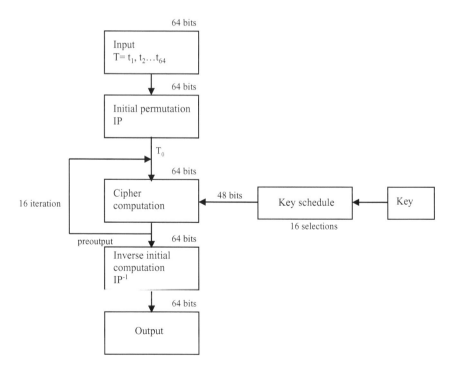

FIGURE 9.14   Basis steps used for enciphering and deciphering [8].

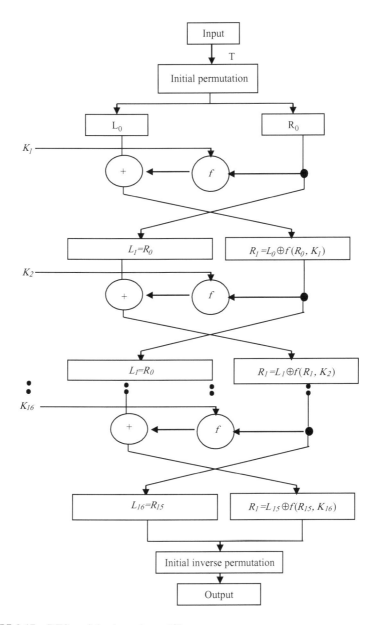

FIGURE 9.15 DES enciphering scheme [2].

where

$$L_0 = t_{58}t_{50}t_{42}\ldots t_{16}t_8$$
$$R_0 = t_{57}t_{49}t_{41}\ldots t_{23}t_{15}t_7$$

If T_i denotes the ith iteration result, and L_i and R_i are the left and right halves of T_i, respectively, then [2]

TABLE 9.1
Initial Permutation [2]

Original Bit Position	New Position of Bits Resulting in Initial Permutation							
1–8	58	50	42	34	26	18	10	2
9–16	60	52	44	36	28	20	12	4
17–24	62	54	46	38	30	22	14	6
25–32	64	56	48	40	32	24	16	8
33–40	57	49	41	33	25	17	9	1
41–48	59	51	43	35	27	19	11	3
49–56	61	53	45	37	29	21	13	5
57–64	63	55	47	39	31	23	15	7

$$L_i = R_{i-1}$$

$$R_i = L_{i-1} \oplus f(R_{i-1}, K_i) \qquad (9.58)$$

where each K_i has a different 48-bit key estimated from a 64-bit starting key [2].

The operation of $f(R_{i-1}, K_i)$ in Figure 9.16 [28] provides the following: first, R_{i-1} is prolonged to a 48-bit block $E(R_{i-1})$ by shuffling the bits according to the bit-selection method pattern E as shown in Table 9.2, where the repetition of various bit positions (such as 1, 4, 5, 8, 9) makes the 32-bit expanded to 48-bit. The first three bits of $E(R)$

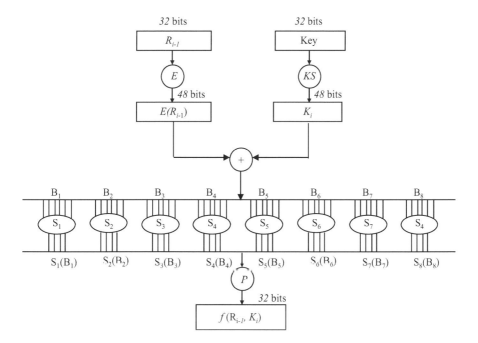

FIGURE 9.16 Estimation of the function $f(R_{i-1}, K_i)$ [2].

TABLE 9.2

E **Bit-Selection Table** [2]

Original Bit Position		E Bit-Selection Table				
1–6	32	1	2	3	4	5
7–12	4	5	6	7	8	9
13–18	8	9	10	11	12	13
19–24	12	13	14	15	16	17
25–30	16	17	18	19	20	21
31–36	20	21	22	23	24	25
37–42	24	25	26	27	28	29
43–48	28	29	30	31	32	1

are those in positions 32, 1 and 2 of R, and the last three bits of $E(R)$ are those in positions 31, 32 and 1.

Next $E(R_{i-1})$ and K_i are modulo-2 added, and the result is written as eight 6-bit blocks $B_1, B_2, \ldots, B_8$, where

$$E(R_{i-1}) \oplus K_i = B_1, B_2, \ldots, B_8 \tag{9.59}$$

Each 6-bit block B_j is transformed into a 4-bit block using the selection functions S_j, as shown in Table 9.3. The transformations are made as given below:

- Considering the 6-bit block $B_1 = b_1 b_2 b_3 b_4 b_5 b_6$ used for the selection box S_1. The first and last bits of B_1 (i.e., $b_1 b_6$) represent a base-2 number in the base-10 range 0–3.
- The number r selects the row of S_1. The middle four bits of B_1 (i.e., $b_2 b_3 b_4 b_5$) show a base-2 number in the base-10 range 0–15.
- This number, q, selects the column of the table. Thus, looking up the number in the rth row and qth column yields a base-10 number in the range 0–15. This number is uniquely represented by a 4-bit base-2 block. We can let each entry $S_j(D_j)$ of S_j be represented by the function [2]:

$$S_j(B_j) = S_J^{b_1 b_6}(b_2 b_3 b_4 b_5) \tag{9.60}$$

For the input block

$$B_j = b_1 b_2 b_3 b_4 b_5 b_6.$$

- Using the transformation, the eight 4-bit blocks $S_j(B_j)$ are then concatenated and the resulting 32-bit block is transposed by the permutation pattern P shown in Table 9.4. This permutation pattern P is used 16 times for every input block.

Inverse initial permutation: Following the 16th iteration, we have 64 output bits $L_{16} R_{16}$, which are known as the preoutput as shown in Figure 9.16. These 64 bits are

TABLE 9.3
Selection Function (S-box) [2]

Function	Row	Column															
		0	1	2	3	4	5	6	7	8	9	10	11	12	13	14	15
S_1	0	14	4	13	1	2	15	11	8	3	10	6	12	5	9	0	7
	1	0	15	7	4	14	2	13	1	10	6	12	11	9	5	3	8
	2	4	1	14	8	13	6	2	11	15	12	9	7	3	10	5	0
	3	15	12	8	2	4	9	1	7	5	11	3	14	10	0	6	13
S_2	0	15	1	8	14	6	11	3	4	9	7	2	13	12	0	5	10
	1	3	13	4	7	15	2	8	14	12	0	1	10	6	9	11	5
	2	0	14	7	11	10	4	13	1	5	8	12	6	9	13	2	15
	3	13	8	10	1	3	15	4	2	11	6	7	12	0	5	14	9
S_3	0	10	0	9	14	6	3	15	5	1	13	12	7	11	14	2	8
	1	13	7	0	9	3	4	6	10	8	2	5	14	12	11	15	1
	2	13	6	4	9	8	15	3	0	11	1	2	12	5	10	14	7
	3	1	10	13	0	6	9	8	7	4	15	14	3	11	5	2	12
S_4	0	7	13	14	3	0	6	9	10	1	2	8	5	11	12	4	15
	1	13	8	11	5	6	15	0	3	4	7	2	12	1	10	14	9
	2	10	6	9	0	12	11	7	13	15	1	3	14	5	2	8	4
	3	3	15	0	6	10	1	13	8	9	4	5	11	12	7	2	14
S_5	0	2	12	4	1	7	10	11	6	8	5	3	15	13	0	14	9
	1	14	11	2	12	4	7	13	1	5	0	15	10	3	9	8	6
	2	4	2	1	11	10	13	7	8	15	9	12	5	6	3	0	14
	3	11	8	12	7	1	14	2	13	6	15	0	9	10	4	5	3
S_6	0	12	1	10	15	9	2	6	8	0	13	3	4	14	7	5	11
	1	10	15	4	2	7	12	9	5	6	1	13	14	0	11	3	8
	2	9	14	15	4	2	8	12	3	7	0	4	10	1	13	11	6
	3	4	3	2	12	9	5	15	10	11	4	1	7	6	0	8	13
S_7	0	4	11	2	14	15	0	8	13	3	12	9	7	5	10	6	1
	1	13	0	11	7	4	9	1	10	14	3	5	12	2	15	8	6
	2	1	4	11	13	12	3	7	14	10	15	6	8	0	5	9	2
	3	6	11	13	8	1	4	10	7	9	5	0	15	14	2	3	12
S_8	0	13	2	8	4	6	15	11	1	10	9	3	14	5	0	12	7
	1	1	15	13	8	10	3	7	4	12	5	6	11	0	14	9	2
	2	7	11	4	1	9	12	4	2	0	6	10	13	15	3	5	8
	3	2	1	14	7	4	10	8	13	15	12	9	0	3	5	6	11

then transposed by the inverse initial permutation $1P^{-1}$ according to the bit position reordering shown in Table 9.5. The output of the algorithm IP^{-1} has bit 40 of the preoutput block as its first bit, bit 8 of the preoutput block as its second bit, and so on until bit 25 of the preoutput block is the last bit of the ciphertext output.

Key calculation: The 48-bit keys K_i used in each iteration are derived from a 56-bit key K, which is made secret by the sender and the receiver. The 56-bit key is sufficiently strong to withstand cryptanalysis attacks [19,20]. The function KS is used to make the different keys K_i as per key schedule. The concept used in the

TABLE 9.4

Permutation _P_ [2]

Bit Position	Permutation Function _P_			
1–4	16	7	20	21
5–8	29	12	28	17
9–12	1	15	23	26
13–16	5	18	31	10
17–20	2	8	24	14
21–24	32	27	3	9
24–28	19	13	30	6
29–32	22	11	4	25

TABLE 9.5

Permutation _IP_$^{-1}$ [2]

Bit Position	Inverse Permutation Function _IP_$^{-1}$							
1–8	40	8	48	16	56	24	64	32
9–16	39	7	47	17	55	23	63	31
17–24	37	6	46	15	54	22	62	30
25–32	36	5	45	14	53	21	61	29
33–40	35	4	44	13	52	20	60	28
41–48	34	3	42	12	51	19	59	27
49–56	33	2	41	11	50	18	58	26
57–64	32	1	40	10	49	17	57	25

DES is shown in Figure 9.17. First K is the input as 64-bit block, designated as KEY, which has eight parity bits in positions 8, 16,…, 64. The function KS uses an integer n between 1 and 16 and the 64-bit block KEY provides a 48-bit output block K_n as a permuted selection of bits from KE, which is written as [2]:

$$K_n = KS(n, \text{KEY}) \tag{9.61}$$

The bit selection is done as follows [2]:

- First, a permutation PC-1 (permuted choice 1) rejects the parity bits and shifts the 56 key bits to the positions indicated in Table 9.6.
- The result is then divided into two equal 28-bit halves: C_0 and D_0. Each of the 28-bit blocks is shifted to the left by one bit as indicated by the key schedule of left shifts i.e., in each block, bit 2 moves to position 1, bit 3 moves to position 2 and so on with bit 1 shifting to position 28. The result is the two blocks C_1 and D_1.
- The 56-bit block $C_1 D_1$ then gets shuffled and reduced to a 48-bit block by means of the permutation PC-2 shown in Table 9.7 which is known as permuted choice 2. The result is the key K_1.

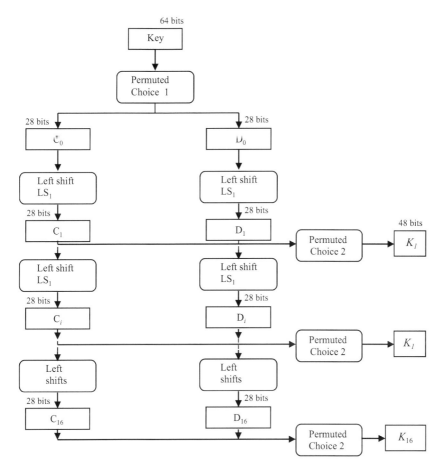

FIGURE 9.17 Key scheduling in DES algorithm [2].

TABLE 9.6
Permutation Choice (PC-1) [2]

Bit Position	PC-1						
	Left Block C_0						
1–7	56	49	41	33	25	17	9
8–14	1	58	50	42	34	26	18
15–21	10	2	59	51	43	35	27
22–28	19	11	3	60	52	44	36
	Right Block D_0						
1–7	63	55	47	39	31	23	15
8–14	7	62	54	46	38	30	22
15–21	14	6	61	53	45	37	29
22–28	21	13	5	28	20	12	4

TABLE 9.7
Permutation Choice 2 (PC-2) [2]

Bit Position			PC-2			
1–6	14	17	11	24	1	5
7–12	3	28	15	6	21	10
13–18	23	19	12	4	26	8
19–24	16	7	27	20	13	2
25–30	41	52	31	37	47	55
31–36	30	40	51	45	33	48
37–42	44	49	39	56	34	53
43–48	46	42	50	36	29	32

For the ith iteration, the blocks C_{i-1} and D_{i-1} are then successively shifted to the left as designated by the left-shift functions shown in Table 9.6 to derive each key K_1. This is represented as [2]

$$C_i = LS_i(C_{i-1})$$

$$D_i = LS_i(D_{i-1}) \tag{9.62}$$

The key K_i is given by

$$K_i = PC - 2(C_i D_i) \tag{9.63}$$

The following algorithm has been adapted in DES scheme [2]:
   Initialization

1. Specify the 64-bit cryptographic key $K = k_1 k_2 k_3, \ldots, k_{63} k_{64}$, which consists of 56 key bits and 8 non-key parity bits.
2. Using K, construct sixteen 48-bit key vectors $k_1 k_2 k_3, \ldots, k_{16}$, which are used for iteration rounds 1 through 16, respectively (Figure 9.17).
3. Select the 64-bit, externally supplied plaintext input $X = x_1 x_2 x_3, \ldots, x_{63} x_{64}$.
4. As shown in Figure 9.15, construct the 32-bit vectors L_0 and R_0 using the initial permutation IP.
5. Set iteration counter to $i = 1$.
      Iterations $i = 1$ through 16
6. For each iteration i, expand R_{i-1} to a 48-bit block $E(R_{i-1})$ according to the bit-selection method E shown in Table 9.2.
7. If enciphering is being done, choose the key K_i for deciphering, select K_{17-i}.
8. Modulo-2 add $E(R_{i-1})$ and K_i. Let the resulting 48-bit vector be $B = b_1 b_2, \ldots, b_{47}, b_{48}$.
9. The results of $E(R_{i-1}) \oplus K_i$ are the inputs to the selection functions S_i through S_8 given in Table 9.3. This then defines the 32-bit vector $A = a_1 a_2, \ldots, a_{31}, a_{32}$.

10. Derive $P(A)$ by applying the permutation function P shown in Table 9.4 to the vector A.
11. Modulo-2 add $P(A)$ to L_{i-1}, and let the result be R_i.
12. Define $L_i = R_{i-1}$.
13. Increment the iteration counter i by 1. If $i < 16$, repeat steps 6 through 13. Otherwise, go to step 14.
14. Using the inverse initial permutation IP^{-1} shown in Table 9.5, transpose the 64-bit preoutput $L_{16}R_{16}$ to yield the final 64-bit output.

9.5.4 PUBLIC KEY CRYPTOGRAPHY

Conventional cryptographic systems employ a single key for enciphering and deciphering a message, so that both the sender and the receiver should know the key keeping it confidential [2]. A major problem with these techniques is distribution of the keys where key distribution is slow for implementation to secure communications. To deal with this problem, various researchers [2,25] devised an approach to the key distribution problem. It is called as public key cryptography having two separate keys: one key for message enciphering and made publicly known and the other key for message deciphering and remains secret. There must be a complex computed method for deriving the deciphering key from the encryption key. This can be made through the use of special mathematical formulations known as trapdoor one-way functions in which the inverse of these functions cannot be derived solely from a description of the function. The knowledge regarding the method of construction of key makes the inverse easy to compute results in the trapdoor [2].

Before discussing public key cryptography schemes [2], we need to know some mathematical concepts and definitions. The principal mathematical concepts relating to public key cryptography are prime and composite numbers, factoring and modular arithmetic [2].

Prime and composite numbers: If an integer is not a prime, it is a composite number representing "product of primes". If an integer is very large, it is difficult to determine whether it is a prime. This is particularly true for the approximately 100-digit-long numbers used in public key cryptography [2].

Euler function: The Euler function $\phi(m)$ represents the number of positive integers, which is less than that of equal to m, which are relatively prime to m having no common factor with m. For a prime p, the number of integers less than p, which are relatively prime to p, is simply given by [2]

$$\phi(p) = p - 1 \tag{9.64}$$

For $m = 31$, we have $\phi(31) = 30$ since every integer less than 31 is relatively prime to 31. Now we consider composite numbers. Given two prime numbers p and q, then for their product $n = pq$, we have

$$\phi(n) = \phi(p)\phi(q) = (p-1)(q-1) \tag{9.65}$$

For $p = 3$ and $q = 5$,

$$\phi(15) = \phi(3)\phi(5) = (3-1)(5-1) = 8$$

Thus, there are eight numbers that are relatively prime to 15. These are the set {1, 2, 4, 7, 8, 11, 13, 14}.

In general, for an arbitrary composite number n, $\phi(n)$ is given by [2]

$$\phi(n) = \prod_{i=1}^{r} P_i^{\alpha_i - 1} \qquad (9.66)$$

where $n = p_1^{\alpha_1} p_2^{\alpha_2} p_r^{\alpha_r}$, which is the prime factorization of n with the positive integers α_i representing the number of occurrences of the prime p_i.

Greatest common divisor: The greatest common divisor (gcd) of a pair of integers is the greater integer which divides both numbers of the given pair. Thus, 3 is the greatest common divisor of the pair (12,15). This is written as gcd (12,15) = 3.

To find the gcd, we list all the divisors of each number and pick out the largest one common to both lists. We can write [30,31]

Divisors of 12 : 1, 2, 3, 4, 6, 12

Divisors of 15 : 1, 3, 5, 15

gcd(12,15) = 3

Euclid's algorithm eliminates listing of all the divisors and is therefore particularly useful for large numbers [32]. Considering the numbers 34 and 704, divide the smaller (34 in this case) by the larger. Doing so yields a remainder of 24. We write this as gcd (34, 704) = gcd (24, 34).

This process is then repeated until we arrive at the step before a divisor comes out even with no remainder. In this case,

$$\text{gcd}\,(34,704) = \text{gcd}(24,34) = \text{gcd}(10,24) = \text{gcd}(4,10) = 2$$

Finally, gcd (4, 10) = 2, since in the next step 4 divided by 2 = 2 with no remainder. Thus, the gcd of 34 and 704 is 2.

9.5.5 CONGRUENCES: MODULAR ARITHMETIC

We consider two integers that are congruent modulo 12 if they differ only by an integer multiple of 12. The numbers 7 and 19 are congruent modulo 12, which is written as 7 = 19 modulo 12 or 7 = 19 mod 12. We thus have the following definition of congruences: for given integers a, b and $n \neq 0$, a is congruent to b modulo n if the difference $(a - b)$ is an integer multiple of n. So we can write [30,31]

$$a = b \bmod n \qquad (9.67)$$

The advantage of using modular arithmetic is that it restricts the size of the numbers occurring at intermediate steps when carrying out an arithmetic operation. This results in the estimation of modular arithmetic (wherein one reduces each intermediate result modulo n) that yields as same as results computing in ordinary arithmetic and reducing the final answer modulo n.

We consider that op denote one of the arithmetic functions $+$, $-$ or $\times$ as addition, subtraction, or multiplication. Then, for integers a_1 and a_2

$$(a_1 \quad or \quad a_2)\,\mathrm{mod}\,n = [(a_1\,\mathrm{mod}\,n)\,op\,(a_2\,\mathrm{mod}\,n)]\,\mathrm{mod}\,n \qquad (9.68)$$

This principle is well explained in Figure 9.18.

In applying this theorem to repeated multiplication, each intermediate step must be reduced modulo n. Thus, for example, to calculate the expression $4^5 \bmod 7$, we repeatedly apply to get [2,30,31]

$$4^5 \bmod 7 = \left\{ \left[\left(4^2 \bmod 7 \right)\left(4^2 \bmod 7 \right) \right] \bmod 7 \left(4 \bmod 7 \right) \right\} \bmod 7$$

To evaluate this, we would perform the following steps:

1. Square 4 $(4 \times 4) \bmod 7 = 2$.
2. Square the result $(2 \times 2) \bmod 7 = 2$.
3. Multiply by 4 $(4 \times 4) \bmod 7 = 2$.

Using ordinary arithmetic and carrying out the same sequence of repeated squaring and multiplication followed by mod 7 reduction at the end, we have to

1. Square 4 $4 \times 4 = 16$.
2. Square the result $16 \times 16 = 256$.
3. Multiply by 4 $256 \times 4 = 1024$.
4. Reduce the answer mod 7 $1024 \bmod 7 = 2$.

The reduction in size of the intermediate numbers is made using the modular arithmetic approach [2]. We can also use modular arithmetic to easily calculate the expressions of the form $a\,x \bmod n$ for large values of a, x and n without creating enormous intermediate numbers [2]. In particular, public key cryptosystems use exponentials of this form where $0 < x < n - 1$.

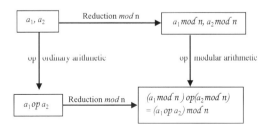

FIGURE 9.18   Modular and ordinary arithmetic [2].

Test for primality: To execute a public key cryptosystem, it needs to generate large prime numbers, which are ~ 100 digits long [32]. A major factor is how to verify the primality of such a large randomly selected number whether a number is prime or composite [33]. For large number verification, the approaches are not computationally feasible since too many multiplications are required. Thus, these approaches are less reliable but highly probable approaches should be used [32]. Here, a trade-off is required to be made between computation time and the risk of assuming that a number is prime when it is really composite. The algorithm of Solovay and Strassen is one of the important techniques [37], where, to test a number p for primarily, one picks a random number b from a uniform distribution $[1, p - 1]$ and checks whether it satisfies the conditions [2]

$$\gcd(b, p) = 1$$

$$J(b, p) = b^{(p-1)/2} \bmod b \tag{9.69}$$

where $J(b, p)$ is the Jacobi symbol,[33,34] which is defined recursively by [2]

$$J(b, p) = \begin{cases} 1 & \text{if} \quad b = 1 \\ J\left(\dfrac{b}{2}, p\right)(-1)^{(p^2-1)/p} & \text{if} \quad b \text{ is even} \\ J(p \bmod b, b)(-1)^{\frac{(p-1)(b-1)}{4}} & \text{otherwise} \end{cases} \tag{9.70}$$

If p is prime, then equation (9.25) always holds. If p is composite, equation (9.25) will be false with a probability of at least 50% for each b selected.

Multiplicative inverses: The modular arithmetic computes multiplicative inverses. For an integer a in the range $[0, n - 1]$, it may be able to find a unique integer x in the range $[0, n - 1]$ such that [2]

$$a x \bmod n = 1 \tag{9.71}$$

We consider 3 and 11 which are multiplicative inverses mod 8 since 33 mod 8 = 1. This characteristic makes modular arithmetic very useful and appealing in cryptographic applications. When a and n are relatively prime, i.e., gcd $(a, n) = 1$, we can write

$$x = a^{\phi(n)-1} \bmod n \tag{9.72}$$

where $\varphi(n)$ is Euler's function. Solutions can also be found for the general equation:

$$a x \bmod n = b \tag{9.73}$$

when gcd $(a, n) = 1$. The solution is

$$x = b x_0 \bmod n \tag{9.74}$$

where x_0 is the solution to

$$ax \bmod n = 1 \tag{9.75}$$

If gcd $(a, n) \neq 1$, equation (9.73) will either have no solution or have more than one solution in the range $[1, n - 1]$. First, we consider $g = \gcd(a, n)$.

If $b \bmod g = 0$ (i.e., if g divides b exactly), then equation (9.73) will have g solutions of the form:

$$x = \left(\frac{b}{g} x_0 + t \frac{n}{g} \right) \bmod n \tag{9.76}$$

For $t = 0, 1, \dots, g - 1$, where x_0 is the solution of the following:

$$\frac{a}{g} x \bmod \frac{n}{g} = 1 \tag{9.77}$$

Otherwise, it will have no solution.

9.5.6 THE RIVEST–SHAMIR–ADLEMAN (RSA) ALGORITHM

Rivest, Shamir and Adleman [34] have developed a method for realizing public key encrypting. In RSA algorithm, two large prime numbers are used to multiply them and to factor the product. The product of two large secret numbers is made public without giving a clue to factors which effectively constitute the deciphering key. Each station independently and randomly selects two primes p and q and also the product $n = pq$, the modulus used in the arithmetic calculations of the algorithm. By making each of the primes of 100 decimal digits long, the product can be determined in a fraction of a second. However, for factoring n with 200 digits, have billions of years with one step per microsecond using the fastest known factoring algorithm [34]. Then,

$$S = \exp\left[(\ln n)\ln(inn) \right]^{1/2} \tag{9.78}$$

Step S to factor n into p and q. Then, Euler's function is written as

$$\phi(n) = (p - 1)(q - 1) \tag{9.79}$$

For $\varphi(n)$, the station randomly selects an integer K between 3 and $\varphi(n) - 1$ so that the greatest common divisor is one

$$\gcd\left[K, \varphi(n) \right] = 1 \tag{9.80}$$

The parameter K is written as either the secret key D or the key E depending on the desired usage in the network. Here, we consider that it is the public key.

The multiplicative inverse of K modulo $\varphi(n)$ is then calculated using Euclid's algorithm [2]. The secret key D is an integer D that is a unique number which satisfies the condition:

$$ED \bmod \varphi(n) = 1 \tag{9.81}$$

The public information has E and n. All other quantities remain secret. To make the ciphertext C showing a plaintext message is first indicated as a sequence of integers with each between 0 and $n - 1$. Denoting this sequence by P, the ciphertext C is made by raising P to the power of E modulo n, i.e.,

$$C = P^E \bmod n \tag{9.82}$$

where C is also in the range from 0 to $n - 1$. To recover the plaintext at the receiving end, the parameters D and n are used as the deciphering key via the relationship:

$$P = C^D \bmod n \tag{9.83}$$

The procedures for selecting the keys and performing the steps of encipherment and decipherment with an example are given below. The various quantities used are summarized in Table 9.8 [2].

Step 1 Choose $p = 5$ and $q = 7$
Step 2 Find $n = pq = 35$
Step 3 Find $\varphi(n) = (p - 1)(q - 1) = 24$
Step 4 Choose $E = 5$ since gcd $(5, 24) = 1$
Step 5 Then $D = E^{-1} \bmod 24 = 5$ since ED mod $\varphi(n) = 5 \times 5 \bmod 24 = 1$
Step 6 The plaintext $P = 2$, calculate C:

$$C = P^E \bmod n = 2^5 \bmod 35 = 32$$

Step 7 Use equation (9.83) to recover the plaintext P from C:

$$P = C^D \bmod n = 32^5 \bmod 35 = 2$$

TABLE 9.8
Quantities in RSA Algorithm [2]

Quantity	Status
Primes p and q	Secret
$n = p \cdot q$	Noncorrect
$\phi(n) = (p - 1)(q - 1)$	Secret
Public key E	Noncorrect
Public key D	Secret
Plaintext P	Secret
Ciphertext C	Noncorrect

which is the original message to encipher an alphanumeric message in practice; the plaintext message characters would be indicated by their 8-bit ASCII codes [2].

9.5.7 COMPARISON OF CRYPTOGRAPHIC TECHNIQUES

Although the public key cryptographic scheme of RSA makes the key distribution problem simpler than that in the DES method, the RSA technique is costly and relatively slow. The encryption times of a few microseconds are needed for the DES, whereas the encryption times of a few milliseconds are needed for the RSA, which restrict the throughput rate to about 50 kb/s using current technology. Although faster public key cryptographic methodologies are being investigated [35], they still need to be implemented and scrutinized for resistance against attack.

9.6 OPTICAL CRYPTOGRAPHY

Due to the tremendous increase in network users and services, the demand of increased accessibility of optical networks having communications through these networks requires eligible security. In this direction, securing of optical communications needs cryptographic operation at optical layers of the protocol. It is enviable to secure against threats in the lowest layer of an optical network. An optical network is faced with a variety of attacks such as jamming, physical infrastructure attacks, eavesdropping and interception [36]. There is a unique challenge for achieving security in optical domain just like a conventional electronic security. By using optical signal processing, the optical communication is made secure at the optical domain [37]. There are two ways: all-optical logic for encryption [38] and optical steganography [39]. Optical logic gates make signals to be encrypted with low latency and high speed, whereas optical steganography gives an additional layer of privacy supplementing data encryption. There are different threat categories:

- Threats by a rival using communications (confidentiality).
- Threats by an unauthorized entity (authentication).
- Threats by an entity manipulating communication (integrity).
- Threats by a rival trying to undermine the successful delivery.
- Privacy risks linked with an adversary having the existence of communications (privacy and traffic analysis).

So we must know survey confidentiality, authentication, privacy and availability threats and solutions at the optical domain.

9.6.1 CONFIDENTIALITY

Optical networks have no electromagnetic signature in which an attacker can spy on an optical system using a variety of approaches, such as tapping into the optical fiber [14], listening to the residual crosstalk from an adjacent channel while impersonating a legitimate subscriber [15]. Tapping optical fiber is common where fiber is exposed without physical protection. There arc two types of fiber tapping:

- Removing cladding of the fiber to leak portion of the light from the optical fiber and coupling of signals by a second fiber placed adjacent to the place where light escapes from the fiber.
- Eavesdropping that is to eavesdrop to the residual adjacent channel cross-talk while impersonating one of the subscribers. Normally in wavelength division multiplexing (WDM) networks, different wavelengths are used by different subscribers, and a desired signal is dropped at its destination using a wavelength demultiplexer.

There are two kinds of eavesdropping which are difficult to realize, but the eaves-droppers with specialized optical equipment are made it possible. Further, the confidentiality in an optical network is improved through the use of optical encryption and optical coding [37], as shown in Figure 9.19.

Due to having high data rates of optical networks, there is a difficulty to implement compatible optical encryption. The development of architectures requiring encryption functions in the optical domain is a challenge to the scientist. Optical encryption also has the benefits from not generating an electromagnetic signature, making it immune to electromagnetic-based attacks [36]. Even if eavesdroppers obtain a small portion of signal by tapping into the optical fiber or listening to a residue adjacent channel, no useful information is recovered without the knowledge of the encryption key in optical domain [37]. Optical code division multiple access (OCDMA) is used to provide confidentially. In an OCDMA system, each data stream is coded with a specific code, which is sensed with the corresponding decoder. Moreover, in a multiple-access system, a plurality of CDMA codes simultaneously existed in the transmission channel, which superposes both in time and in optical spectrum. Although optical coding does not give confidentiality that is as strong as optical encryption, it introduces an additional layer of protection from eavesdropping [36].

The configuration of optical encryption and decryption in Figure 9.20 has two channels: a single WDM channel with OOK (on–off keying) for host of the public network and a secure M-ary ($M = 2$) channel for the secure user [36,37]. For simplicity, one WDM channel is used in the public network accessed by many users through time division multiple access (TDMA).

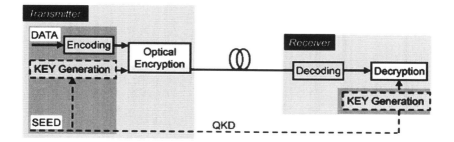

FIGURE 9.19 A simple optical encryption and decryption scheme.

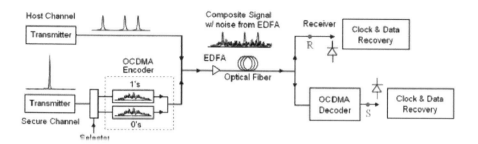

FIGURE 9.20 A optical encryption and decryption scheme for WDM with 00K modulation [36].

9.6.2 OCDMA-BASED ENCODER/DECODER

The secured signal is coded by using time spread with an OCDMA encoder [36]. The signals from both channels are independently united into the optical fiber–optical link having dispersion compensation (using, e.g., dispersion compensating fiber and/or other dispersion compensators). At the destination, a photoreceiver is used for the collection of optical signals for the user in the host channel and a relevant decoder is additionally required for the secure user to recover the secure data. The OCDMA encoder/decoder has a coherent spectral phase encoder/decoder shown in Figure 9.21a, which uses pulse shaping technique [40]. The decoder function is made with the phase mask replaced with its conjugate [41].

Optical architectures show the realization of fast encryption functions in the optical domain. It provides less side-channel risk than their electrical counterparts. Optical XOR logic has been carried out by optical logic gates used for optical encryption algorithms [42–44]. Optical XOR gates have an electromagnetic signal that is supervised by an eavesdropper. The XOR logic is one of the key elements for building optical layer encryption, and combining XOR with feedback generates long key streams from smaller keys used in the process of enciphering [45]. So the block of optical ciphers has XOR, feedback and feed-forward capabilities. These building blocks having operation in the optical domain perform a high-speed, electromagnetic wave immunity and all-optical means for encryption.

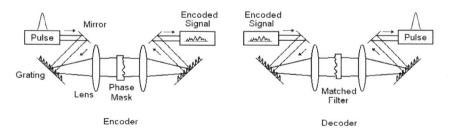

FIGURE 9.21a Spectral phase encoder and decoder.

(Continued)

9.6.3 DSP-Based Approach

Digital signal processing (DSP) technique is employed in optical network security [46,47]. The encrypted wireless CDMA codes using a single-user secure optical transmission link [48] are shown in Figure 9.21b, where basic principle is to break the original data stream into multiple sub-streams which are encoded by one encrypted CDMA code. The multiple encoded sub-streams are joined together to modulate the optical carrier. Since the CDMA code is encrypted off-line by advanced encryption methods, the rival cannot find the CDMA code in use for decoding the transmitted signal [36]. The receiver collects the signal waveform and decodes the signal with the same encrypted CDMA codes which are known to the receiver. The data encoding/decoding and code encryption are realized through DSP techniques [36].

9.6.4 Spread Spectrum-Based Approach

The spread spectrum-based approach provides cryptographic with steganographic security where both encryption of the signal (frequency hopping encoding) and the hiding of it under the noise floor (time spreading) are employed [36,49]. In the receiver end, a photoreceiver and WDM filters are used for data recovery of the individual WDM channels (Figure 9.22). The incoherent decoder contains a dispersion compensator and delay lines. A random frequency hopping assignment is used for providing more security to the system [49]. The WDM channels give an *ad hoc* security enhancement and make more difficult for an eavesdropper to intercept and decode the secure signal, because any slight imperfection in the decoding scheme with complementary dispersion compensation and delay lines makes the WDM signals swarm out the secure signal. An analytical description is made for the BER performance analysis having the following assumptions [49]:

- The dispersion is properly compensated during propagation.
- The nonlinearity effect is very small and negligible in contributing small peak powers in the signals.

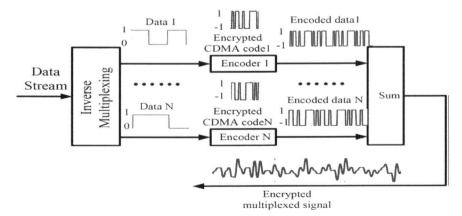

FIGURE 9.21b (CONTINUED)   CDMA-based *n* encoder [36].

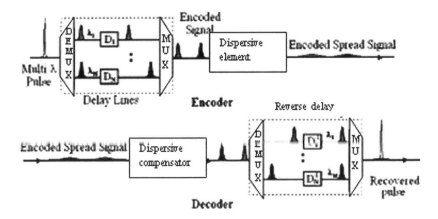

FIGURE 9.22 WDM-based approach for encoding and decoding of encryption scheme [51].

The performance of both the host WDM and secure CDMA channels is affected by noise, contributions from amplified spontaneous emission and multi-access interference. The former is in Gaussian white noise form [50]. For a large number of hopping frequencies, the multi-access interference noise distribution (in the secure CDMA channel) shows a zero mean Gaussian. Considering Gaussian-distributed effective noise in the secure channel, the probability density functions of the intensity of the decoded bits (I_0, I_1) for incorrect and correct decoding are denoted by $\overline{N_0}$ and $\overline{N_1}$ as the corresponding average noise intensities and P is the signal power of the secure signal [49]:

$$f_{I_0}(I) = \frac{1}{N_0} e^{-1/\overline{N_0}} \tag{9.84}$$

$$f_{I_1}(I) = \int_{-\sqrt{I}}^{\sqrt{I}} \frac{\exp\left(-\frac{I + P - 2\sqrt{P}y}{\overline{N_1}}\right)}{\overline{N_1}\pi\sqrt{1-y^2}} dy \tag{9.85}$$

Equation (9.84) represents the distribution of the bit intensity of any code word which is not easy to reconstruct due to unmatched decoder and applies to eavesdropper, whereas equation (9.85) represents the distribution of the bit intensity of any code word which makes it easy to construct (i.e., bit 1 and bit 0 for secure channel) due to the presence of matched decoding. The components of the average noise intensities are as follows:

$$\overline{N_0} = \left\langle N^{\text{WDM}}(t) \right\rangle + \left\langle N^{\text{secure}}(t) \right\rangle + P_{\text{noise}} \tag{9.86}$$

$$\overline{N_1} = \left\langle N^{\text{WDM}}(t) \right\rangle + P_{\text{noise}} \tag{9.87}$$

where P_{noise} represents the average Gaussian noise power originated from the optical amplifier, $\left\langle N^{\text{WDM}}(t) \right\rangle$ is the ensemble average of the spread and delayed WDM signals after the incoherent decoder, and $\langle N^{\text{secure}}(t)\rangle$ is the ensemble average of noise intensity in which secure pulse is not reconstructed. The ensemble average of the sum of Gaussian-spread pulses are taken in the WDM channels separated by a frequency spacing of ω_0. $\langle N^{\text{WDM}}(t)\rangle$ and is written as [49]

$$\langle N^{\text{WDM}}(t)\rangle = \frac{1}{2m} \sum_{y=-(m-1/2)}^{(m-1)/2} \sum_x \sum_N \frac{P_0 \exp\left[-\dfrac{\left(t - xT_{\text{WDM}} - yT_{\text{chip}} + S\tau_0^2 N\omega_0\right)}{\left(1 + S^2\right)\tau_0^2}\right]}{\sqrt{1 + S^2}}$$

(9.88)

where τ_0 is the half-width at the $1/e$th intensity point, P_0 is the initial peak power of the WDM pulses, S is the spreading factor, T_{WDM} is the bit period of the WDM signal, T_{chip} is the bit period of a time chip in the encoder [51] and m stands for the number of time chips within the bit period of secure user. $\langle N^{\text{secure}}(t)\rangle$ is written as

$$\langle N^{\text{secure}}(t)\rangle = \frac{1}{2m} \sum_{x=0}^{N} x^2 P_{ind} e^{-t^2/\tau_0^2} \frac{N!}{x!(N-x)!} \cdot \frac{1}{m^x}\left(\frac{m-1}{m}\right)^{N-x}$$

(9.89)

where $P_{ind} = P/N^2$ representing the initial peak power of WDM wavelengths in secure CDMA signal having N pulses. BER is derived from equations (9.84) and (9.85) as [49]

$$\text{BER} = \frac{1}{2} \int_0^{I_{th}} dI. \int_{-\sqrt{I}}^{\sqrt{I}} \frac{\exp\left(-\dfrac{I + P - 2\sqrt{P}y}{\overline{N_1}}\right)}{\overline{N_1}\pi\sqrt{1 - y^2}} dy + \frac{1}{N_0} e^{-1/\overline{N_0}}$$

(9.90)

where I_{th} is the threshold power.

SUMMARY

Network security is one of the important issues for protecting the confidentiality and integrity of signal transmission in the networks. One uses a finite-bit encryption key to perform some mathematical operations on the plaintext to transform it into ciphertext. At the receiving end, a decryption algorithm is used for encryption transformation with the appropriate decryption key. The effectiveness of a particular encryption technique is judged for the confidentiality of the encryption algorithm. In a conventional (secret) cipher, the same key is used to encrypt and decrypt a message. A general implementation of a conventional one-key cryptosystem is DES. In a public-key cryptography, one key is for encryption and a different key is for decryption. The encryption key is made publicly known, whereas the corresponding decryption key is kept secret. Such a product can thus be made publicly known without giving a clue to its factors that contain the required information for the deciphering key.

We have also discussed the optical signal security schemes used in optical network having WDM and no WDM system.

EXERCISES

9.1. One way to tap optical power off the line (without breaking it and inserting a repeater and/or tap-off device) is to bend the fiber, thereby inducing higher order modes to scatter out of the core. A sensitive optical receiver can then monitor the information on the line. Based on a simple analysis,[44,45] the fraction F of modes lost in a bent graded-index fiber is

$$F = \frac{2a}{R\Delta}$$

Here a is the core radius, R is the bending radius, and $\Delta = 1 - \left(n_2/n_1\right)^2$ is the core-to-cladding index difference where n_1 and n_2 are the refractive indices of the fiber core center and cladding, respectively. Given that $n_1 = 1.005n_2$, what bending radii are needed to scatter 1%, 5% and 10% of the light of a 50-micrometer core diameter fiber?

9.2. Decipher the following cryptogram which was enciphered by geometrical pattern encoding using a 5-by-5 matrix with taken off in the columnar order 2 3 1 5 4:

LATESSSOTUEHRHAEATEEIPFRR

9.3. Using a permutation cipher with a fixed period $d = 5$ which permutes the sequence i : 1 2 3 4 5 into $f(i)$; 2 4 5 3 1, showing that message "why do I have so many homework problems now" becomes

HDOYWHVEAIOANMSHMEOYOKPOWOLEBR SOWNM

9.4. Decrypt the following message which was encrypted using the shifted alphabet cipher $f(a) = (a+4) \bmod 26$:

YLMATVSFPQKEZIQIELIEHEGLI

9.5. We consider $R_i = 101010101010101010101100011001$

Using the bit-selection matrix E given in Table 9.2, find the expanded block $E(Ri)$. For simplicity, write the result as an 8-row, 6-column matrix.

9.6. We consider $Y = 11001001011010111010100110000101$ be the 32-bit input block to the permutation matrix P given in Table 9.4. Find rearranging of Y according to P.

9.7. Use Euclid's algorithm to find the greatest common divisor of
(a) (14, 35) (d) (2873, 6643)
(b) (11, 12) (e) (4148, 7684)
(c) (180, 252) (f) (1001, 7655)

9.8. Write a program to solve for a when $a = b \bmod n$. Solve the following expressions using this program:
(a) 13 mod 11 (c) 2594 mod 48
(b) 87 mod 9 (d) $2^7 \bmod 21$

9.9. First solve the expression 3^7 mod 5 using ordinary arithmetic followed by mod 5 reduction at the end. Next solve 3^7 mod 5 using modular arithmetic. Compare the number sizes used in the two methods.

9.10. Solve the following congruence of x in the range $[0, n-1]$:
(a) $3x$ mod 7 = 1 (c) $3x$ mod 7 = 2
(b) $5x$ mod 17 = 1 (d) $5x$ mod 8 = 3

9.11. Consider a linear substitution cipher that uses the transformation $f(a) = a$ k mod 26.
(a) Find k if the plaintext letter G (letter number 7) corresponds to the ciphertext letter Q (letter number 16). (b) Find k if the plaintext letter D becomes the ciphertext letter S.

9.12. Find all solutions to the equation $6x$ mod 10 = 4 in the range $[0, 9]$.

9.13. For the RSA algorithm, let us choose $p = 47$ and $q = 61$. Then, $n = pq = 2867$ and $\varphi(n) = 2760$. Select a secret key $D = 167$, which is relatively prime to $\varphi(n)$.
(a) Calculate the multiplicative inverse of D mod $\varphi(n)$ to derive the public key $E = 1223$.
(b) To encipher a message first divide it into a series of blocks such that the value of each block does not exceed $n - 1$. One way to do this is to substitute the following two-digit numbers for each letter in the alphabet: bland = 00, A = 01, B = 02, ..., Y = 25, Z = 26. Using this code, write the message "RSA ALGORITHM" in blocks of 4 bits:
$b_1 \, b_2 \, b_3 \, b_4 \, b_5 \, b_6 \, b_7 \, b_8 \, \ldots \, b_{25} \, b_{26} \, b_{27} \, b_{28}$
(c) Encipher each 4-bit block by raising it to the power D, dividing by n, and taking the remainder as the ciphertext. Show that the ciphertext becomes
2756 2001 0542 0669 2347 0408 1815.

9.14. Suppose a system having two identical units connected in parallel. We consider each having the reliability of 95%.
(a) What is the reliability of the system?
(b) What is the reliability when a third identical unit is added in parallel?

9.15. Suppose we have two nonidentical units connected in parallel which have the failure rates of 3×10^{-4}/h and 2×10^{-4}/h.
(a) What is the component reliability at 1000 hour?
(b) What is the system reliability at 1000 hour?

9.16. A LAN operates 24 hours a day over average 20 days in a month. If there is a break down (once in a month) it takes 4 hours to repair
(a) What is the MTTF?
(b) What is the steady-state availability?
(c) What is the probability of failure in 12-hour shift?

9.17. Given that $D_T = 10^{-4}$, find the values of D_s for which L_{opt} is equal to m when there are (a) 100 terminals and (b) 256 terminals.

9.18. Make data security in optical network using CDMA based on DSP.

9.19. Make data security in optical network using spectral phase controller.

9.20. Design an optical network security system using 10 encoder/decoder systems based on DSP.

REFERENCES

1. Y. F. Lam and V. O. K. Li, "Reliability modeling and analysis of computer networks with dependent reliability," *IEEE Transaction on Communications*, vol. 34, pp. 82–84, January 1986.
2. G. E. Keiser, *Local Area Networks*, New York: Tata McGraw Hill, 1997.
3. R. Boorstyn and H. Frank, "Large scale network topological optimization," *IEEE Transaction on Communication*, vol. 25, pp. 29–47, 1977.
4. R. S. Wilkov "Analysis and design of reliable computer networks," *IEEE Transaction on Communication*, vol. 20, pp 660–673, 1972.
5. J. F Hyness, *Modeling and Analysis of Computer Communication Network*, New York: Plenum, 1984.
6. E. Haensler "A fast recursive algorithm to calculate the reliability of a communication network," *IEEE Transaction on Communication*, vol. 20, pp. 637–640, 1972.
7. E. Haensler, G. K. McAuliffe and R. Wilkov "Exact calculation of computer network reliability," *Networks*, vol. 4, pp. 95–112, 1974.
8. V. O. K. Li and J. A. Silvester, "Performance analysis of networks with unreliable components", *IEEE Transaction on Communication*, vol. 32, pp. 1105–1110, 1984.
9. M. J. Johnson, "Proof that the timing requirements of the FDDI token ring protocol are satisfied," *IEEE Transaction on Communication*, vol. 35, pp. 620–625, 1987.
10. B. S. Dhillon, *Reliability Engineering in System Design and Operation*, New York: Van Nostrand Reinhold, 1983.
11. C. S. Raghabendra and S. V. Makam, "Reliability modeling and analysis of computer networks", *IEEE Transaction on Reliability*, vol. 35, pp. 156–160, 1986.
12. P. Zafiropulo. "Performance evaluation of reliability improvement techniques for ingle loop communication systems," *IEEE Transaction on Reliability*, vol. 22, pp. 742–751, 1974.
13. A. Albanese, "Fall-safe nodes for lightguide digital networks," *Bell System Technical Journal*, Vol. 61, pp. 247–256, 1982.
14. J.H. Saltzer and M.D. Schroeder, "The protection of information in computer systems," *Proceedings of the IEEE*, vol. 63, pp. 1278–1308, September 1975.
15. S.T. Kent, "Security in computer networks," in *Protocols and Techniques for Data Communication Networks*, (F.F. Kuo ed.), Englewood Cliffs, NJ: Prentice-Hall, 1981.
16. W. Diffie and M.E. Hellman, "Privacy and authentication: An introduction to cryptography," *Proceedings of the IEEE*, vol. 67, pp. 397–427, March 1979.
17. M.D. Abrams and H.J. Podell (eds.), "Computer and network security," *IEEE Computer Society,* DC, 1987.
18. V.L. Voydock and S.T. Kent, "Security in high-level network protocols," *IEEE Communications Magazine*, vol. 23, pp. 12–24, July 1985.
19. B. Bosworth, *Codes, Ciphers, and Computers*, Rochelle Park, NJ: Hayden, 1982.
20. D.E. Denning, *Cryptography and Data Security*, Reading, MA: Addison-Wesley, 1982.
21. A.G. Konheim, *Cryptography: A Primer*, New York: Wiley, 1981.
22. H.F. Gaines, *Cryptanalysis*, New York: Dover, 1956.
23. A. Sinkov, *Elementary Cryptanalysis: A Mathematical Approach*, New York: Random House, 1968.
24. G.J. Simmons, "Symmetric and asymmetric encryption," *Computing Surveys*, vol. 11, pp. 305–330, December 1979.
25. W. Diffie and M.E. Hellman, "New directions in cryptography," *IEEE Transactions on Information Theory*, vol. IT-22, pp. 644–654, November 1976.
26. D.M. Balenson, "Automated distribution of cryptographic keys using the Financial Institution Key Management Standard," *IEEE Communications Magazine*, vol. 23, pp. 41–46, September 1985.

27. D.E. Denning, "Protecting public keys and key signatures," *Computer*, vol. 16, pp. 27–35, February 1983.

28. Data encryption standard, Federal information processing standards publication, 46, National Bureau standard, Washington DC, 1977.

29. H. Feistel, "Cryptography and computer privacy," *Scientific American*, vol. 228, pp. 15–23, 1973.

30. W. J. Leveque, *Fundamentals of Number Theory*. Reading, MA: Addison-Westley, 1977.

31. O. Ore, *Number Theory and Its History*, New York: McGraw Hill, 1988.

32. R. Solovay and V. Strasson, "A fast Monte-Carlo test for primality," *SiAM Journal on Computing*, vol. 6, pp. 84–85, 1977.

33. C.H. Meyer and S.M. Matyas. *Cryptography: A New Dimension in Computer Data Security*, New York: Wiley, 1982.

34. R. L. Rivest, A. Shamir and L. Aldeman, "On digital signature and public key crypto-system," *Communication ACM*, vol. 21, pp. 121–126, 1978.

35. S. C. Kak, "Secret-hardware public key cryptograph," *Proceedings of the IEEE*, vol. 133, pp 94–96, 1986.

36. Mable P. Fok, Zhexing Wang, Yanhua Deng, and P. R. Prucnal, "Optical layer security in fiber-optic networks," *IEEE Transaction on Information Forensics and Security*, vol. 6, 725–735, 2011.

37. Y. K. Huang, B. Wu, I. Glesk, E. E. Narimanov, T. Wang, and P. R. Prucnal, "Combining cryptographic and steganographic security with self-wrapped optical code division multiplexing techniques," *Electronics Letters*, vol. 43, no. 25, pp. 1449–1451, December 2007.

38. K. Vahala, R. Paiella, and G. Hunziker, "Ultrafast WDM logic," *IEEE Journal of Selected Topics in Quantum Electronics*, vol. 3, pp. 698–701, April 1997.

39. B. Wu and E. E. Narimanov, "A method for secure communications over a public fiber-optical network," *Optics Express*, vol. 14, pp. 3738–3751, 2006.

40. M. P. Fok and P. R. Prucnal, "A compact and low-latency scheme for optical stegan-ography using chirped fiber Bragg gratings," *Electronics Letters*, vol. 45, pp. 179–180, 2009.

41. Y. K. Huang, B. Wu, I. Glesk, E. E. Narimanov, T. Wang, and P. R. Prucnal, "Combining cryptographic and steganographic security with self-wrapped optical code division multiplexing techniques," *Electronics Letters*, vol. 43, no. 25, pp. 1449–1451, December 2007.

42. N. Gogoi and P. P. Sahu, "All-optical surface plasmonic universal logic gate devices," *Plasmonics*, vol. 11, pp. 1537–1542, 2016.

43. P. P. Sahu, "Optical pulse controlled two mode interference coupler based logic gates," *Optik-International Journal for Light and Electron Optics*, Vol. 126, no. 4, 404–407, 2015.

44. N. Gogoi and P. P. Sahu, "All optical compact surface plasmonic two mode interference device for optical logic gates," *Applied Optics* 54, no. 5, 1051–1057, 2015.

45. W. Trappe and L. C. Washington, *Introduction to Cryptography with Coding Theory*, 2nd ed. Englewood Cliffs, NJ: Prentice-Hall, July 2005.

46. E. Ip, A. P. Lau, D. J. Barros, and J. M. Kahn, "Coherent detection in optical fiber sys-tems: Erratum," *Optics Express*, vol. 16, pp. 21943–21943, 2008.

47. S. J. Savory, G. Gavioli, R. I. Killey, and P. Bayvel, "Electronic compensation of chromatic dispersion using a digital coherent receiver," *Optics Express*, vol. 15, pp. 2120–2126, 2007.

48. Z. Wang, L. Xu, T. Wang, and P. R. Prucnal, "Secure optical transmission in a point-to-point link with encrypted wireless CDMA codes," *IEEE Photonics Technology Letters*, vol. 22, no. 19, pp. 1410–1412, 2010.

49. B. B. Wu, P. R. Prucnal, and E. E. Narimanov, "Secure transmission over an existing public WDM lightwave network," *IEEE Photonics Technology Letters*, vol. 18, 1870–1872, 2006.
50. B. B. Wu and E. E. Narimanov, "A method for secure communications over a public fiber-optical network," *Optics Express*, vol. 14, pp. 3738–3751, 2006.
51. V. Baby, et al., "Experimental demonstration and scalability analysis of a four-node 102-Gchips/s fast frequency-hopping time-spreading optical CDMA network," *IEEE Photonics Technology Letters*, vol. 17, no. 1, pp. 1870–1872, January 2005.

10 FTTH Standards, Deployments, and Issues

Fiber-to-the-X (FTTX, X is home/curb/building) is used for broadband access. There are two services: cable TV [1] and plain old telephone service (POTS) using FTTX equipment [2]. "Fiber-to-the-Home" (FTTH) is a communication architecture in which a communication path is established from optical fiber cables to an optical line terminal (OLT) unit located in telecommunications operator's switching setup connecting to optical network terminal (ONT). The above two services were provided by two separate architectures. The arrival of high-bandwidth WWW using FTTX has made easy home data communication services along with TV and telephone [3]. The diverse type of traffic flowing through the Internet demands constant bandwidth for quality-of-service (QoS) of transmission. The transmission system is made to give a best-effort service by statistical multiplexing. The data rates range from 100 Mbit/s to 10 Gbit/s in LANs, but MANs and WANs need broadband access having a range in 100 Kbit/s to a few Mbit/s. It is evident from literature that no other technologies other than optical fiber can provide very high bandwidth and optical fiber becomes the ideal medium for high-bandwidth access to integrated services in the networks. In this direction, the FTTX technology such as FTTH has been focused for accessing high-bandwidth services.

FTTX model has FTTH [4], Fiber-to-the-Curb (FTTC) [5], and Fiber-to-the-Building (FTTB) [5]. These models offer fiber directly to the home, or nearby home. The FTTX models consist of passive optical network (PON).

The installation and maintenance costs of access networks are mainly contributed by active elements. Here, PON has been focused with the use of passive elements and becomes alternative for FTTX [4], where the operating costs are mainly due to copper network and its maintenance.

10.1 PONs

PON is a tree architecture to increase coverage area having a smaller number of splits providing reduction of optical power loss. PON is Ethernet PON (EPON) [5], Broadband PON (BPON) [6], and generalized framing procedure PON (GPON) [6], which uses two wavelengths for two-directional data transmission. The signals are multiplexed with time division multiplexer (TDM) to accommodate a great number of services and users in both ways. Wavelength division multiplexing (WDM) is required in PONs to increase the transmission capacity. The cost is increased due to having more tunable and wavelength-sensitive optical components used in the network. WDM-PONs are required after TDM-PONs [4, 5]. There are common components used in all types of PONs.

Optical line terminal (OLT)

OLT is a system equipment at office end providing multiservice for accommodating IP service and the traditional TDM services at the same time, ascending to
each ONU bandwidth allocation in accordance with the time division multiplexing,
downward to the ONU through broadcast Ethernet [7].

Optical network unit (ONU)

ONU consists of mainly active optical network unit and PON unit. The optical
node includes optical pickup and monitoring equipment, and its main function is to
receive the broadcast Ethernet data from OLT [7]. In response to the distance and
power control, commands are issued by the OLT, while the corresponding adjustment.

ODN

Optical distribution network (ODU) [7] consists of fiber and passive optical splitter having passive optical devices, and optical channel between the OLT and ONU,
and it also provides the link between the OLT and ONU with very high reliability.

10.1.1 STANDARDS OF DIFFERENT PON TECHNOLOGY

The major PONs are EPON, ATM PON (APON), GPON, and WDM-PON, which
are discussed in this section. Figure 10.1 shows FTTH access network having three
architectures: (a) direct to subscriber homes: an architecture having a fiber set from
the central office (CO) to each end-user subscriber for delivering bandwidth-intensive
integrated voice, data, and video services; (b) via curb switch: simple architecture
having connector termination space curb switch connected to the local exchange
(called as central office) with bidirectional fiber in which N subscribers are connected with curb switch; and (c) via passive optical switch: an architecture consisting
of passive optical power splitter connected to the local exchange (called as central
office) with bidirectional fiber in which N subscribers are connected with curb switch
at an average distance L km from the CO.

Due to high cost, active curb-side switch is replaced with passive optical split.
PON [8] requires a smaller number of optical transceivers, CO terminations, and
fiber deployment. Point to multipoint (POM) optical network has no active elements
in the signal path from source to destination. The interior elements used in a POM
are passive optical components, such as optical fiber, splices, and splitters. Access
networks based on a single-fiber PON only require (N + 1) transceivers and L km of
fiber.

10.1.2 EPON

Considering the IEEE Ethernet protocol, PON has been made and its standardization was developed using inexpensive Ethernet in PONs. This PON system is EPON.
Ethernet is mostly required for busty data and not for constant bit rate (CBR) or TDM
services. EPON was mainly based on IEEE 802.3 standards. The maximum rate for
EPON is 1.25 Gbit/s. If it uses 8B/10B coding, it becomes 1 Gbit/s.

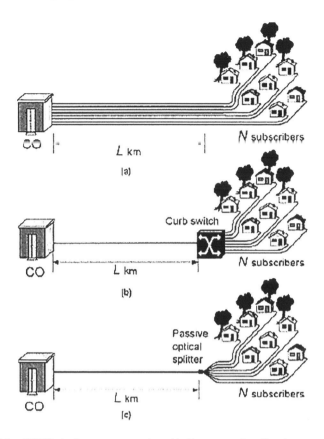

FIGURE 10.1 FTTH deployment scenarios: (a) direct to subscriber homes, (b) via curb switch, and (c) via passive optical switch [5].

Deployments

Nippon Telegraph and Telephone (NTT) produces currently 1M FTTX subscribers (including PON and P2P), which are also used in EPON. There are two data streams: upstream and downstream having OLT and ONU. In the downstream direction (OLT to ONU), Ethernet frames transmitted by the OLT pass via a 1:N passive splitter and reach each ONU. Typical values of N are between 8 and 32. The packets are transmitted by the OLT to their destination (ONU) based on a Logical Link Identifier (LLID), in which the ONU is assigned where it registers with the network.

In the upstream direction, data frames are originated from an ONU transmitting to the OLT and not send to any other ONU due to the directional properties of a passive optical combiner. In the upstream direction, the function of EPON is similar to that of a point-to-point architecture. In upstream direction, a contention-media-access control is used (like carrier sense medium access/collision detection). An OLT faces a collision and informs ONUs by sending a jam signal. Figure 10.2b shows downstream operation in EPON.

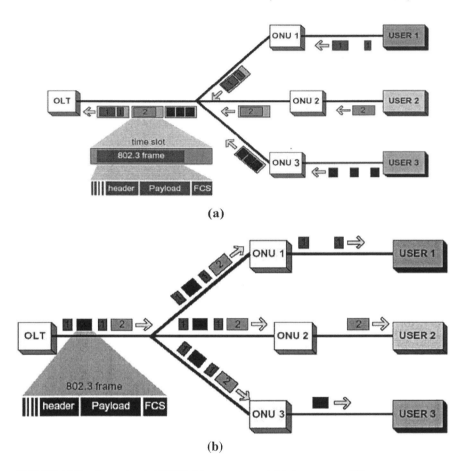

FIGURE 10.2   Operation in EPON: (a) upstream and (b) downstream [5].

10.1.3   APON

ATM PON (APON) is supported by Asynchronous Transfer Mode (ATM) as the medium access control (MAC) layer [9]. Figure 10.3 shows downstream frames consisting of 56 ATM cells having 53 bytes each for the basic rate of 155 Mbps, scaling up to 224 cells for 622 Mbps. There are Physical Layer Operation, Administration, and Maintenance (PLOAM) cells, starting one at the beginning of the frame and one in the middle. The remaining 54 cells are data ATM cells.

APON provides ATM-based services, and it is also named as Broadband PON (BPON). BPON is also standardized by Internal Telecommunication Union (ITU) specification G.983.1. BPON provides over lay capabilities for services such as video and Ethernet traffic [27]. Burst-mode receivers are used in OLT for synchronizing different ONUs at a different distance from the OLT; hence, the power received at the OLT from different ONUs is different. The ATM cell is an ATM data cell. In downstream direction, the ATM cells carry permission from the OLT

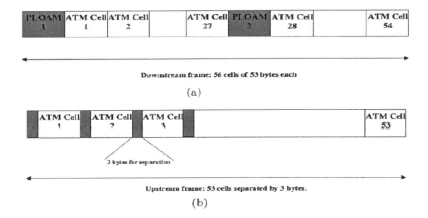

FIGURE 10.3 Frame formats of APON: (a) upstream and (b) downstream.

to the ONU. One-time permission is used for the ONU to transmit payload data in an ATM cell. During its operation, OLT transmits a continuous stream of grants to all the ONUs in PON. Thus, the OLT has upstream bandwidth assigned to each ONU. In upstream direction, the ATM cells are taken by ONUs for transmission of these queue sizes to the OLT. Table 10.1 shows the comparison of different PONs standards.

The upstream transmission is made in burst forms of ATM cells, with a 3-byte physical overhead added to each 53-byte ATM cell for burst-mode receivers.

10.1.4 GENERALIZED FRAMING PROCEDURE PON (GPON)

The GPON based on protocol of general framing procedure (GFP) is described by the ITU G.984.x. It provides maximum bit rates of 2.5 Gbps for hierarchical TDM STS-48. Higher efficiency is obtained for multiple services over PON. GFP uses a generic system to accommodate traffic from higher layers over a transport layer such as synchronous optical network (SONET) or synchronous digital hierarchy (SDH). The dynamic bandwidth assignment, operation, and maintenance are made by APON.

TABLE 10.1
The Key Characteristics of Different PONs with Its Standards

	EPON	APON	GPON	WDM-PON
Framing	Ethernet	ATM	GFP/ATM	Protocol independent
Maximum bandwidth	1 Gbit/s	622 Mbit/s	2.488 Gbit/s	1–10 Gbit/s/channel
Average bandwidth	60×106 bit/s	20×106 bit/s	40×106 bit/s	$1–10 \times 109$ bit/s
Estimated cost	Inexpensive	Inexpensive	Expensive	Highly expensive
Standard	IEEE 802.3	ITU G.983	ITU G.984	None

10.1.5   WDM-PON

For the basic form of PON using only a single optical channel, the available bandwidth is restricted to the maximum bit rate of 10 Gbps using STS-192 at optical transceivers. The splitting restriction is due to attenuation, and the maximum number of ONUs of 64 kbps is also restricted. This restricts the network's scalability. Although PON has higher bandwidth than metallic access network, an increase in the bandwidth of the PON is provided by WDM, where multiple wavelengths are used in both upstream and downstream directions for the increase in bandwidth. This type of PON is a WDM-PON. In WDM-PON, each ONU requires its own unique dedicated wavelength. The mode of operation of WDM-PONs provides multiple ONUs using a shared wavelength or an ONU using multiple wavelengths for different ONUs as destinations/sink different traffic. A WDM-PON provides a point-to-point access, whereas PON provides POM access having one wavelength for OLT and each ONU.

10.2   HYBRID PON

For an efficient use of bandwidth capacity of WDM-PON, hybrid models of TDM-PONs are used at user interface for flexible and cost-effective use of PON. This technique is called as hybrid PON (HPON). Some Hybrid TDM/WDM-PONs such as Samsung's HPON [5] and Stanford's SUCCESS-DWA [6] and SUCCESS-HPON [7] have already been reported.

10.2.1   SUCCESS-HPON

Figure 10.4 shows SUCCESS HPON having a smooth path from TDM to WDM-PONs. It gives the bandwidth as same as that of regular WDM-PON I in expensive manner where we avoid lasers at the customer side and tunable components at CO. Here, fixed or tunable lasers (TLs) at the customer's WDM-ONUs are also avoided by sending a continuous wave (CW) from the CO modulating the upstream data, as shown in Figure 10.4.

There are additional options for modulation at the ONUs. Semiconductor optical amplifiers (SOAs) with electronic devices reduce the costs, and TLs and receivers at the CO provide the services to all the network users for both downstream and upstream. This decreases the number of components in the system. But for the addition of new customers, the sharing of network is difficult, and for that, innovative scheduling algorithms can be used [8].

10.2.2   SUCCESS DRA

Figure 10.5 shows optical access architecture for SUCCESS (Stanford University Access) network with dynamic wavelength allocation (DWA). The SUCCESS-DWA PON in the network builds a bridge between TDM and WDM PONs. Cost-effective network performance is obtained using bandwidth for multiple physical PONs. The existing PON infrastructures work together with a SUCCESS-DWA PON.

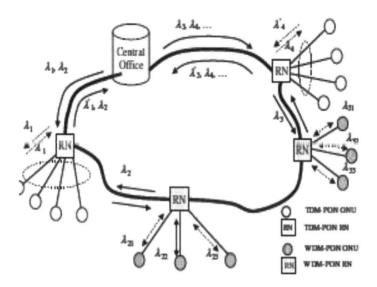

FIGURE 10.4 Architecture of success-HPON.

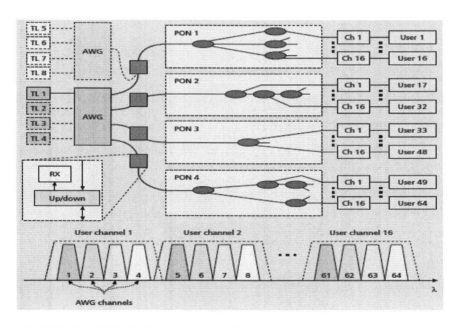

FIGURE 10.5 SUCCESS DRA architecture [8].

Figure 10.5 shows the proposed network architecture having TLs, an array waveguide grating (AWG), and thin-film WDM filter as required components of this network. TLs and the AWG are placed in the CO, whereas the WDM filters are at ONUs. The AWG with TLs is cyclic or near-cyclic.

10.3   OPEN RESEARCH ISSUES

In FTTX PONs, the main issues include QoSs, signal security, and transmission reliability. There are other issues especially in each access network.

10.3.1   ISSUES IN EPON

The main issues of EPON standards are proper specification of hardware devices and protocols used for assigning bandwidth to the users providing fairness and QoS. Another issue is how to provide open access for multiple ISPs for sharing same fiber infrastructure when the local legislation requires it.

10.3.2   ISSUES IN LARGE-SCALE IP VIDEO NETWORKS

There are issues of implementation of IP video network in PONs and for its dependence on legislation and copyright issues. The deployment of scalable multicasting protocols, mapping multicast groups to conventional video channels, and development of inexpensive integrated set-top boxes are other difficulties.

10.3.3   ISSUES IN INTEGRATED ONU/WIRELESS BASE STATION/HOME GATEWAY/DSLAM

Other issues are integration of various components, and development of combined scheduling algorithms and MAC protocols for multiple interfaces and users to optimize performance.

10.3.4   ISSUES IN HYBRID TDM/WDM-PON ARCHITECTURES

For a smooth transition from TDM to WDM, the proposed network architectures require to have knowledge how to share efficient and fast tunable components. In SUCCESS-HPON, there are issues related to sharing the most expensive components for the reduction in the overall cost of the network but they enhance computational complexity of the system. For making trade-off between the number of components and computational complexity, research has to be done.

For SUCCESS-DWA protocols, issues are related to laser tuning time, burst-mode preamble sequence, and additional guard time. The degradation of network efficiency increases the overhead of the frames. The proper design of queuing schedule reduces the overhead size.

10.3.5   ISSUES IN WDM-PON

Issues are expensive cost of devices that are used for passive and WDM functional. There are issues related to the integration of these devices with electronics (e.g., SOAs in an ONU).

EXERCISES

10.1. In an EPON, the downstream delay is equal to the upstream delay; when a REPORT message arrives from ONUi, the OLT's local time is 12. The time stamp on the REPORT message is 3. What is the current local time at ONUi assuming negligible processing time at the OLT?

10.2. Consider the PON network architecture in which we modify the PON architecture to build a (1 + 1) protected PON, so that if any component, namely, the OLT or ONU transceivers, fiber, or splitter, fails, the PON network is not disrupted at all. Design a suitable architecture for such a (1 + 1) protected PON.

10.3. Consider a PON which has 32 ONUs. Let the transmission power of the OLT transceiver be 0.02 mW, and the receiver be able to detect signals of at least 0.00015 mW for a reasonable bit-error rate. Assume that the splitter is 9% lossy. The optical fiber has an attenuation of. 0.12 dB/ltm. What is the maximum possible distance of an ONU from the OLT? What is the RTT corresponding to this distance?

REFERENCES

1. F. Wiener, "Examining video architectures which optimize the network infrastructure for video delivery over DSL and PON," *Proceedings of OFC*, NTuB2, March 2005.
2. "FTTP: Revolutionizing the Bells' TelecomNetworks," Bernstein Research – Telcordia Technologies Report, May 2004.
3. N.J. Frigo et al., "A view of fiber to the home economics," *IEEE Communications Magazine*, August 2004.
4. B. Lung, "PON architecture future proofs FTTH," *Lightwave, PennWell*, vol. 16, no. 10, pp. 104–107, September 1999.
5. B. Mukherjee, *Optical WDM Networks*, Springer-Verlag, New York, 2006.
6. "PON IC Opportunities Expand as Market Ram," RHK Report, June 2004.
7. Q. Han and X. Huan, "Design of GPON network based on FTTH, 2nd international conference on Automation, Mechanical Control and Computational Engineering (AMCCE 2017)," *Advances in Engineering Research*, vol. 118, pp. 148–151, 2017.
8. D.J. Shin et al., "Hybrid WDM/TDM-PON with wavelength-selection-free transmitters," *IEEE/OSA Journal of Lightwave Technology*, vol. 23, pp. 187–195, January 2005.
9. F.-T. An, et al., "SUCCESS: A next-generation hybrid WDM/TDM optical access network architecture," *IEEE/OSA JLT*, vol. 22, pp. 2557–2569, November 2004.

11 Math Lab Codes for Optical Fiber Communication System

An engineering approach based on an analytical model is used to design long-haul optical fiber communication link by the optimization of parameters, especially in optical fiber, which can constrain the number of optical channels as well as distances [1]. Considering industrial data such as ITU-G-652 standard (Corning R SMF-28™ [1]) and ITU-G-655 nonzero dispersion-shifted fiber (Corning LEAFR – large effective area fiber [1]), one can design the optical fiber transmission system in an optical network. Since the interaction of these parameters gives a large number of optimization/combination, computer-aided engineering approach is required to get accurate design [1–3] with less computation time in a precise engineering point of view. There are a number of computer-based Math lab programs and simulation tools for optical system design modeling. MATLAB and Simulink are packages in which simulation programs can be written by using different building blocks, tools, and functions. The simulation written by appropriate techniques/functions makes the integration of optical modules modeled at physical operating principle such as optical modulator with mathematical model that is required to solve fiber propagation equations. In this chapter, we have mentioned math lab codes used for the design of optical transmission for optical networks.

11.1 SPECIFICATION OF DESIGN OF OPTICAL FIBERS

For the design of geometrical and index profile of silica optical fibers, the fiber performance of dispersion depends on the fiber core radius and the relative refractive index difference. A dispersion-compensated fiber, in this case, the SSMF of total length of transmission ~10,000 km, the length of your designed fiber and the SMF so that the average fiber dispersion is 0.01 PS/(km) [1]. Normally, the spacing of optical amplifiers is 100 km.

11.1.1 MATERIAL SPECIFICATION

The refractive index $n(\lambda)$ is written using Sellmeier's formula for dispersion as [1,4,5]

$$n^2(\lambda) = 1 + \sum_k \frac{G_k \lambda^2}{(\lambda^2 - \lambda_k^2)} \tag{1}$$

where G_k is Sellmeier's coefficient and $k = 1, 2, 3$.

The material parameters such as Sellmeier's coefficients are given in Tables 11.1 and 11.2 [1].

TABLE 11.1
Sellmeier's Coefficients for Ge-Doped Optical Fiber Silica-Based Materials (G_k in μm^{-2} (where $k = 1, 2, 3$)) [1]

Sellmeier's Constants	Germanium Concentration, C (mol%)			
	A	B	C	D
Types	0	3.1%	5.8%	7.9%
G_1	0.6961663	0.7028554	0.7088876	0.7136824
G_2	0.4079426	0.4146307	0.4206803	0.4254807
G_3	0.8974794	0.8974540	0.8956551	0.8964226
λ_1	0.6840432	0.0727723	0.0609053	0.0617167
λ_2	0.1162414	0.1143085	0.1254514	0.1270814
λ_3	9.896161	9.896161	9.896162	9.896161

TABLE 11.2
Sellmeier's Coefficients of Ge Depend on Optical Fiber Silica Materials in the Core Region

Sellmeier's Constants	Concentration Composition			
	E	F	G	H
Types	Quenched SiO_2	13.5GeO_2:86.5 SiO_2	9.1 P_2O_5:90.0 SiO_2	13.3B_2O_3:86.7 SiO_2
G_1	0.696750	0.711040	0.695790	0.690618
G_2	0.408218	0.408218	0.452497	0.401996
G_3	0.890815	0.704048	0.712513	0.898817
λ_1	0.069066	0.064270	0.061568	0.061900
λ_2	0.115662	0.129408	0.119921	0.123662
λ_3	9.900559	9.425478	8.656641	9.098960

Source: J.W. Fleming, "Material dispersion in lightguide glasses", *Electronic Letters*, vol. 14, p. 326, 1978. G_k in μm^{-2}.

11.1.2 TRANSMISSION SPECIFICATIONS

Attenuation [1,5] is less than 0.25 dB/km at the wavelength range 1.55–1.625 μm.

Attenuation vs. Wavelength		
Wavelength range (nm)	Central wavelength	Propagation loss (dB/km)
1525–1575	1550 nm	0.05
1525–1625	-	0.05

Bending Loss for Single Turn of Optical Fiber Bending		
Diameter (mm)	Wavelength (nm)	Loss (dB)
32	1550 and 1625	≤0.50
75	1310	≤0.005
50	1550	≤0.001
75	1625	≤0.005
Mode-field diameter is written as 9.20–10.00 μm at 1550 nm		

Dispersion [1–5]

Overall dispersion: 2.0 to 6.0 PS/(nm.km) for 1.530–1.565 μm wavelength range
4.5 to 11.2 PS/(km) for 1.565–1.625 μm wavelength range
Dispersion formula is written as [1]

$$D(\lambda) \approx \frac{S_9}{4}\left[\lambda - \frac{\lambda_0^4}{\lambda}\right] \text{ ps/nm km}$$

where S_0 represents the zero dispersion slope.

Fiber polarization mode dispersion (PMD) [1,6] is written as

$$\text{PMD link value} \qquad \leq 0.4 \text{ ps}/\sqrt{\text{km}}$$

$$\text{Maximum individual fiber} \qquad \leq 0.1 \text{ ps}/\sqrt{\text{km}}$$

These values represent the upper limit of PMD of fiber (also known as PMD_Q).

11.1.3 ENVIRONMENTAL SPECIFICATIONS

Attenuations in optical fiber depend on environmental conditions. The specifications for different conditions are given below [1]:

Environmental Test Condition	Induced Attenuation (dB/km) at 1550 nm
Temperature dependence: −60°C to +85°C	≤0.05
Temperature–humidity cycling up to 98% Relative humidity: −10°C to +85°C	≤0.05
Water immersion: 23°C	≤0.05
Heat aging: 85°C	≤0.05
Operating temperature range: −60°C to +85°C.	

Dimensional specifications of optical fiber are normally fixed as follows [1].

Standard Length (km/reel): 4.4–25.2
Glass Geometry
    Fiber curl: ≥4.0 m radius of curvature
    Cladding diameter: 125.0 ± 1.0 μm
    Core/clad concentricity: ≤0.5 μm
    Cladding noncircularity: ≤1.0%
Coating Geometry
    Coating diameter: 245 ± 5 μm
    Coating/cladding concentricity: <12 μm

Mechanical specifications of optical fiber direct transmission characteristics restrict the fiber length with a tensile stress ≥100 kpsi. The transmission performance characteristics are important while designing the optical fiber. Typical values are written as [1]

Wavelength range: 1.469 at 1550 nm
Single-mode core diameter: 8.2 μm
Numerical aperture (NA): 0.14
NA is measured at the 1% power level of a one-dimensional far-field scan at 1310 nm.
Zero dispersion wavelength (λ_0): 1313 nm
Zero dispersion slope (S_0): 0.086 ps/(nm^{2}km)
Refractive index difference: 0.36%
Fatigue resistance parameter (ns): 20

11.2 MATH LAB CODES FOR DESIGN OF OPTICAL FIBERS

Practical limits for the fiber core radius and the relative index difference must be taken into account. A set of curves must be obtained with the core radius or the relative index difference as a parameter. Make sure that the material dispersion factors are correctly modeled.

Optical Fibers
Signal attenuation and dispersion.

11.2.1 CODES FOR PROGRAM OF THE DESIGN OF OPTICAL FIBERS

```
Math lab program of the Design of Optical Fibers is given
  below[1]
c = 2.997925e8;
G1 = 0.711040;
G2 = 0.408218;
G3 = 0.704048;
lambda1 = 0.064270e-6;
lambda2 = 0.129408e-6;
lambda3 = 9.425478e-6;
a = 4.1e-6;
```

```
delta = 0.0025;
start = input ('Enter lambda start point (nm) —: ');
finish = input ('Enter lambda end point (nm) —-: ');
resolution = input ('Enter lambda resolution (nm) —-: ');
disp ('');
lambda = start*1e-9;
lambdavector (1,1) = lambda;
for (i = 1:(((finish-start)/resolution) + 1))
n1squared = 1 + ((G1*power(lambda,2))/
(power(lambda,2) -power(lambda1,2))) + ((G2*power(lambda,2))/
(power(lambda,2) -power(lambda2,2))) + ((G3*power(lambda,2))/
  (power(lambda,2)
-power(lambda3,2)));
n1 = sqrt(n1squared);
n1vector(1, i) = n1;
n2 = n1*(1 + delta);
n2vector(1, i) = n2;
V = (2*pi/lambda) *a*n1*sqrt(2*delta);
Vvector(1, i) = V;
dy1dx = (-2*G1*power(lambda1,2) *lambda)/
  (power(power(lambda,2)-power (lambda1,2),2));
dy2dx = (-2*G2*power(lambda2,2) *lambda)/
  (power(power(lambda,2)-power (lambda2,2),2));
dy3dx = (-2*G3*power(lambda3,2) *lambda)/
  (power(power(lambda,2)-power
(lambda3,2),2));
d2y1dx2 = (2*G1*power(lambda1,2) *(3*power(lambda,2) +
  power(lambda1,2)))/(power(power
  (lambda,2)-power(lambda1,2),3));
d2y2dx2 = (2*G2*power(lambda2,2) *(3*power(lambda,2) +
  power(lambda2,2)))/
  (power(power(lambda,2)-power(lambda2,2),3));
d2y3dx2 = (2*G3*power(lambda3,2) *(3*power(lambda,2) +
  power(lambda3,2)))/(power(power(lambda,2)
  -power(lambda3,2),3));
d2ndx2 = 0.5*(((d2y1dx2 + d2y2dx2 +
  d2y3dx2)*power(n1,2)-0.5*(power (dy1dx + dy2dx + dy3dx,2))
  )/power(n1,3));
M = (-d2ndx2/c) *lambda;
Mvector(1, i) = M; %row vector
Dw = (-n2*delta)/c*(0.080 + 0.549*power(2.834-V,2)) *(1/lambda);
Dwvector(1, i) = Dw;
if (i < (( (finish-start)/resolution) + 1)) lambdavector(1,i +
  1) = lambdavector(1,i) + (resolution*1e-9);
lambda = lambdavector (1,i + 1);
end
end
plot (lambdavector, Mvector,lambdavector,Dwvector,lambdavector,
  Mvector+ Dwvector);
grid;
```

11.2.2 CODES FOR DESIGN OF STANDARD SINGLE-MODE FIBERS

Math lab program for design of standard single-mode fibers is given below [1]:

```
% totdisp_SMF.m
% Matlab script for calculating of total dispersion for
% Non-Zero Dispersion Shifted Fiber
% Design of Optical Fiber
lambda = [1.1:0.01:1.700] *1e-6;
G1 = 0.7028554; %Sellmeier's coefficients for germanium
G2 = 0.4146307; %doped silica (concentration B in table)
G3 = 0.8974540;
lambda1 = 0.0727723e-6; %Wavelengths for germanium doped
  silica
lambda2 = 0.1143085e-6;
lambda3 = 9.896161e-6;
c = 299792458; %Speed of light
pi = 3.1415926; %Greek letter pi
a = 4.1e-6; %Core radius
delta = 0.003; %Greek letter delta (ref. index difference
%between core and cladding)
% Calculating the refractive index
npow2oflambda = 1 + (G1. *lambda.^2. /(lambda.^2.-
  lambda1*lambda1)) …
+ (G2. *lambda.^2. /(lambda.^2.-lambda2*lambda2)) …
+ (G3. *lambda.^2. /(lambda.^2.-lambda3*lambda3));
noflambda = sqrt(npow2oflambda);
pointer = find (lambda = = 1.550e-6);
n1 = noflambda(pointer) % core index
% material dispersion calculation
t1 = diff(noflambda);
t2 = diff(lambda);
t3 = t1. /t2;
t4 = diff(t3);
t5 = diff(lambda);
t5 = adjmat(t5);
lambda = adjmat(lambda);
lambda = adjmat(lambda);
%Material dispersion
Matdisp = - (lambda. /c).* (t4./t5);
% Converting to ps/nm.km
Matdisp = Matdisp. *1e6;
figure (1)
clf
hold xlabel('nm')
ylabel('ps/nm.km')
title('Standard Single Mode Fiber')
plot(lambda, Matdisp, '. -')
grid on
% waveguide dispersion calculation
V = (2 * pi * a * n1 * sqrt (2 * delta)) ./(lambda);
```

```
Dlambda1 = - (n1 * delta). / (c * lambda);
%plot (lambda,Dlambda2)
Dlambda2 = 0.080 + 0.549 * (2.834 − V).∧2;
%plot(lambda, Dlambda2)
Dlambda = Dlambda1.* Dlambda2;
% Converting to ps/nm.km
Dlambda = Dlambda. *1e6;
plot (lambda,Dlambda, '-')
% Calculating total dispersion
TotDisp = Matdisp + Dlambda;
plot (lambda, TotDisp, ':')
legend('Material Dispersion', 'Waveguide Dispersion', 'Total
  Dispersion',0)
% Finding the dispersion at 1460, 1550 and 1625 nm
pointer = find (lambda = = 1.460e-6);
Disp1460 = TotDisp(pointer)
pointer = find (lambda = = 1.550e-6);
Disp1550 = TotDisp(pointer)
pointer = find (lambda = = 1.6250e-6);
Disp1625 = TotDisp(pointer)
% refind_SMF.m
% PROJECT DESIGN: Optical Fiber Design
lambda = [1.1:0.01:1.700] *1e-6;
n1 = [1.0487:0.001:1.8587]; % core index
n2 = [1.0435:0.001:1.8535]; % claddingx index
% Determining the index profile
delta = (n1 - n2)./ n1;
deltap = delta * 100;
plot (n1, deltap)
grid on
xlabel('n1')
ylabel('Delta - Relative Refractive Index (%)')
title('Refractive Index')
```

11.2.3 CODES OF NONZERO DISPERSION-SHIFTED FIBERS

Math lab codes of nonzero dispersion-shifted fibers are given below [1,7]:

```
% total disp_NZDSF.m for Nonzero Dispersion-Shifted Fibers
% Math Labcodes for total dispersion of Non-Zero Dispersion
  Shifted Fiber
% material dispersion, the waveguide dispersion andthe total
  dispersion of the designed fiber.
%dispersion shitted Optical Fiber design
lambda = [1.1:0.001:1.700]*1e-6;
G1 = 0.7028554; %Sellmeier's coefficients for germanium
G2 = 0.4146307; %doped silica (concentration B in table)
G3 = 0.8974540;
lambda1 = 0.0727723e-6; %Wavelengths taken for germanium doped
  silica
```

```
lambda2 = 0.1143085e-6;
lambda3 = 9.896161e-6;
c = 299792458; %Speed of light
pi = 3.1415926; %Greek letter pi
a = 2.4e-6; %Core radius
delta = 0.0043; % ref. index difference between core and
  cladding)
% Estimating the refractive index
npow2oflambda = 1 + (G1. *lambda.∧2. /(lambda.∧2.-
  lambda1*lambda1)) …
+ (G2. *lambda.∧2. /(lambda.∧2.-lambda2*lambda2)) …
+ (G3. *lambda.∧2. /(lambda.∧2.-lambda3*lambda3));
noflambda = sqrt(npow2oflambda);
pointer = find (lambda = = 1.550e-6);
n1 = noflambda(pointer) %Refractive index in the core @1550 nm
% Calculating the material dispersion
t1 = diff(noflambda);
t2 = diff(lambda);
t3 = t1./t2;
t4 = diff(t3);
t5 = diff(lambda);
t5 = adjmat(t5);
lambda = adjmat(lambda);
lambda = adjmat(lambda);
%Material dispersion
Matdisp = −(lambda. /c).* (t4./t5);
% Converting to ps/nm.km
Matdisp = Matdisp. *1e6;
Figure(1)
clf
hold
xlabel('nm')
ylabel('ps/nm.km')
title('Non-Zero Dispersion Shifted Fiber')
plot(lambda, Matdisp, '.')
grid on
% Calculating waveguide dispersion
V = (2*pi*a*n1*sqrt(2*delta)). /(lambda);
pointer = find (lambda = = 1.550e −6);
V1550 = V(pointer)
Dlambda1 = −(n1 * delta). / (c * lambda);
Dlambda2 = 0.080 + 0.549 * (2.834 - V).∧2;
Dlambda = Dlambda1.* Dlambda2;
% Converting to ps/nm.km
Dlambda = Dlambda. *1e6;
plot (lambda,Dlambda, '-')
% Estimation of total dispersion
TotDisp = Matdisp + Dlambda;
Plot(lambda, TotDisp, '+')
Legend('Material Dispersion', 'Waveguide Dispersion', 'Total
  Dispersion', 0)
```

```
% Estimatingthe dispersion for wavelengths 1460, 1550 and
  1625 nm
pointer = find(lambda = = 1.460e-6);
Disp1460 = TotDisp(pointer)
pointer = find(lambda = = 1.550e-6);
Disp1550 = TotDisp(pointer)
pointer = find(lambda = = 1.6250e-6);
Disp1625 = TotDisp(pointer)
% PerihenZDSP.m
% Math Lab script for calculating of possible values for %
  refractive index in the core, n1, and cladding, n2,
% and its relative refractive index.
lambda = [1.1:0.01:1.700]*1e-6;
n1 = [1.0487:0.001:1.8587]; % core index
n2 = [1.0324:0.001:1.8424]; % cladding index
% Estimating index profile
delta = (n1 - n2)./ n1;
deltap = delta * 100;
plot (n1, deltap)
grid on
xlabel('n1')
ylabel('Delta - Relative Refractive Index (%)')
title('Refractive Index')
```

11.2.4   CODES OF SPLIT-STEP FOURIER METHOD (SSFM)

These codes perform the propagation of the optical signals a long fiber link distance. Figure 11.1 shows schematics of optical fiber transmission link system composed of transmitter producing RZ-DQPSK modulation format [1,8], fiber propagation model including optical amplifiers and DCFs, and optical receiver at the transmission end and monitoring at the output of the transmitter.

An initialization program must also be included to set the data and parameters required for the Simulink model and subroutines. Codes of the SSMF, including Raman gain amplification effects are shown below.

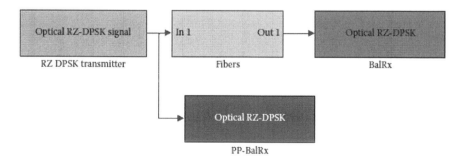

FIGURE 11.1   Simulink model of the SSMF [1].

11.2.5 CODES FOR OPTICAL FIBER TRANSMISSION SYSTEM

Math lab programs for optical fiber transmission system composed of an optical transmitter generating RZ-DQPSK modulation format, fiber propagation model including optical amplifiers and DCFs, and optical receiver are given below [1,8,9].

```
function output = ssprop_matlabfunction_modified(input)
nt = input(1);
u0 = input(2:nt + 1);
dt = input(nt + 2);
dz = input(nt + 3);
nz = input(nt + 4);
alpha_indB = input(nt + 5);
betap = input(nt + 6:nt + 9);
gamma = input(nt + 10);
P_non_thres = input(nt + 11)
maxiter = input(nt + 12);
tol = input(nt + 13);
tic;
alpha = log(10)*alpha_indB/10; % alpha (1/km)
ntt = length(u0);
w = 2*pi*[(0:ntt/2-1),(−ntt/2:-1)]'/(dt*nt);
%w = 2*pi*[(ntt/2:ntt-1),(1:ntt/2)]'/(dt*ntt);
clear halfstep
halfstep = −alpha/2;
for ii = 0:length(betap)−1;
halfstep = halfstep-j*betap(ii + 1)*(w.^ii)/factorial(ii);
end
clear LinearOperator
% Linear Operator in Split Step method
LinearOperator = halfstep;
% pause
halfstep = exp(halfstep*dz/2);
u1 = u0;
ufft = fft(u0);
% Nonlinear operator will be added if the peak power is
  greater than the
% Nonlinear threshold
iz = 0;
while (iz <nz) & (max((abs(u1).^2 + abs(u0).^2)) >P_non_thres)
iz = iz + 1;
uhalf = ifft(halfstep.*ufft);
for ii = 1:maxiter,
uv = uhalf .* exp(−j*gamma*(abs(u1).^2 + abs(u0).^2)*dz/2);
ufft = halfstep.*fft(uv);
uv = ifft(ufft);
%fprintf('You are using SSFM\n');
if (max(uv-u1)/max(u1) <tol)
u1 = uv;
break;
else
```

```
u1 = uv;
end
end
if (ii = = maxiter)
warning(sprintf('Failed to converge to %f in %d iterations',…
tol,maxiter));
end
u0 = u1;
end
if (iz <nz) & (max((abs(u1).∧2 + abs(u0).∧2)) <P_non_thres)
% u1 = u1.*rectwin(ntt);
ufft == fft(u1);
ufft = ufft.*exp(LinearOperator*(nz-iz)*dz);
u1 = ifft(ufft);
%fprintf('Implementing Linear Transfer Function of the Fibre
Propagation');
end
toc;
output = u1;
```

11.3 MATLAB CODES FOR OPTICAL TRANSMISSION SYSTEM WITH MUX AND DEMUX

Simulation package can be developed using MATLAB and Simulink platforms for modeling advanced digital optical transmission systems. A digital photonic transmission system has three subsystems: optical transmitter, optical fiber channel, and optical receiver, as demonstrated in Figure 11.2 [1]. The generic model is also shown in Figure 11.3 [1].

In Figure 11.2, an optical transmitter normally consists of a narrow line width laser source, external optical modulators, a bit pattern data generator, and optionally, an electrical precoder or an electrical shaping filter. The noise-equivalent current, as seen from the input of the electronic preamplifier, is modeled as a random noise source superimposed on the signal input to the receiver; relevant quantum shot noise is also calculated by the model at each sampled numerical instant. Balanced optical receiver with a phase comparison at the input to obtain constructive and destructive

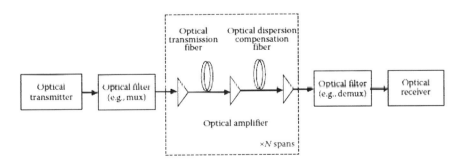

FIGURE 11.2 Optical transmission system [1].

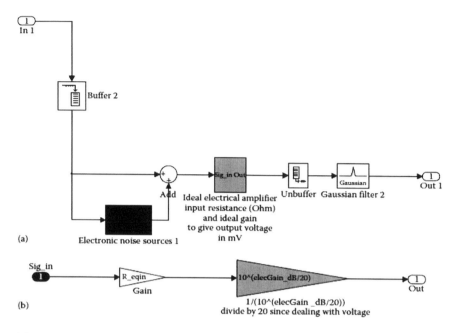

FIGURE 11.3 Generic model of Simulink model of transmitter (a) noise and (b) gain section [1].

interfered modes for the identification of the "1" and "0" of the phase difference of consecutive bits in differential coded sequence is shown in Figure 11.3 [1,10].

```
% Initialization file for data and parameters transfer for
  Simulink model [1]
% Inclusion of parameters for solving the NLSE
% Implementing the split-step propagation
clear all
close all
% Constants
c = 3e8; % speed of light (m/s)
% Numerical values
numbitspersymbol = 1;
P0 = 0.003162; % peak power (W)
FWHM = 25,% pulse width FWHM (ps)
halfwidth = FWHM% for square pulse
bitrate = 1/halfwidth; % THz
baudrate = bitrate/numbitspersymbol;
signalbandwidth = baudrate;
PRBSlength = 2^5;
num_samplesperbit = 64; % should be 2^n
dt = FWHM/num_samplesperbit ; % sampling time(ps); % time
  step (ps)
nt = PRBSlength*num_samplesperbit; % FFT length
```

```
dz = 0.1; % distance step-size (km)
nz = 10; % number of z-steps
maxiter = 20; % max # of iterations
tol = 1e-5; % error tolerance
% OPTICAL PARAMETERS
nonlinearthreshold = 0.005; % 5mW - % Nonlinear Threshold Peak
   Power for silica
% core fibre
lambda = 1550;
lambda_carrier = 14648.4375; % wavelength (nm)with Level 4
   group the carrier freq
optical_carrier = c/(lambda_carrier*1e-9); %carrier freq
alpha_indB = 0.001; % loss (dB/km)
D = 17; % NZDSF GVD (ps/nm.km); if anomalous
beta3 = 0.3; % GVD slope (ps^3/km)
ng = 1.46; % group index
n2 = 2.6e-20; % nonlinear index (m^2/W)
Aeff = 47; % effective area (um^2)
% CALCULATED QUANTITIES
T = nt*dt; % FFT window size (ps) -Agrawal:
gamma = 2e24*pi*n2/(lambda*Aeff); % nonlinearity coef (km^-1.
   W^-1)
t = ( (1:nt)'-(nt+1)/2)*dt; % vector of t values (ps)
t1 = [(-nt/2+1:0)]'*dt; % vector of t values (ps)
t2 = [(1:nt/2)]'*dt; % vector of t values (ps)
w = 2*pi*[(0:nt/2-1),(-nt/2:-1)]'/T; % vector of w values
   (rad/ps)
v = 1000*[(0:nt/2-1),(-nt/2:-1)]'/T; % vector of v values
   (GHz)
vs = fftshift(v); % swap halves for plotting
v_tmp = 1000*[(-nt/2:nt/2-1)]'/T;
L = nz*dz
Lnl = 1/(P0*gamma) % nonlinear length (km)
Ld = halfwidth^2/abs(beta2) % dispersion length (km)
N = sqrt(abs(Ld./Lnl) ) % governing the which one is
   dominating:
dispersion or Non-linearities
ratio_LandLd = L/Ld % if L << Ld -> NO Dispersion Effect
ratio_LandLnl = L/Lnl % if L << Lnl -> NO Nonlinear Effect
```

11.3.1 MODELING OF NONLINEAR OPTICAL FIBER TRANSMISSION SYSTEMS

Modeling of critical subsystem of digital photonic transmission system considers optical transmitter, properties of optical fibers and receiver. Here, other fiber transmission performances are chromatic dispersion (CD), PMD, and Kerr-effect nonlinearities, which are included for Math lab codes. Fiber core and cladding refractive indices depend on both operating wavelengths and the intensity of the guided mode distributed across the area of the fiber core and cladding. The refractive index as function of power is written as [1,3]

$$n' = n + \overline{n_2}\left(\frac{P}{A_{\mathrm{eff}}}\right)$$

where P = average optical intensity in fiber
$\overline{n_2}$ = non-linear index coefficient
A_{eff} = effective area of the fiber
The maximum transmission limit $L_{\max}$ considering PMD effect is given below [1,11]:

$$L_{\max} = \frac{0.02}{\langle \Delta t \rangle^2 \bullet R^2}$$

where R is the bit rate.
At $\langle \Delta t \rangle$ = 1 ps/km, for R = 40 Gb/s and 10 Gbps, $L_{\max}$ is estimated as 12.5 km and 200 km, respectively.
At $\langle \Delta t \rangle$ = 0.1 ps/km for R = 40 Gb/s and 0 Gbps, $L_{\max}$ is determined as 1,250 km and 20,000 km, respectively.
For high-bandwidth long-haul transmission employing external modulation, L_D is estimated due to CD as [1]

$$L_D = \frac{10^5}{D.R^2}$$

where R is the bit rate (Gb/s), D is in ps/(nm km), and L_D is in kilometers. In case of 10 Gb/s OC-192 OOK systems ($D = \pm 17$ ps/(nm km)), the dispersion length L_D is estimated as ~60 km SSMF. In the case of 40 Gb/s OC-768 systems, L_D is estimated as ~4 km due to D of ±60 ps/nm. The math lab codes are written as follows:

```
% Initialization file for data and parameters transfer to
   Simulink model [1]
% Implementing the split-step propagation
clear all
close all
% CONSTANTS
c = 3e8; % speed of light (m/s)
% PARAMETERS
numbitspersymbol = 1;
P0 = 0.003162; % peak power (W)
FWHM = 25 % width of FWHM (ps)
halfwidth = FWHM
bitrate = 1/halfwidth; % THz
baudrate = bitrate/numbitspersymbol;
signalbandwidth = baudrate;
PRBSlength = 2^5;
num_samplesperbit = 64; % should be 2^n
dt = FWHM/num_samplesperbit ; % sampling time(ps); % time
   step (ps)
nt = PRBSlength*num_samplesperbit; h
```

```
dz = 0.1; % distance step-size (km)
nz = 10; % number of z-steps
maxiter = 20; % max # of iterations
tol = 1e-5; % error tolerance
% OPTICAL PARAMETERS
nonlinearthreshold = 0.005; % Nonlinear Threshold Peak Power
  for silica
% core fibre
lambda = 1550;
lambda_carrier = 14648.4375;
optical_carrier = c/(lambda_carrier*1e-9); %carrier freq
alpha_indB = 0.001; % loss (dB/km)
D = 17; % NZDSF GVD (ps/nm.km); if anomalous
beta3 = 0.3; % GVD slope (ps^3/km)
ng = 1.46; % group index
n2 = 2.6e-20; % nonlinear index (m^2/W)
Aeff = 47; % effective area (um^2)
% Estimatedquantities
T = nt*dt;
alpha_loss = log(10)*alpha_indB/10; % alpha (1/km)
beta2 = -1000*D*lambda^2/(2*pi*c); % beta2 (ps^2/km);
gamma = 2e24*pi*n2/(lambda*Aeff); % nonlinearity coef (km^-1.
  W^-1)
t = ( (1:nt)'-(nt+1)/2)*dt; % vector of t values (ps)
t1 = [(-nt/2+1:0)]'*dt; % vector of t values (ps)
t2 = [(1:nt/2)]'*dt; % vector of t values (ps)
w = 2*pi*[(0:nt/2-1),(-nt/2:-1)]'/T; % vector of w values
  (rad/ps)
v = 1000*[(0:nt/2-1),(-nt/2:-1)]'/T; % vector of v values (GHz)
vs = fftshift(v); % swap halves for plotting
v_tmp = 1000*[(-nt/2:nt/2-1)]'/T;
L = nz*dz
Lnl = 1/(P0*gamma) % nonlinear length (km)
Ld = halfwidth^2/abs(beta2) ;
N = sqrt(abs(Ld./Lnl) ) ;
ratio_LandLd = L/Ld ;
ratio_LandLnl = L/Lnl ;
```

11.3.2 Phase Modulation Model and Intensity Modulation

There are a number of blocks such as communication blocks, control system blocks, and signal-processing blocks. A phase modulation of communication blocks can be implemented by using the phase shift block as shown in Figure 11.4. Different formats, the minimum shift keying, and differential phase shift keying (DPSK) transmitters are employed in this block. There should be set sufficiently wide to accommodate the spectrum of all multiplexed channels. The integration of phase shift blocks forms an intensity modulator. The overall transmitter is shown in Figure 11.4. In addition to the optical modulation, for DPSK, there must be a differential coder. The math lab codes are given below.

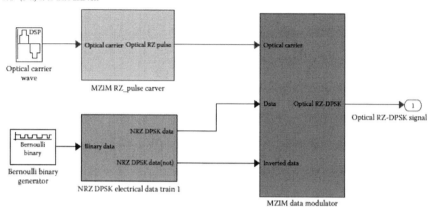

FIGURE 11.4 Photonic transmitters for carrier-suppressed RZ and data modulation using DPSK modulation scheme [1].

```
Program for Solving the Laser Rate Equation is given
  below[1,10].
% DFB Laser diode parameters for wavelength 1.55micron ;
%InGaAsP/InP ; 2 Gbit/s ; 40 Km
global q
global Vg
global Beta
global Imax
global Mune
global Gamma
global Anull
global Nnull
global Epsil
global Tphot
global Tcarr
global Vactv
global Alpha
global Trise
global Ibias
global MidLamda
global SpecWidth
Gamma = 0.8;
Anull = 3.2e-20;
Vg = 7.5e+7;
Nnull = 1.0e+12;
Epsil = 2.5e-23;
Tphot = 3.0e-12;
Beta = 3.0e-5;
Tcarr = 0.3e-9;
```

```
q = 1.6e-19;
Vactv = 1.5e-16;
Alpha = 5;
Trise = 1e-12;
Ith = q/Tcarr*Vactv*(Nnull+1/Gamma/Anull/Vg/Tphot);
Ibias = 7.1*Ith;
Imax = 9.0*Ith;
Mune = 0.042;
% change from 1553.8nm to 1553.3nm to match with ITU-grid AC
  and GA
% else
MidLamda = OpLamda(Channel);
% end
SpecWidth = 40e-15;
set
% Solving The Rate equations
function rate=rate_equ(t,y)
global q
global Vg
global Nsd
global Ssd
global Beta
global Gamma
global Anull
global Nnull
global Epsil
global Tphot
global Tcarr
global Vactv
global Alpha
global Trise
global Ibias
global Sigma
global Imax
global dummy
global Initial
global Impulse
global BitPeriod
Nsd=1/(Gamma*Anull*Vg*Tphot)+Nnull ;
Ssd=(Tphot/Tcarr)*Nsd*(Imax/Ibias-1) ;
Sigma=2e-20;
%---- Computing Rate Equations ------------
rate=zeros (3,1);
rate (1) =(Gamma*Anull*Vg*( (y(2)-Nnull)/...
(1+Epsil*y(1) ) )-1/Tphot)*y(1)+Beta*Gamma*y(2)/Tcarr + ...
randn*sqrt(2*Beta*Vactv*Nsd*(Vactv*Ssd+1)^3/Tcarr) ;
rate(2)=dgcoding(t,InPulse)/(q*Vactv)-y(2)/...
Tcarr-Vg*Anull*y(1)*(y(2)-Nnull)/(1+Epsil*y(1) ) + ...
randn*sqrt(2*Vactv*Nsd/Tcarr*(Beta*Vactv*Ssd+1) ) ;
rate(3)=Alpha*(Gamma*Vg*Anull*(y(2)-Nnull)-1/Tphot)/2
  +randn*sqrt(Gamma*Vg*Sigma*(y(2)-Nnull)/y(1) ) ;
```

```
if t<0
rate=Initial;
end
dummy=dummy+1;
if dummy>=50
% waitbar(t/(BitPeriod*size(InPulse,2))) ;
dummy=0;
end
```

11.3.3 MATH LAB CODES FOR RAMAN AMPLIFICATION AND SPLIT-STEP FOURIER METHOD

Math lab codes for Raman amplification and SSFM are given below [1,12,5].

```
nt = input(1);
u0 = input(2:nt + 1);
dt = input(nt + 2);
dz = input(nt + 3);
nz = input(nt + 4);
alpha_indB = input(nt + 5);
betap = input(nt + 6:nt + 9);
gamma = input(nt + 10);
P_non_thres = input(nt + 11);
maxiter = input(nt + 12);
tol = input(nt + 13);
tic;
u1 = ssprop(u0,dt,dz,nz,alpha,betap,gamma);
INPUT
u0 - starting field amplitude (vector)
to have P = P0*exp(-alpha*z)
alpha = 1e-3*log(10)*alpha_indB/10;
ntt = length(u0);
w = 2*pi*[(0:ntt/2-1),(-ntt/2:-1)]'/(dt*nt);
gain = numerical_gain_hybrid(dz,nz);
for array_counter = 2:nz + 1
grad_gain(1) = gain(1)/dz;
grad_gain(array_counter) =
   (gain(array_counter)-gain(array_counter-1))/
dz;
end
gain_lin = log(10)*grad_gain/(10*2);
clear halfstep
halfstep = -alpha/2;
for ii = 0:length(betap)-1
halfstep = halfstep - j*betap(ii + 1)*(w.^ii)/factorial(ii);
end
square_mat = repmat(halfstep, 1, nz + 1);
square_mat2 = repmat(gain_lin, ntt, 1);
size(square_mat);
size(square_mat2);
```

```
total = square_mat + square_mat2;
clear LinearOperator
LinearOperator = halfstep;
halfstep = exp(total*dz/2);
u1 = u0;
ufft = fft(u0);
iz = 0;
while (iz <nz) && (max( (gamma*abs(u1).^2 + gamma*abs(u0).^2))
  >P_non_thres)
iz = iz + 1;
uhalf = ifft(halfstep(:,iz).*ufft);
for ii = 1:maxiter,
uv = uhalf.*exp((-j*(gamma)*abs(u1).^2 + (gamma)*abs(u0).
  ^2)*dz/2);
ufft = halfstep(:,iz).*fft(uv);
uv = ifft(ufft);
if (max(uv-u1)/max(u1) <tol)
u1 = uv;
break;
else
u1 = uv;
end
if (ii = = maxiter)
fprintf('Failed to converge to %f in %d
  iterations',tol,maxiter);
end
u0 = u1;
end
if (iz <nz) && (max((gamma*abs(u1).^2 + gamma*abs(u0).^2))
  <P_non_thres)
ufft = fft(u1);
ufft = ufft.*exp(LinearOperator*(nz-iz)*dz);
u1 = ifft(ufft);
end
output = u1;
```

11.4 MODELING OF OPTICALLY AMPLIFIED
TRANSMISSION SYSTEM AND BER

Optical receiver is represented by square-law detection, i.e., taking the absolute value
of the amplitude and the square into account. This is the photo-detection process [1].
This means that the photodetector absorbs the optical power and then converts it into
electronic current, which is then further amplified in the electronic domain and then
displayed on an oscilloscope. At this point, the eye diagram is achieved and the bit
error rate (BER) can be evaluated by getting the raw amplified data from optically
amplified transmission model. An error detection and calculation of BER is obtained
in the models for DPSK modulation format transmission system [5,9]. Monitoring of
data received at the output of the receiver is required to obtain the raw data for the
evaluation of the BER [1].

```
Initialization codes[1]
close all
clear all
% CONSTANTS
c = 3e8; % speed of light (m/s)
% NUMERICAL PARAMETERS
numbitspersymbol = 1;
P0 = 0.001; % peak power (W)
FWHM = 25e-12 ; % pulse width FWHM (ps)
FWHMps = 25e-12;
% halfwidth = FWHM/1.6651 % for Gaussian pulse
halfwidth = FWHM ; % for square pulse
bitrate = 1/halfwidth; % THz
bitrateG = 1/FWHMps;
baudrate = bitrate/numbitspersymbol;
signalbandwidth = baudrate;
PRBSlength = 2^8;
num_samplesperbit = 32; % should be 2^n
dt = FWHMps/num_samplesperbit ; % sampling time(ps);%time
  step(ps)
nt = PRBSlength*num_samplesperbit; % FFT length
dz = 500; % distance stepsize (m)
nz = 100; % number of z-steps
maxiter = 20; % max # of iterations
tol = 1e-5; % error tolerance
% OPTICAL PARAMETERS
nonlinearthreshold = 0.005; % 5mW -- % Nonlinear Threshold
  Peak Power for
silica core fiber
lambda = 1550e-9;
optical_carrier = c/lambda; %carrier freq
num_samplesperperiod = 8;
sampling_fac = 16;
carrier_freq =
  num_samplesperperiod*num_samplesperbit*bitrateG;
sampling time(ps); % time step (ps)
FFT length
alpha_indB = 0.2; % loss (dB/km)
D = 17e-6; % NZDSF GVD (s/m^2); if anomalous dispersion (for
compensation),D is negative
compensation),D is negative
ng = 1.46; % group index
n2 = 2.6e-20; % nonlinear index (m^2/W)
Aeff = 80e-12; % effective area (um^2)
% Estimated quantities
T = nt*dt;
 % FFT window size in sec. (ps) -Agrawal: should be about
  10-20 times of the pulse width
alpha_loss = log(10)*alpha_indB/10^4; % alpha (1/m)
beta2 = D*lambda^2/(2*pi*c); % beta2 (s^2/m);
beta3 = 0.3e-39; % GVD slope (s^3/m)
```

```
gamma = 2*pi*n2/(lambda*Aeff); % nonlinearity coef (m^-1.W^-1)
% STARTING FIELD
L = nz*dz
Lnl = 1/(P0*gamma) % nonlinear length (m)
Ld = halfwidth^2/abs(beta2) % dispersion length (m)
N = sqrt(abs(Ld./Lnl)) % governing the which one is
  dominating:
dispersion or Non-linearities
ratio_LandLd = L/Ld % if L << Ld  -> NO Dispersion Effect
ratio_LandLnl = L/Lnl % if L << Lnl -> NO Nonlinear Effect
% Monitor the broadening of the pulse with relative the
  Dispersion Length
% Calculate the expected pulsewidth of the output pulse
% Eq 3.2.10 in Agrawal "Nonlinear Fiber Optics" 2001 pp67
FWHM_new = FWHM*sqrt(1+(L/Ld)^2);
% N<<1 -> GVD ; N >>1 -> SPM
Leff = (1-exp(-alpha_loss*L))/alpha_loss;
expected_normPout = exp(-alpha_loss*2*L);
NlnPhaseshiftmax = gamma*P0*Leff ;
betap = [0 0 beta2 beta3]';
% Constants for ASE of EDFA
```

11.4.1 PROPAGATION OF OPTICAL SIGNALS OVER A SINGLE-MODE OPTICAL FIBER–SSMF

Matlab program for the propagation of optical signals over a single-mode optical fiber–SSMF is given below [1,5].

```
function output = ssprop_matlabfunction_modified(input)
nt = input(1);
u0 = input(2:nt+1);
dt = input(nt+2);
dz = input(nt+3);
nz = input(nt+4);
alpha_indB = input(nt+5);
betap = input(nt+6:nt+9);
gamma = input(nt+10);
P_non_thres = input(nt+11);
maxiter = input(nt+12);
tol = input(nt+13);
tic;
% pulse propagation in an optical fiber using the split-step
% GVD, higher order dispersion, loss, and self-phase
  modulation (gamma).
alpha = 1e-3*log(10)*alpha_indB/10;
ntt = length(u0);
w = 2*pi*[(0:ntt/2-1),(-ntt/2:-1)]'/(dt*nt);
%w = 2*pi*[(ntt/2:ntt-1),(1:ntt/2)]'/(dt*ntt);
clear halfstep
halfstep = -alpha/2;
```

```
for ii = 0:length(betap)-1;
halfstep = halfstep - j*betap(ii + 1)*(w.^ii)/factorial(ii);
end
clear LinearOperator
% Linear Operator in Split Step method
LinearOperator = halfstep;
% pause
halfstep = exp(halfstep*dz/2);
u1 = u0;
ufft = fft(u0);
iz = 0;
while (iz <nz) & (max( (abs(u1).^2 + abs(u0).^2))
  >P_non_thres)
iz = iz+1;
uhalf = ifft(halfstep.*ufft);
for ii = 1:maxiter,
uv = uhalf .* exp(-j*gamma*(abs(u1).^2 + abs(u0).^2)*dz/2);
ufft = halfstep.*fft(uv);
uv = ifft(ufft);
fprintf('You are using SSFMn');
if (max(uv-u1)/max(u1) <tol)
u1 = uv;
break;
else
u1 = uv;
end
if (ii == maxiter)
warning(sprintf('Failed to converge to %f in %d
  iterations',tol,maxiter));
end
u0 = u1;
end
if (iz <nz) & (max( (abs(u1).^2 + abs(u0).^2)) <P_non_thres)
% u1 = u1.*rectwin(ntt);
ufft == fft(u1);
ufft = ufft.*exp(LinearOperator*(nz-iz)*dz);
u1 = ifft(ufft);
fprintf('Implementing Linear Transfer Function of the Fiber
Propagation');
end
toc;
output = u1;
```

11.4.2   BER EVALUATION

BER is formulated using Gaussian distribution which follows the transmission of optical signals/photons [1]

$$\text{BER} = \frac{1}{2}\left[Q\left(\frac{d}{\upsilon_{NT0}}\right) + Q\left(\frac{\upsilon_{01} - d}{\upsilon_{NT1}}\right)\right]$$

where d is the decision voltage detecting bit error, v_{ol} is the data bit pattern (either 0 or 1) transmitted in optical transmission media, v_{NT0} represents 0 bits, v_{NT1} represents 1 bits, and $Q(\alpha)$ is a standard function, which is written as [1]

$$\frac{1}{\sqrt{2\pi}} \int_{\alpha}^{\infty} e^{-x^2/2} \, dx$$

Math lab codes for BER calculation are given below [13,14]:

```
clear Q_Gaussian BER_simple varied_threshold index_bit1 index_
   bit0 …
mean_bit1 mean_bit0 std_bit1 std_bit0;
%-------------------------------------------------------------
Q_Gaussian = 0; % Initialise Q value
% Simple Q-factor calculation with assumption:
% Gaussian Noise distribution
% Assign variables from SimuLink to Matlab variables
%************************************************************
delay = 32;
% To determine received "1"
index1_temp = find(Bernoulli==0);
tmp1 = find(index1_temp >length(demodsignal)- delay);
index1 = index1_temp(2:tmp1(1)-1);
for i = 1 : length(index1)
var_1(i) = demodsignal(delay+index1(i));
end
% To determine received "0"
index0_temp = find(Bernoulli==1);
tmp0 = find(index0_temp >length(demodsignal)-delay);
index0 = index0_temp(2:tmp0(1)-1);
```

*Simulink® Models of Optically Amplified Digital Transmission Systems*511

```
for ii = 1 : length(index0)
var_0(ii) = demodsignal(delay+index0(ii));
end
% End
%*****************
Thres_Volt = 0;
%*****************
% % BER
% % Assume bit 1 and 0 have a Gaussian normal distribution
% mean_bit1 = mean(var_1); % I1
% std_bit1 = std (var_1) , % sigma1
%
% mean_bit0 = mean(var_0); % I0
% std_bit0 = std(var_0) ; % sigma2
% % Q-factor
% Q = (mean_bit1 - mean_bit0) / (std_bit1 + std_bit0) ;
%
```

```
% BER = 1/2 * (erfc(Q/sqrt(2))) ;
% Determine BER
% Determine Mean value of samples for bit 1
mean_bit1 = mean(var_1);
std_bit1 = std(var_1);
% Determine Mean value of samples for bit 0
mean_bit0 = mean(var_0);
std_bit0 = std(var_0);
% Calculation of Q-factor and BER
% Assumption: Gaussian approximation for the distributions of
  samples of
% bit 1 and bit 0
% Pe = P0*E(1/2 of erfc( (S-X0)/sqrt(2)*sigma0) + P1*E(1/2 of …
% erfc( (S-X1)/sqrt(2)*sigma0)
BER_Gaussian = length(var_0)/length(Bernoulli(delay+1:length
  (Bernoulli)))*
mean(1/2.*erfc(abs(Thres_Volt - var_0)./(sqrt(2)*std_bit0))) +
  …
length(var_1)/length(Bernoulli(delay+1:length(Bernoulli)))*
mean(1/2.*erfc(abs(var_1 - Thres_Volt)./(sqrt(2)*std_bit1)));
% clc
mean_bit1
mean_bit0
std_bit1
std_bit0
Q_Gaussian = sqrt(2)*erfcinv(BER_Gaussian*2)
BER_Gaussian
BER_Gaussian_Agrawal = 1/2*(erfc(Q_Gaussian/sqrt(2)))
Q_Gaussian = abs(mean_bit1 - mean_bit0)/(std_bit1 + std_bit0)
```

SUMMARY

In this chapter, we have discussed math lab codes used for the design of optical fiber transmission system having transmitter module, receiver module, optical fiber module, optical amplification module, and optical modulation module, and their performance models. Math lab codes are linked with MATLAB and Simulink module and packages.

EXERCISES

11.1. Write down matlab codes for the design of optical transmission link using graded index fiber.
11.2. Write down matlab codes for the design of graded index optical fiber considering dispersion reduction.
11.3. Write down matlab codes for the design of two modes step index optical fiber considering dispersion reduction.

11.4. Write down matlab codes for the design of long-distance optical transmission link using single-step index optical fiber considering dispersion reduction and placement of optical amplifiers.

11.5. Write down matlab codes for the design of long-distance optical transmission link using graded index optical fiber considering dispersion reduction and placement of optical amplifiers.

REFERENCES

1. L. N. Binh, *Optical Fiber Communication System: Theory and Practice with MATLAB and Simulink® Models*, CRC Press, London, 2010.
2. G. Mahlkc and P. Gossing, *Fiber Optic Cables*. Siemens A. G. and J. Wiley, Chichester, 1987.
3. G. P. Agrawal, *Fiber Optic Communications Systems*, 3rd edn. Academic Press, New York, 2002.
4. J. A. Buck, *Fundamentals of Optical Fibers*, Wiley, New York, 1995.
5. G. P. Agrawal, *Nonlinear Fiber Optics*, 3rd edn. Academic Press, San Diego, CA, 2001.
6. IEC SC 86A/WGI, Method1, September 1997.
7. L. N. Binh, T. L. Huynh, K. Y. Chin, and D. Sharma, "Design of dispersion flattened and compensating fibers for dispersion-managed optical communications systems," *International Journal of Wireless and Optical Communications*, vol. 2, no. 1, pp. 63–82, June 2004.
8. K. P. Ho, *Phase-Modulated Optical Communication Systems*, Springer, New York, 2005.
9. A. H. Gnauck and P. J. Winzer, "Optical phase-shift-keyed transmission," *IEEE Journal of Lightwave Technology*, vol. 23, no. 1, pp. 115–130, 2005.
10. M. Lax, "Rate equations and amplitude noise," *IEEE Journal of Quantum Electronics*, vol. 3, no. 2, pp. 37–46, 1967.
11. J. P. Gordon and H. Kogelnik, "PMD fundamentals: Polarization mode dispersion in optical fibers," *PNAS*, vol. 97, no. 9, pp. 4541–4550, April 2000.
12. W. D. Jones, *Optical Fiber Communications Systems*, Holt, Rinhart and Winston, New York, 1988.
13. D. Ye and W. D. Zhong, "Improved BER monitoring based on amplitude histogram and multi-Gaussian curve fitting," *Journal of Optical Networking*, vol. 6, no. 6, pp. 584–598, 2007.
14. L. Ding, W.-D. Zhong, C. Lu, and Y. Wang, "New bit-error-rate monitoring technique based on histograms and curve fitting," *Optics Express*, vol. 12, no. 11, pp. 2507–2511, 2004.

Index

A

Adaptive grooming policy (AGP), 230
Add drop multiplexer (ADM), 2, 120, 202
Algorithm
    heuristic greedy, 10
    one-step, 259
    sequential coloring, 88
    simulated annealing, 105
    two-step, 258
All-optical waveband switching, 232–234
Application layer, 72
Array waveguide grating, 409
Asynchronous transfer mode, 1
Automatically-switched optical networks
    (ASON), 222
Autonomous system, 250
Average (hop), 150

B

Bandwidth provisioning, 225
Best effort (BE), 403
Bidirectional line-switched ring (BLSR),
    252–254, 322–323
Bit error rate (BER), 76, 231, 250, 351
Bridge services, 76
Blocking probability, 30, 84, 125, 135, 189,
    274, 275, 277, 280–282
Broadcast layer, 291
BLSR, *see* Bidirectional line-switched
    ring (BLSR)

C

Call admission control (CAC), 346
    centralized control, 266, 322
    channel capacity, 8–9, 15, 17, 129, 149, 169,
        228, 237, 298, 300, 310
Carrier sense multiple access (CSMA)/CD, 405
Channel-capacity constraint, 8
Circuit switching, 233
Connection, 4–6, 8–7, 28–31, 222–228, 238–242,
    268, 346
Constraints
    traffic flow constraints, 124, 130
    wavelength constraints, 125, 128, 153,
        161, 276
Control information (CI), 26, 27, 56, 205
CPLEX, 10, 18, 170, 215, 291, 300
CR-LDP, 225

Cross-Connect (OXC), 2, 12, 13–18, 33, 83, 141,
    162, 166, 204, 207–209, 222–226

D

Data link layer, 37, 370
Deadlock, 332, 334
Digital crossconnect (DXC), 5, 13, 18, 207,
    257, 298
Digital subscriber line (DSL), 1
Directed Steiner minimum tree (DSMT), 287
Directed-link-disjointness, 301–302
Direction, wavelength, and ROAD assignment
    (DWRA), 30
Distance matrix, 123, 146, 151
Distributed control, 222, 225, 265–266

E

Encryption
    Block Ciphers, 375
    Data encryption standards (DES), 374
    Product cipher, 375
    Public key cryptography, 385
    Rivest-Shamir-Adleman (RSA), 389
    Substitution Ciphers, 372–373
    Transposition Ciphers, 372

F

Fault management, 250–254, 310, 323
Finite queue
    balanced traffic, 21, 23
    unbalanced traffic, 21, 23
First come first served (FCFS), 37
First fit (FF), 206
First-in-first-out (FIFO), 23, 25
FTTx
    FTTB, 403
    FTTC, 403
    FTTH, 403
    FTTP, 403
Framing protocol (FP), 3, 238

G

GMPLS, 222
Graph
    coloring, 87
    in degree, 142
    out-degree, 142

439

Printed and bound by CPI Group (UK) Ltd, Croydon, CR0 4YY

23/10/2024

01778223-0017